D0438019

The Book of

Strange but True Science

Publications International, Ltd.

Images from Shutterstock.com

Copyright © 2019 Publications International, Ltd. All rights reserved. This book may not be reproduced or quoted in whole or in part by any means whatsoever without written permission from:

Louis Weber, CEO
Publications International, Ltd.
8140 Lehigh Avenue
Morton Grove, IL 60053

Permission is never granted for commercial purposes.

ISBN: 978-1-64030-833-6

Manufactured in China.

8 7 6 5 4 3 2 1

Contents

❋ ❋ ❋ ❋

How to Beat a Lie Detector Test • The Route to China Via a Hole • Running Through the Rain vs. Walking Through the Rain • It's Iron-IC • The Mysterious Blue Hole • Forensic Dentist • Exposed to Poison • Color Me Toxic • Get Savvy on the Seas • Anatomical Anomalies

There's a Stranger Side to Science

❋ ❋ ❋ ❋

Welcome to *The Book of Strange but True Science*. If you're one of those people who find themselves wondering whether a crazy physics theory might really be true, how long people have been making mummies, if quantum teleportation is actually possible, or how flak jackets stop bullets, then you're sure to find lots of satisfying answers in this book. In these pages we'll investigate the possibility of elevators to outer space, self-driving cars, and the science of invisibility. We'll unravel the mysteries behind fireworks, X-rays, microwaves, electric chairs, and indoor plumbing. We'll find out why skunks are stinky, beans cause gas, and doctors have such awful handwriting. But these articles don't limit themselves to just the dry and dusty facts and figures: We'll also investigate the ideas behind the inventions, the histories leading up to the discoveries, and the scientists behind the science.

Each chapter takes an intriguing subject—medical science, pets and animals, consumer gadgets, astronomy, food, mad scientists, the human body—and ferrets out the strange stories and lesser known truths. Whether it's a world-changing event or a curious anecdote of relative obscurity, every story is sure to pique your interest. How do bananas ripen so conveniently? Read about it on page 609. How did Silly Putty get invented? Find out on page 210. Will cockroaches really survive a nuclear war? Turn to page 371 . . .

Chapter 1

Science Fiction to Science Fact

Faster than a Science Fiction Plot Device

Why does lightning-quick travel in science fiction happen at warp speed? What exactly is undergoing the "warping"?

* * * *

THE ANSWER IS, basically, the fabric of space and time. If a spaceship were able to use antimatter to generate a special magnetic field around itself, it could essentially slip through the seams of the space-time continuum and travel light-years in an instant. Most famously, the fictional starship *Enterprise* from *Star Trek* was able to go into "warp drive" when it had to go somewhere far and fast.

So far the concept is just fiction. Matter that travels faster than light theoretically might be converted to pure energy, ceasing to exist as matter. Warp speed takes antimatter, and that's not something you can pick up at the gas station. But it's probably not mere fantasy. William Shatner claims that when *Star Trek* fan Stephen Hawking toured the set of *Star Trek: The Next Generation,* he paused at the warp engines and said, "I'm working on that." NASA researchers have indeed worked on designs for an engine partly propelled by tiny amounts of antimatter.

By the time NASA actually launches something at warp speed, the phrase may have lost its scientific meaning through widespread general use of the words to simply mean "fast." An old headline about the release of the Apple iPad referred to "Warp Speed Publishing." An article about the Arizona State football team's fast-paced offense said the team would play at warp speed. A headline about faster broadband read, "Intel pushes warp-speed wireless broadband."

This follows a common pattern of specific terms entering general use by extension or association of meaning. But it raises the question of the limits of our exaggeration. What hyperbole can we turn to next when *warp speed* loses its power, when we need something faster than the starship *Enterprise* just to brag about broadband?

An Elevator to Outer Space

Rockets are for suckers. Looking for a cheaper way to build space stations and launch satellites, NASA scientists have an idea that sounds ludicrous but that they swear is feasible: an elevator that reaches from Earth's surface to outer space.

* * * *

How could an elevator to outer space be possible? you wonder. The answer is nanotubes. Discovered in 1991, nanotubes are cylindrical carbon molecules that make steel look like a 98-pound weakling. A space elevator's main component would be a 60,000-odd-mile nanotube ribbon, measuring about as thin as a sheet of paper and about three feet wide.

It gets weirder. That ribbon would require a counterweight up at the top to keep it in place. The counterweight, hooked to the nanotube ribbon, would be an asteroid pulled into Earth's orbit or a satellite. Once secured, the ribbon would have moving platforms attached to it. Each platform would be powered by solar-energy-reflecting lasers and could carry several thousand tons of cargo up to the top. The trip would take about a week. Transporting materials to outer space in this fashion would supposedly reduce the cost of, say, putting a satellite into orbit from about $10,000 a pound to about $100 a pound.

The elevator base would be a platform situated in the eastern Pacific Ocean, near the equator, safe from hurricanes and many miles clear of commercial airline routes. The base would be mobile so that the whole thing could be moved out of the path of potentially damaging space junk orbiting Earth. Although there are a lot of theoretical kinks to work out, the more optimistic of the scientists who have hatched this scheme believe the whole thing could be a reality within a couple of decades.

It's a Bird! It's a Plane! It's . . . Avrocar?!?

Not all UFOs are alien spaceships. One top-secret program was contracted out by the U.S. military to an aircraft company in Canada.

* * * *

Oh, the 1950s—a time of sock hops, drive-in movies, and the Cold War between America and the Soviet Union, when each superpower waged war against the other in the arenas of scientific technology, astronomy, and politics. It was also a time when discussion of life on other planets was rampant, fueled in part by the alleged crash of an alien spaceship near Roswell, New Mexico, in 1947.

Watch the Skies

Speculation abounded about the unidentified flying objects spotted nearly every week by everyone from farmers to pilots. As time passed, authorities began to wonder if the flying saucers were, in fact, part of a secret Russian program. Fearful that such a craft would upset the existing balance of power, the U.S. Air Force decided to produce its own saucer-shape ship.

In 1953, the military contacted Avro Aircraft Limited of Canada, an aircraft manufacturing company that operated in Malton, Ontario, between 1945 and 1962. Project Silverbug was initially proposed simply because the government wanted to find out if UFOs could be manufactured by humans. But before long, both the military and the scientific community were speculating about its potential. Intrigued by the idea, designers at Avro—led by British aeronautical engineer John Frost—began working on the VZ-9-AV Avrocar. The round craft would have been right at home in a scene from the classic science fiction film *The Day the Earth Stood Still*. Security for the project was so tight that it probably generated rumors that America was actually testing a captured alien spacecraft—speculation that remains alive and well even today.

Of This Earth

By 1958, the company had produced two prototypes, which were 18 feet in diameter and 3.5 feet tall. Constructed around a large triangle, the Avrocar was shaped like a disk, with a curved upper surface. It included an enclosed 124-blade turbo-rotor at the center of the triangle, which provided lifting power through an opening in the bottom of the craft. The turbo also powered the craft's controls. Although conceived as being able to carry two passengers, in reality a single pilot could barely fit inside the cramped space. The Avrocar was operated with a single control stick, which activated different panels around the ship. Airflow issued from a large center ring, which was controlled by the pilot to guide the craft either vertically or horizontally.

The military envisioned using the craft as "flying Jeeps" that would hover close to the ground and move at a maximum speed of 40 mph. But that, apparently, was only going to be the beginning. Avro had its own plans, which included not just commercial Avrocars, but also a family-size Avrowagon, an Avrotruck for larger loads, Avroangel to rush people to the hospital, and a military Avropelican, which, like a pelican hunting for fish, would conduct surveillance for submarines.

But Does It Fly?

The prototypes impressed the U.S. Army enough to award Avro a $2 million contract. Unfortunately, the Avrocar project was canceled when an economic downturn forced the company to temporarily close and restructure. When Avro Aircraft reopened, the original team of designers had dispersed. Further efforts to revive the project were unsuccessful, and repeated testing proved that the craft was inherently unstable. It soon became apparent that whatever UFOs were spotted overhead, it was unlikely that they came from this planet. Project Silverbug was abandoned when funding ran out in March 1961, but one of the two Avrocar prototypes is housed at the U.S. Army Transportation Museum in Fort Eustis, Virginia.

Quantum Teleportation

Another idea popularized on Star Trek, *teleportation involves travel by "beaming" from one location to another. The idea of teleporting from one location to another is not as unbelievable as it sounds. In fact, some scientists say that there is nothing in the laws of physics that would render it impossible.*

* * * *

On *Star Trek*, teleportation is made possible by converting the atoms that make up a human into energy, beaming that energy to another location, and then reassembling the atoms. Voila! A reconstructed human. We are definitely not at a point where we can break down humans to an atomic level and reassemble them in another location; for now, that will have to remain science fiction. However, scientists have been able to accomplish something else that, at first glance, seems like it must be fiction. But like so many other amazing things in the scientific world, this is reality.

It's called "quantum teleportation," but unlike *Star Trek*, this is not teleportation of matter, but rather information. Quantum teleportation can become possible due to an interesting, and somewhat strange, property of quantum mechanics called "quantum entanglement." This occurs when particles are linked together in such a way that the state of one particle determines the states of the linked particles, even if they are separated by vast distances.

The classic example used to describe quantum entanglement is called the EPR Paradox, or Einstein-Podolsky-Rosen Paradox, named after physicists Albert Einstein, Boris Podolsky, and Nathan Rosen, who first described it in the 1930s. In the paradox, two entangled particles have an uncertain state until one of them is measured, at which point the state of the other particle is immediately certain. But this would suggest that there is communication between the two particles that is faster than the speed of light, conflicting with Einstein's theory that nothing can travel faster than the speed of light; hence, the paradox.

This paradox was confusing, even for Einstein. So a few years later, physicist David Bohm suggested a more simplified example of the EPR Paradox. In Bohm's example, an unstable particle with a spin of 0 decays into two new particles, Particle A and Particle B, which then travel in different directions. Since the original particle had a spin of 0, the two new particles have a spin of + and –, giving them an entangled property.

The famous "Schroedinger's Cat" thought experiment, dreamed up by physicist Erwin Schroedinger, describes more of the paradoxes of quantum mechanics. Schroedinger envisioned a cat that was placed in a steel box with a vial of poison, a Geiger counter, a hammer, and a radioactive substance. When the radioactive substance decays, it triggers the Geiger counter, the hammer hits the vial of poison, and the cat dies. But radioactive decay is a random process, and there is no way to predict when it will occur. So, according to physicists, the radioactive substance exists in a state known as "superposition," or in a state that is both decayed and not decayed at the same time. Therefore, until an observer actually opens the box, there is no way to know whether or not the cat is alive. It exists, according to Schroedinger, as both "living and dead . . . in equal parts."

Now, setting aside the fact that Schroedinger seemed to really hate cats, this same paradox can be applied to quantum entanglement. Before observing the spins of Particle A and Particle B, they don't have a definite state; rather, they are in a state of superposition, just like Schroedinger's poor cat. Therefore, the spin of Particle A and Particle B is both + and –. But if you measure the spin of Particle A and find it to be +, then you immediately know that the spin of Particle B is –, without even having to measure.

Scientists are already finding practical ways to use quantum entanglement, such as to send and receive information between spacecraft and ground-based receivers, and to create entanglement-enhanced microscopes. But back to quantum teleportation: in order to accomplish this feat, scientists must begin with entangled particles. They must then take whatever information they wish to teleport and include it in the entanglement. The entanglement then becomes the teleportation channel between particles. So far, physicists have been successful in teleporting what is called a spin state, also known as a "qubit," across a quantum communication channel. And in 2017, Chinese scientists successfully teleported the quantum state of a photon from Earth to an orbiting satellite more than 800 miles away.

When the process was first theorized in 1993 by physicists Asher Peres, William Wootters, and Charles Bennett, Peres and Wootters suggested calling it "telepheresis." But Bennett proposed the catchier, more impressive sounding "quantum teleportation," which has perhaps led to greater interest in the subject. Because who doesn't want teleportation to be possible? While we may not be anywhere close to being able to actually teleport objects, that doesn't mean that quantum teleportation may not have some amazing uses. Quantum mechanics have the potential to make all kinds of new technology possible, such as quantum computers, which could make even today's lightning-fast computers look slow, or a "quantum internet" which can securely send information over long distances without actually traveling those distances.

Teleporting from place to place like a *Star Trek* space explorer may remain fiction for the foreseeable future, but we still have much to learn about the way matter, space, and time work. Properties of the universe that were mysteries a century ago are now explainable, and researchers are continually finding new and amazing things throughout the cosmos. Perhaps we simply haven't found the exact mechanisms to make our science fiction dreams a reality, but one thing is certain: scientists will continue to search the universe and run experiments until we've reached the limits of what's possible. But with an infinite amount of space to study, perhaps those possibilities are as limitless as our universe itself.

Losing Philip K. Dick's Head

Consider the sorts of things an air traveler might accidentally leave in an overhead bin: a novel, a jacket, a laptop computer . . . or perhaps the $750,000 android head of author Philip K. Dick.

* * * *

Meet David Hanson

IN THE WORLD of robotics, David Hanson is known as the genius inventor of "frubber"—an uncannily authentic-looking synthetic skin. With this breakthrough, Hanson has created robots modeled after popular figures, ranging from Albert Einstein to rocker David Byrne. In 2004, his firm, Hanson Robotics, led a team of artists, scientists, and literary scholars to create an android modeled after science fiction writer Philip K. Dick, who died in 1982. To build the android's body of synthetic knowledge, a team led by artificial intelligence expert Andrew Olney scanned 20 of Dick's novels as well as interviews, speeches, and other information into its computer brain.

After six months of work, the completed android was an impressive achievement. Using a camera for its eyes, the android could follow movement, make eye contact, and recognize familiar faces in a crowd. The android's cutting-edge AI applications allowed it to respond to queries using its inputted source material (though its replies were often off-topic and bordered on the surreal—not unlike the real-life Dick, according to friends). The team called him "Phil."

Meet Philip K. Dick

Hanson and his team chose an apt subject for their robot. Many of Dick's stories are peopled with synthetic humans who call into question what qualifies as human. This theme is evident in several of the films adapted from his works, including *Blade Runner* (based on the novel *Do Androids Dream of Electric Sheep?*) and *Total Recall* (based on *We Can Remember It for You Wholesale*). Hanson's team was particularly inspired by Dick's novel *We Can Build You*, in which two unsuccessful electric-organ salesmen construct a lifelike robot of Edward M. Stanton, Secretary of War under Abraham Lincoln. The competing needs of the human characters, coupled with the android's increasing humanity, play out in a rich tale that challenges the reader's imagination. Little did Hanson's team realize that their robot would also become part of its own odd legend.

David Hanson Loses His Head

In June 2005, Phil made his debut at Chicago's NextFest technology exhibition. He was a sensation. The technically minded marveled at the feat of engineering, the sci-fi fans thrilled at the wonderful irony of interacting with an android replication of their long-dead hero, and the merely curious were rewarded by the eerie sense of humanity that Phil inspired.

On the heels of this success, Hanson, juggling time between his growing firm and his doctoral work, carted Phil around the country. In Pittsburgh, Hanson received the Open Interaction Award for Robotics from the American Association for

Artificial Intelligence. At the San Diego Comic-Con, Phil took part in a panel discussing the upcoming release of *A Scanner Darkly* (based on the Dick novel of the same name). Phil was a hit: Hanson and Phil made appearances in Memphis, New Orleans, and Dallas. There was talk of a tour to promote the movie, an appearance on the *Late Show with David Letterman*, and a stint in the Smithsonian Institute's traveling collection. It was awfully heady stuff for a head.

In early 2006, Phil was on his way to Mountain View, California. He was stored face down in Styrofoam, bundled in a gym bag, and stowed in the overhead bin on the plane. Hanson, bleary from sleep, changed flights in Las Vegas. Soon afterward, he realized that Phil was still on the plane. The airline confirmed that Phil had traveled with the plane to Orange County. The android was supposed to be put on a flight to San Francisco but never arrived. Phil has not been seen since.

All Is Not Lost

Though the unique head has vanished, Hanson's laptop, containing Phil's brain, is safe. Despite the heartbreaking loss, Hanson eventually built another head for Phil. A failed lawsuit against the airline, however, means that, ironically, like the real-life Philip K. Dick, Phil must wait for the two things most precious to a struggling writer: time and money.

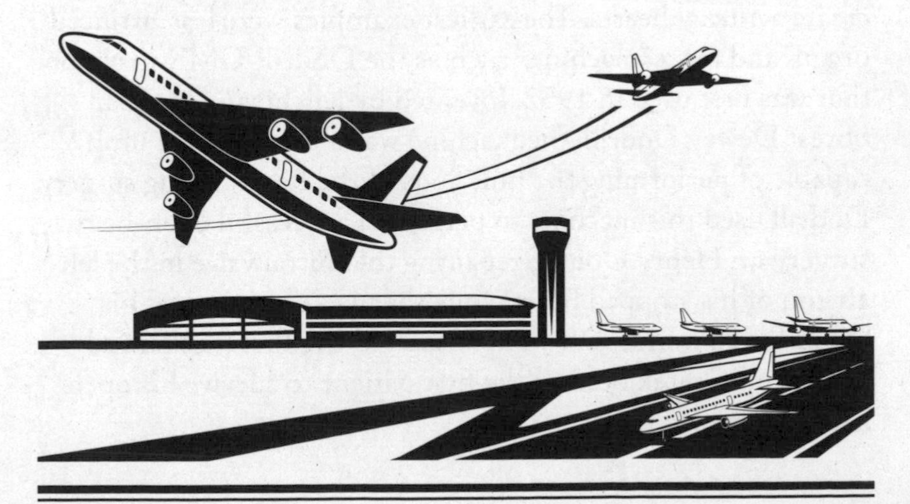

The Artificial Heart

In Mary Shelley's 1818 classic novel Frankenstein, *scientist Victor Frankenstein creates a creature by cobbling together body parts and using science to animate the once-dead tissue. The work of fiction has sent chills down the spines of readers for two hundred years, but some also see it as a cautionary tale about what can go wrong when man tries to play "God." And on a cold December day in 1982, some were convinced that humans had gone too far with their application of science. On that day, a man with no heart was brought back from the dead.*

* * * *

THERE ARE SOME organs that humans can live without, like the spleen or gallbladder. We can also lose one lung or kidney, and even part of our liver, and still live a relatively normal life. But there is no way to live without a heart. The blood that the heart pumps through our bodies provides absolutely vital oxygen and nutrients to our brains and every other part of us. There's simply no way to survive without one. Heart transplants are a possibility for some who suffer from disease or injury, but these require waiting for donors (who, themselves, must die in order to provide the life-saving organ).

So for decades, doctors and scientists have searched for ways to create artificial hearts. The earliest examples were less "artificial organ" and more "machine," such as the Dodrill-GMR machine that was first used in 1952. Invented by Michigan physician Forest Dewey Dodrill, the machine was a mechanical pump capable of performing the functions of the heart during surgery. Dodrill used this machine to perform a successful open-heart surgery on Henry Opitek, repairing the mitral valve in the left atrium of his stopped heart while the machine pumped his blood for 50 minutes. It was considered the first mechanical heart, and Opitek became the first patient to survive his open-heart surgery.

While Dodrill's machine was a noteworthy innovation, it could only be used during surgery. This made it ideal for someone like Opitek, who simply needed a valve repair in an otherwise healthy heart; but for those who suffered from conditions that surgery couldn't fix, another revolutionary idea was needed. So in 1964, the National Institutes of Health implemented the Artificial Heart Program, with the goal of creating an implantable artificial heart. Physician Willem Johan Kolff took up the challenge, starting the Division of Artificial Organs at the University of Utah in 1967.

Over the next fifteen years, Kolff, along with a team of about 200 doctors, engineers, and students, designed and tested several versions of artificial hearts. Each new version was implanted in a laboratory animal, with each new heart functioning just a little bit longer than the last. Kolff assigned project managers for each different venture under his supervision, naming each project after its manager. Graduate student Robert Jarvik, who would go on to receive his M.D. in 1976, joined the Division of Artificial Organs in 1971, and was immediately assigned the task of improving an artificial heart that had kept a lab animal alive for 10 days.

Progress was slow, but a decade later, Jarvik had created an artificial heart that was able to keep a laboratory calf, named Alfred Lord Tennyson, alive for 268 days. After a few more tweaks, the heart, called the Jarvik-7, was as ready as it would ever be for a human subject.

So, on December 2, 1982, a Seattle dentist named Barney Clark volunteered to be the first recipient of the Jarvik-7 artificial heart. Clark was suffering from severe congestive heart failure with no hope of recovery, and although he really didn't expect to live more than a few days with his artificial heart, he wanted to undergo the procedure in the hopes that he could contribute to medical science. During a seven-hour operation performed by surgeon William DeVries, reporters from around the world camped out at the hospital to await news of this groundbreaking procedure, which some felt crossed ethical boundaries, just like Victor Frankenstein's monster.

When Clark survived the surgery, interest in the Jarvik-7 heart skyrocketed, with reporters asking for daily updates and asking endless questions. Some criticized the doctors and university, saying they were using Clark as a "human guinea pig," even though Clark had willingly volunteered for the surgery. It was an experience that Jarvik described as "a medical experiment in a fishbowl." The world watched as Clark survived with his artificial heart for 112 days, allowing him to celebrate his 39th anniversary with his wife, Una Loy. He died suddenly on March 23, 1983, after multi-organ failure. But his wife worked to continue his legacy, wanting people to know that he was a selfless pioneer in the research of artificial heart advancement.

The second recipient of the Jarvik-7, William Schroeder, who received his heart on November 25, 1984, lived for 620 days, and was able to travel, go to ball games, and go fishing, although a series of strokes left him limited in his abilities. After he died on August 7, 1986, his headstone was engraved with an image of the Jarvik-7, a tribute to the extra days of life he was given.

These first artificial hearts were powered by huge, dishwasher-sized pneumatic machines, but could be made portable with a fifteen-pound compressed air system. These limitations made the Jarvik-7 more practical for use as a stopgap measure to be used while waiting for a heart transplant. This often gave patients several more years of life, with some of the earliest artificial heart recipients living for more than a decade after having subsequent heart transplants.

Today, patients suffering from heart failure can often extend their lives with the help of a ventricular assist device, or VAD. These small mechanical devices, which weigh only ten ounces and are less than three inches in diameter, can be implanted in the heart and take over the pumping functions of one or both ventricles. They can be used for just a few days or weeks, or even a few years, while patients wait for heart transplants, allowing them to resume a more normal life, even with heart disease. The implantation of a VAD has even been known to help a diseased heart heal, as it helps to take some of the stress off the muscle.

Today, both artificial hearts and VADs are mostly used as a bridge to keep the heart going until a transplant can be performed. But perhaps one day technology will progress to the point that artificial organs are as good as the real ones. With the scientific advancements we've seen just in the first decades of the 21st century, it doesn't seem like an unrealistic goal. And unlike Victor Frankenstein, we don't have to fear scientific advancement. Lucky for us, it doesn't result in monsters; just longer lives.

Edging Closer to Immortality

No, scientists haven't yet unlocked the secret to eternal life, but they do say that we'll soon be living significantly longer than we are now.

✻ ✻ ✻ ✻

SOME SCIENTISTS BELIEVE that within 50 years, people in industrialized nations will routinely live 100 years or longer. When that time comes, perhaps a few healthy, energetic individuals will be pushing 140, 145, 150, and beyond.

The average American these days is expected to live 78 years, and the average life expectancy worldwide has been increasing by about two years every decade since the 1840s. Back then, Sweden boasted the population with the most impressive longevity: Healthy folks lived to the ripe old age of 45.

Increasing life expectancy is attributed to a number of factors, such as vaccinations, antibiotics, better sanitation, and stricter food regulations. Furthermore, improved safety regulations in the workplace and on the road have helped to prevent fatal injuries. Experts such as James Vaupel, director of the laboratory of survival and longevity at the Max Planck Institute in Germany, believe that life expectancy will continue to climb as techniques improve for treating age-related maladies such as heart disease and cancer.

Centennial birthday parties should be commonplace by early in the 22nd century. That may seem like a long way off, but consider that those centigenarians are being born now. The baby that your sister-in-law just brought home from the hospital may come close to reaching 140.

Without a crystal ball to forecast the medical advances we may achieve, it's impossible to say how long people eventually will be living. But as Daniel Perry, executive director of the Alliance for Aging Research, said, "There is no obvious barrier to living well beyond 100."

Satellites

More than 1,000 active artificial satellites currently orbit the earth, along with more than 2,000 that are no longer operable. It's a good thing outer space is a big place! Satellite technology is truly a case of life imitating art. Science, too, of course. Isaac Newton, well ahead of his time in the 17th century, published A Treatise of the System of the World, *in which he wrote the first account of the mathematical possibility of satellites orbiting the earth. Almost 200 years later, the first fictional depiction of a satellite being launched into orbit appeared. It came in* The Brick Moon, *a short story by American author Edward Everett Hale.*

⁂ ⁂ ⁂ ⁂

SEVERAL IMPORTANT 20TH century publications began the tactical march to making satellites a reality. The first was *Exploring Space Using Jet Propulsion Devices* by Russian rocket scientist Konstantin Tsiolkovsky in 1903. Because he identified the speed required for minimal orbit and pointed out that a multi-stage rocket fueled by liquid propellants could get the job done, he holds an esteemed place among the founding fathers of modern rocketry. His work contributed greatly to the Soviet space program. Science fiction author Arthur C. Clarke pushed the discussion further in two important articles that detailed a plan to get satellites into space.

Not surprisingly, it was the Soviet Union that led the charge in the effort to make satellites a reality. Sputnik 1, launched by the Soviet Sputnik program on October 4, 1957, was the first artificial satellite in space. None too pleased about being beaten to the punch by the Soviets, the United States increased its efforts in space innovation. It was the start of the Space Race between the two rival nations.

Sputnik 2 broke more new ground one month later, bringing the first living passenger into orbit. It was not a happy occasion for Laika; the dog died of overheating, although the cause and time of death were kept under wraps for more than 40 years.

The United States' first artificial satellite came about four months after Sputnik 1 made orbit. On January 31, 1958, Explorer 1 was launched. Within three years, more than 100 satellites were orbiting the Earth. The Space Race led to huge and rapid development of technology that allowed the United States, the Soviet Union, and eventually many other nations to put satellites to all kinds of productive uses such as earth observation, space exploration, communication and broadcasting, and navigation. Of course, they have also been key to military strategy through the years.

When it comes to how satellites work, Newton, of course, is the master. His gravitational theories have everything to do with it. It's the gravitational pull of the globe that keeps satellites in orbit. The key is that the satellites need to keep moving at all times. Some circle the planet more than 22,000 miles above the earth's surface. Medium-range satellites are more like

12,000 miles away, while the low-earth orbits used largely for telecommunications circle anywhere from 400 to 1,000 miles from the earth.

Satellites come in virtually all shapes and sizes. They began as vastly different machines crafted by the early pioneers of the technology. The American-made HG-333 GEO Commsat, launched in 1972, became the first standard design in production. It paved the way for several replicas to reach space. The largest artificial satellite is the International Space Station, a joint project among five different space programs (including NASA). It had its first component launched in 1998 and its last module fitted in 2011.

An Ominous Chain Reaction

Few events have changed the course of history as greatly as the movement of a few simple particles at the University of Chicago one sobering December afternoon in 1942.

* * * *

As the United States braced itself for a second year of war both in Europe and in the Pacific, the world's top scientists set off a mesmerizing chain reaction that proved possible the development of nuclear technology. Its consequences were also, however, devastating and final, capable of obliterating civilization with the press of a button. The atomic bomb was born at 3:25 P.M. on December 2, 1942.

The Split Heard 'Round the World

By the close of the 1930s, the Axis powers were said to have begun developing a weapon that would harness atomic energy in order to cause vast, devastating damage. Urged by a letter from physicists Albert Einstein and Leó Szilárd, U.S. President Franklin Roosevelt created a committee of scientists to research the feasibility of such a project, the first in a series of official measures taken by the American government to develop the bomb. After Allied scientists working in England discovered Uranium's fissile properties (or, more specifically, the fact that the isotope Uranium-235 had the ability to split), the project was given the full resources of the military. Physicists and engineers were recruited from throughout the world to work on what became known as the Manhattan Project.

Numerous labs and thousands of personnel throughout the country worked overtime to understand, first, how to enact nuclear fission and, second, how to safely harness its energy. Under the auspices of the now-defunct Office of Scientific Research and Development, the Manhattan Project constituted the most complex integration of science and military technology in modern history.

The Chicago Metallurgical Laboratory, the secret midwestern arm of the project, was overseen largely by Leslie Groves, a meat-and-potatoes general, while Enrico Fermi, an Italian academic and Nobel Laureate—and namesake of Fermilab, which still conducts nuclear research outside Chicago—worked with a team of physicists to create the conditions necessary for nuclear fission and the subsequent chain reaction.

A Most Imposing Pile

Using the work of such theorists as Szilárd, who had discovered the process of nuclear chain reaction, and the expertise of his engineering team, Fermi constructed a block—a "pile"—of Uranium and other materials called Chicago Pile 1. This was the famous reactor, and it stood almost 26 feet high. From the pile, a rod coated in the element Cadmium would be withdrawn, causing neutrons to collide with and split the Uranium isotopes, and by doing so, cause more collisions. Slowed by the non-Uranium materials and quickly shut off by reinserting the Cadmium rod, the reactor showed the potential for larger reactions in an uncontrolled setting.

Though his famous reactor was meant to be constructed outside of the city, at the Argonne National Laboratory in nearby DuPage County, a local labor strike forced Fermi to locate a different space.

Since named a historic landmark by the federal government, the room Fermi chose was a squash court beneath some rusty bleachers at the University of Chicago's long-abandoned Stagg Field. While scores of fellow scientists, officials, and dignitaries looked on, Fermi's team completed a successful self-sustaining nuclear reaction—the world's first. Just three years later, detonated in a split second, the same type of reaction would annihilate a city.

Sources say that the mood that day was both exhilarating and terrifying; after the war, Fermi and Szilárd expressed ambivalence about the consequences of their work, which held immense promise for energy production but could also result in such destruction and utter despair.

Today these themes resonate on the spot of that first detonation. Just above that old squash court stands a sculpture that is, in its raw power and simplicity, as beguiling as the sight of that reactor must have been. A vague, amorphous shape, the bronze statue *Nuclear Energy* by British sculptor Henry Moore is a reminder of the totality of that day.

Cloaking Device: Now You See It, Now You Don't

In 1966, with the United States and the Soviet Union both wondering what the other was capable of in the space race, the first season of Star Trek *introduced a frightening futuristic plot device. A Romulan starship, anathema to Captain Kirk and his cohorts, was on the move and presumably up to no good. Particularly troublesome was its use of a previously unknown technology: a cloaking device, which rendered it invisible to the starship* Enterprise. *This bit of technological whimsy has been a staple of science fiction and video games ever since.*

❋ ❋ ❋ ❋

THE TERM RELIES on the figurative use of the verb *to cloak* (itself deriving from the noun denoting a loose-fitting garment), an application that dates back to the 1500s. But while *cloak* has enjoyed semantic flexibility for centuries, we owe this particular manifestation to the classic science fiction series.

The term *cloaking device* is of relatively recent coinage, but the provocative notion of manufacturing invisibility certainly didn't originate in the mind of *Star Trek* creator Gene Roddenberry. In fact, the idea dates back thousands of years: There is a prototype for the Romulans' technology in the ancient Greek myth of Perseus. In addition to his iconic winged sandals, this son of Zeus employed a helmet of invisibility to aid him in vanquishing the Gorgon Medusa.

Returning to this millennium, we find another fictitious cloaking device—this time, in a clever marrying of the term's literal and figurative meanings, taking the form of an actual cloak. Those familiar with J. K. Rowling's Harry Potter series might recall Harry's invaluable invisibility cloak, a magical garment Harry and his friends employ for missions that are variously mischievous and noble.

But it isn't just Trekkies and wizard wannabes who can vouch for the usefulness of a cloaking device; nor is this seemingly far-fetched technology only the stuff of fantasy. Scientists today are exploring the real-life possibilities surrounding manufactured invisibility—and with encouraging success. So, while there's still a formidable gap between reality and fiction, it may not be light-years before *cloaking device* is as much a part of our daily vocabulary as *Internet* and *cell phone*.

A Weapon for Peace

How an eccentric genius caused a stir over ray guns—30 years after his death.

* * * *

Tech Dissention

THE U.S. GOVERNMENT has never been one to throw away a good idea—after all, it might come in handy someday. In a May 1977 issue of *Aviation Week & Space Technology*, Major General George J. Keegan, retired head of Air Force intelligence, asserted that the Soviet Union was in the final stages of developing a particle beam weapon capable of neutralizing intercontinental ballistic missiles. Yet soon afterward another article appeared in the *Baltimore Sun* entitled, "Moscow Yet to Develop Laser Weapon, [President] Carter Says." It seemed U.S. intelligence needed to get their story straight.

The United States had experimented with developing a particle beam weapon as early as 1958, but they eventually abandoned the device, deeming it unfeasible. Why then, nearly 20 years later, was there a need to deny the existence of such a weapon in the enemy's hands? What possible edge could the Soviets have over the United States that might lend a semblance of credibility to Keegan's claims?

The edge lay in the work of Nikola Tesla, the brilliant scientist and inventor who claimed to have developed the plans for a "teleforce" weapon against which there could be no defense.

Enter Nikola Tesla

Tesla's particular area of specialty was electricity, but unlike Thomas Edison—his contemporary, rival, and onetime employer—Tesla's knowledge of the medium seemed more akin to sorcery than science. He claimed that his discoveries and inventions came in hallucinatory flashes. In his laboratory, he entertained luminaries such as Mark Twain and president Theodore Roosevelt with displays of electrical wonder.

As Tesla grew older his genius led to increasingly bizarre claims and concepts for inventions. Many of these were born from the inventor's hatred of warfare. He was convinced that the future peace and prosperity of mankind lay in the development of weapons so terrible, yet so universally available, that all nations would cease to consider war a means of settling disputes.

Death Rays for Peace

In 1931, Tesla conceived of an invention he later termed a "New Art of Projecting Concentrated Non-Dispersive Energy through Natural Media"—what we now call a particle accelerator. The press, always a big fan of Tesla's dramatics, chose to call the invention a "Death Beam." In a 1934 article in the *New York Times,* Tesla characterized the invention as one that would make war impossible because every nation would possess the ability to send highly concentrated bursts of energy through the air with enough power to "bring down a fleet of 10,000 enemy airplanes at a distance of 250 miles."

Tesla sought funding for the device. With World War II imminent, Tesla sent his proposal to all the Allied nations in the hope that they would cooperate to manufacture the weapon in time. Only the Soviets took an interest and provided Tesla with $25,000 for plans to develop the "teleforce" weapon.

Tesla's increasingly bold assertions led many of his contemporaries to consider him a crackpot. After his death in 1943, it was discovered that certain items, including a notebook, had been removed from his belongings by an unknown agent (suspected of being Russian). The U.S. government quickly moved to confiscate his remaining papers.

From Stalingrad to Star Wars

When reports of Soviet experiments in high-velocity particle beam weapons began to proliferate in the 1970s, the U.S. government definitely paid attention. Many experts already believed that the Soviets, capitalizing on the knowledge gleaned from their dealings with Tesla in the '30s, had successfully tested such a weapon as early as 1968. The furor was sufficient to reinvigorate U.S. interest in particle beam weapon applications, which soon gave rise to the heavily publicized (but ultimately fruitless) Strategic Defense Initiative, or "Star Wars," program of the Reagan administration.

3D Printing

Imagine if doctors could simply print three-dimensional organs for patients in need of a transplant. Or if perfectly fitted, custom-made prosthetics could give people (or even animals) the chance to walk normally. Or how about a custom-printed implant that exactly replicates the body part it is replacing? While it sounds like the work of some sort of futuristic machine, these examples are already in the works, and some have even been accomplished, thanks to 3D printing technology.

* * * *

3D PRINTING, ALSO known as additive manufacturing because it forms objects by adding material together, has been around for decades, but has only recently started earning attention from the general public. Early equipment to enable additive manufacturing was developed in the early 1980s, to make the development of prototypes easier. Up until this point, those who were in the business of manufacturing new products, such as cars, machine parts, or even spacecraft, had to create prototype models by painstakingly creating them from plastic or carved wood—processes which could take days or weeks to complete.

So in the early 80s, the idea of "rapid prototyping" took hold, as a way to complete these models much more quickly. An account of a rapid prototyping system was published in 1981 by Hideo Kodama of the Nagoya Municipal Industrial Research Institute in Japan, which described how three-dimensional plastic models could be constructed with a hardening polymer.

A few years later, in 1984, several researchers, including Chuck Hull, the co-founder of the 3D printer company 3D Systems, invented an imaging process called stereolithography. Stereolithography uses photopolymerization, in which light causes chains of molecules to link and form polymers: If a liquid photopolymer is hit with a beam of ultraviolet light, the portion exposed to the UV light will harden into a plastic. Stereolithography allowed designers to use digital data to then create solid 3D objects out of liquid photopolymer.

In 1992, Hull's company, 3D Systems, created the first stereolithographic apparatus machine, which made it possible to create complex 3D models in a fraction of the time it would normally take. The same year, a startup company called DTM, founded by inventors Carl Deckard and Joe Bearman in Austin, Texas, designed the first selective laser-sintering machine, which used a laser and a powdered nylon/polyamide material to create 3D objects. These first machines were less than perfect, often creating warped models, and they were far too expensive for anyone who wanted one in their home. But the possibilities they hinted at kept engineers and inventors working on them, creating machines that produced more precise objects with cleaner lines.

By the end of the century, medical professionals started to realize the potential of 3D printing machines, and in 1999, the first 3D-printer-assisted organ was implanted in a human. Scientists at Wake Forest Institute for Regenerative Medicine 3D-printed a synthetic model of a human bladder, which was then covered with the cells of a human patient. Once the cells were incubated and functioning together, the organ was implanted in the patient. And since it was made from the patient's own cells, there was little chance of rejection. The next decade saw great progress for 3D printing in the medical field, with researchers printing a miniature kidney prototype, creating custom-made prosthetics, and bioprinting blood vessels using only human cells.

By the mid-2000s, affordable 3D printers became a reality with the RepRap project, started by English engineer Adrian Bowyer. With funding from the British Engineering and Physical Sciences Research Council, RepRap aimed to develop a low-cost 3D printer with an extra-useful property: it can print its own components. So with the right materials, someone could purchase the printer and then print out another printer for a friend. And of course, the printers could also print whatever models consumers dreamed up in their heads, launching the beginning of a new fascination with this technology.

Today, 3D printers are not limited to printing with plastic, with enthusiasts using all sorts of materials to print objects. Fashion designers have been experimenting with 3D printed shoes and clothing, a jeweler can print out a perfect ring with gold or silver, and even chocolate has been used to create 3D-printed (and edible!) objects. But how do 3D printers take someone's imagination and make it reality?

Most 3D printers work very much like an inkjet printer, connected to a computer. A user creates a computer-aided design, or CAD, which is a three-dimensional computer model that can be manipulated on the computer to create just the right design. The 3D printer can then print the object, starting from the bottom and working its way up, layer by layer in a method known as fused depositional modeling. The printer takes the 3D CAD drawing and basically turns it into lots of 2D, cross-sectional drawings, with each 2D section sitting on top of the next one.

Scientists are hopeful that one day 3D printers will be able to print completely functional organs made of biological material. But until then, these machines are still revolutionizing medicine, as well as the automotive and aerospace industries. The future, it seems, will be quite three-dimensional.

Sci-Fi Settings

While science-fiction movies are often set in regions of space entirely alien to our own, a simple fact is consistent with them all—no matter the setting, every one is filmed right here on Earth. The stories and locations may be out of this world, but here are some real places that stood in for a galaxy far, far away.

✻ ✻ ✻ ✻

Ape World

ALTHOUGH CHARLTON HESTON'S character returned home at the end of *Planet of the Apes* (1968), he didn't actually make it to the East Coast of the United States. The crash scene at the beginning of the movie was filmed in Glen Canyon, Utah, while the Statue of Liberty scenes, along with much of the rest of the film, were shot in Malibu, California.

More Monkeys

In Tim Burton's 2001 remake of *Planet of the Apes*, astronaut Mark Wahlberg crash-lands in an unknown time and place that looks remarkably like Hawaii. Actually, it was Hawaii — some of it anyhow. Additional footage was shot at California's Trona Pinnacles and at Lake Powell, which straddles the Utah-Arizona border.

One for the Conspiracy Theorists

Capricorn One (1977) took its story from the conspiracy theory that the 1969 *Apollo* mission to the moon had been faked by NASA and the U.S. government. In Hollywood's version, three astronauts become pawns of the space program when their mission to Mars is canceled due to faulty equipment and lack of funds. They are ordered to fake it in the desert, which was actually Red Rock Canyon State Park in California. This is a clever twist on the use of locations because *Capricorn One*'s fictional American public is fooled into believing they are seeing Mars, just as real-life moviegoers suspend their disbelief regarding locations when they watch sci-fi movies.

Tatooine, Home Planet of Luke Skywalker

Luke Skywalker may have been a poor moisture farmer from a truly backwater planet, but it was actually the upscale Sidi Driss Hotel in Tunisia that served as the backdrop for his boyhood home on the planet Tatooine in the original *Star Wars* (1977), as well as the later prequels. Other North African locations, including Chott el Djerid, were also used, and Death Valley National Park in California doubled for the planet as well.

The Ewoks' Forests of Endor

Whether you love 'em or hate 'em, the Ewoks of Endor did save the day for the Rebel Alliance at the end of *Return of the Jedi* (1983), and the tall trees of the Redwood National and State Parks in northern California served as stand-ins for those of the forest moon.

Chill Out on Ice Planet Hoth

Luke Skywalker and the Rebel Alliance cooled off on the frozen world of Hoth before fighting off an invasion by the Empire's giant AT-AT walkers. The real Hoth locations—Finse and the nearby Hardangerjøkulen, the fifth largest glacier in mainland Norway—were actually part of the Nazi occupation of the Scandinavian nation during World War II.

Dune's Planet Arrakis

The desert world known as Dune in the 1984 film was actually the Samalayuca Dunes in the Mexican state of Chihuahua. Located near the Texas border, they are among the largest and deepest sand dunes in North America, but don't expect to find any giant sand worms there.

Mars Invasion

In one of a few recent films about the colonization of Mars, Val Kilmer led a mission to the *Red Planet* (2000), but instead of training for interplanetary travel, he merely had to travel to Coober Pedy in South Australia, while Gary Sinise's *Mission to Mars* (2000) took its cast on a journey to Jordan.

A World of Aliens

James Cameron's 1986 blockbuster *Aliens* was set on a planet known as LV-426, but most of the film was actually shot at Pinewood Studios in Buckinghamshire, England. The climactic scenes at the atmosphere-processing station were filmed in London at the Acton Lane Power Station. No aliens were harmed during the production of the film.

Total Arnold

A favorite among Arnold Schwarzenegger fans, *Total Recall* (1990) sends Arnold to Mars, but California's future "Governator" didn't have to venture too far from home—most of the film was shot in Mexico.

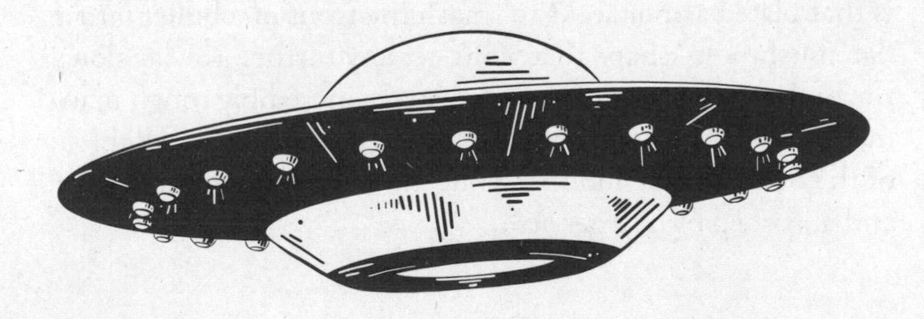

A Tale of Two Armors

Spider silk may not be as strong as steel, but it weighs a fraction as much and works in much different applications. Could spider silk be used to make body armor? Amazingly, could corn starch?!

* * * *

IT'S HARD TO browse the cable guide or Netflix without seeing a handful of crime shows, if not more. Americans can't get enough of watching actors pretend to be law-enforcement agents, detectives, prosecutors, or military versions of any of those jobs. Terms like "armor-piercing round" are floating around in the vernacular now, but what does it really take to pierce the armor worn by law enforcement or private security?

Arming and Armoring

Sorting out the language of the ballistic or "bulletproof" industry and its armor-piercing counterparts is a slippery task. The language isn't standardized, and marketing has affected how makers describe their products. Traditional armor-piercing rounds are meant for use against armored vehicles or ships. Bullets designed for use in rifles, and therefore to shoot individuals, must be manufactured with super rigid materials like tungsten carbide in order to be able to penetrate body armor.

The goal with armor on a ship or person is to slow down a projectile enough to drastically reduce how much harm it does. We might picture bullets bouncing off of armor, but the truth is that plated armor seeks to smash the front of a bullet into a flat "mushroom" shape that won't get any further. Kevlar slows the bullet with layers of densely woven, incredibly tough man-made fibers. The person wearing the armor feels the full force of the bullet's mass and speed but the force is distributed over and absorbed by a larger area.

Sounds great, right? It is. But the wearer can also expect *some* injury, ranging from a severe bruise to internal organ damage. Someone who's shot while wearing a "bulletproof" vest can very easily need to go to the hospital afterward, especially if multiple shots were fired. Armor can warp after impact, even Kevlar, and that warping affects how the armor absorbs subsequent shots.

Bulletproof vests save countless lives with technology that seems like magic to the average person, and scientists spend entire careers experimenting with ways to improve and update this technology. Potential for warping is a great place to begin those experiments.

Mysteries of Oobleck

If you have, know, teach, or exist near one or more children, you almost definitely know what oobleck is, if not by name alone. Oobleck is the term for a mixture of corn starch and water that reacts as a solid sometimes and a liquid sometimes. You can "walk on water" across a kiddie pool of oobleck if you move quickly. It's more like a run on water—check it out on YouTube. But if you stand on the oobleck, you'll sink in, and the oobleck will try to hold you down.

These special qualities make oobleck one of very few known "non-Newtonian fluids," meaning it literally defies the classic laws of physics as revealed in the work of Sir Isaac Newton. And in 2014, an Air Force Academy cadet made the connection between oobleck and the liquid binders or epoxies used in body armor. She wanted to replace traditional gluey fillers that dried into a solid with a fluid that stayed flexible and loose.

The cadet and her advisors tried layers of oobleck sandwiched between layers of Kevlar. They tried different setups and numbers of layers until they found an arrangement that can, in fact, stop bullets. Kevlar is a trademarked product, but oobleck just needs a box of cornstarch. It's flexible until impact and then again immediately afterward. It could be lighter than traditional armor and cost a lot less. The possibilities are endless.

Arach-ing Our Brains

Spiders produce strong natural proteins in the form of silk strands. Maybe a scientist started to wonder about the tensile strength of spider silk after walking through a web and trying to remove all traces of it for the rest of an entire day. This strong, durable, biological product requires no unusual ingredients or energy sources. In strength, flexibility, and other attributes, it could rival Kevlar, without the hangups of patent law or petrochemicals.

The U.S. military loves the idea of spider silk as a part of the armor kit soldiers wear. Its flexibility compared to Kevlar means it could be used to build comfortable underlayers that still offer protection, the same way consumer "worm silk" makes superlative long underwear. Scientists are working with the military to genetically engineer stronger, more versatile, or simply more abundant spider silk. They've bred silkworms that can make spider silk, because worms are much faster at silk production than spiders are.

Scaling to a military application is still a huge problem, literally. The U.S. Department of Defense employs over 2 million active duty or reserve military personnel. Can there ever be enough spider silk to supply these soldiers with even their smallest piece of armor? Will spider silk be used as one component in a much larger application?

So far, the investment in spider silk research is tiny compared to almost any other Department of Defense project. And the military has a terrific track record with technologies that trickle down to the civilian level: GPS navigation, the jet engine, walkie talkies, duct tape, and the Internet were all originally invented for military use.

The Maglev

Short for magnetic levitation, Maglev trains have gone from the stuff of science fiction to the stuff of reality.

* * * *

SOME THREE OR four decades ago, television shows and movies—generally in the science fiction genre—were depicting floating, high-speed trains that were going to render other forms of transportation obsolete, or at least give them a major scare. As with many advancing technologies, sometimes life doesn't keep up with art when it comes to timelines. However, high-speed, magnetic trains have arrived, and it may be a matter of time before they become accessible to the masses.

The principles of magnetism are what drive these frictionless trains. Opposite poles attract. Like poles repel. The trains can work using either of those two dynamics. They do require a large electrical power source to produce magnetic energy, metal coils lining a track, and magnets attached to the underside of a train to guide it along its way. In addition to amazing speed potential that comes with a transportation mode that does not have to deal with friction, another great advantage of a Maglev train is its freedom from engine reliance. No fossil fuels are burned. No pollution is spewed into the air.

James Powell, while stuck in traffic on a New York City bridge in 1968, came up with the concept of magnetically levitated transportation. The Brookhaven National Laboratory researcher, considered the father of Maglev transportation, pulled in colleague Gordon Danby and drew up a concept using static magnets mounted on a moving vehicle that could be levitated by superconducting magnets for frictionless travel. Such concepts had been sketched out long before, dating back to the early 20th century, but Powell and Danby were the first to acquire a patent for the technology.

While the concept was an awe-inspiring one, putting it into practice would be the challenging part. "A Maglev train car is just a box with magnets on the four corners," says Jesse Powell, son of the Maglev inventor who went into business with his father. That's making it sound easy, of course, but it is also completely true.

Superconducting magnets, when cooled to less than 450 degrees (Fahrenheit) below zero, can generate magnetic fields 10 times more powerful than normal magnets can. As it turns out, they generate enough to lift and propel a train. Conductive materials such as aluminum are used to make the loops that are set at regular intervals into the concrete walls of the guideway. When a magnetic field moves past the loop, an electric current is created that in turn generates another magnetic field. Typically, two types of loops are set into the guideway—one to keep the train hovering a few inches above the "track" and another to keep the train in place horizontally. When a train begins to stray from the optimal position, the resistance is increased to keep it in line.

A third loop then propels the train forward via alternating current. Electrifying these propulsion loops can generate fields that pull the train from the front (opposite poles attract) or push it from the back (like poles repel). And just like that, magnetic trains can attain speeds up to 375 miles per hour on a smooth ride that features no friction at all. Controlling the speed and positioning with magnets on a single track produces many benefits. Magnetic fields prevent derailment. And because trains on the same route are powered to run at the same speed, the risk of a collision is minimal. "With Maglev, there is no driver," Jesse Powell said. "The vehicles have to move where the network sends them. That's basic physics. So now that we have computer algorithms for routing things very efficiently, we could change the scheduling of the entire network on the fly. It leads to a much more flexible transportation system."

While the elder Powell concocted the idea in the United States, it was Great Britain that ran the first one: the Birmingham airport AirLink shuttle that debuted in 1984. While it was inexpensive to get started and popular enough with passengers, AirLink was expensive to maintain. It ran at slow speed between the airport terminal of Birmingham International Airport and the nearby Birmingham International railway Station through 1995.

The next big debut came in the form of Germany's M-Bahn in the late 1980s. It opened to passenger traffic in 1989 in West Berlin, connecting just three stations over the course of a mile. The line was closed the following year after the fall of the Berlin Wall. Its manufacturer, TransRapid, went on to launch the next big project in Shanghai in 2002. The fastest high-speed Maglev train in the world, the Shanghai TransRapid can zip along at 268 miles per hour. A 19-mile trip from the airport to Shanghai's business district that takes an hour by car requires just eight minutes on the train!

A slower-speed Maglev also runs in China, which has embraced the technology and has said Maglev transportation could be available in as many as 12 cities by 2020.

In the United States, leaders at Boeing and NASA came together to form Hyperloop Transportation Technologies and announced in 2016 it was pursuing plans to connect San Francisco and Los Angeles with a Maglev train that would aim for speeds of a whopping 750 miles per hour. If that were to come to fruition, a six-hour drive (on a good traffic day) or all-day train trip from San Francisco to L.A. might require less than an hour.

Building a Structure of Science

"Science is built up of facts, as a house is built of stones; but an accumulation of facts is no more a science than a heap of stones is a house," noted Henri Poincare. One of the great polymaths and scientific thinkers of the 19th century Poincare's contributions to modern mathematics and physics are astonishing, and he single-handedly laid the groundwork for many of the discoveries that were made after his death.

* * * *

BORN IN NANCY, France in 1862, Poincare was an avid reader, and began reading popular science books at a young age before moving to more advanced texts. He had an incredible memory, and preferred linking the ideas he was reading in his head rather than learning by rote memorization. This skill would prove useful when he attended university lectures, as his poor eyesight kept him from being able to see what the professors were writing on the blackboard.

Upon graduation, he became a professor at the University of Paris, where he served for many years. He held positions in multiple departments, including astronomy, physics, and mathematics. In 1880 he made the discovery that elliptic and automorphic functions were related to the same set of algebraic equations. He continued to do fundamental work through the 1880s in celestial mechanics and physics.

Poincare also helped to discover the theory of special relativity with Hendrik Lorentz and Albert Einstein. Einstein would later say that Poincare was one of the pioneers of the theory of relativity. In 1905 Poincare proposed the theory that gravitational waves emanated from bodies with mass, and propagated outward at the speed of light; this was later also predicted by Einstein based on general relativity.

While working in the field of mining he was also making breakthrough discoveries in science and mathematics; in 1893 he became a chief engineer in the French Mining Corps, and was promoted to Inspector in 1910. He also worked to coordinate worldwide time in the French Bureau of Longitudes. Poincare's contributions to modern science and mathematics were numerous and varied. He contributed to the special theory of relativity and quantum mechanics, Algebraic topology and geometry, electromagnetism, differential equations, to name just a few. He died on July 12, 1912, at the age of 58.

The Three-Body Problem

Predicting the motions of a group of orbiting celestial bodies has been a problem since Sir Isaac Newton first published the *Principia*. The problem concerns how to predict the individual motions of three or more bodies that are acting on one another gravitationally. Solving the problem would allow physicists to better understand the stability of a solar system. In 1887, the King of Sweden established a prize for anyone who could solve the problem. Although Poincare was unable to solve the problem, he was still given the award for his work, which had considerably advanced understanding of celestial mechanics.

The Poincare Conjecture

Poincare is perhaps most famous for the conjecture named for him. It proposes that any circle made on a surface can be contracted to a single point: for example, a rubber band wrapped around the sphere can be slid down to a single point. This property is unique in 3D space to the sphere; it is not true for a disk (which has an edge) or a donut-shaped object, for example.

Poincare asked whether the same holds true for a sphere in 4D space. The conjecture directed much exploration in mathematics, and in particular topology, which studies the properties of continuous space. The conjecture, proposed in 1904, was not solved until 2003, when Grigori Perelman showed that the same is true for a sphere in 4-dimensional space.

R U a Robot?

The word robot *was coined in 1920 in the play* R.U.R. (Rossum's Universal Robots)*, written by Czech writer Karel Čapek. It comes from the Czech* robota*, meaning "forced labor" or, figuratively, "drudgery, hard work," referring to the central European system of serfdom that existed before 1848. Čapek's dystopian play, which premiered on Broadway in October 1922, featured a race of enslaved "Robots," humanlike creatures artificially assembled from organic materials.*

❋ ❋ ❋ ❋

FAR FROM THE images of mechanized metal that the word *robot* typically conjures today, these flesh-and-blood beings more closely resembled what we would now call *androids*. But, perhaps because of the relevance of Čapek's theme—the play came about in a time when industrialization and factory labor were combining with rising totalitarianism (the Bolsheviks revolted in Russia in 1917, and Mussolini came to power in Italy in 1922)—the word *robot* caught on, even as the popular conception of one diverged from Čapek's dramatic vision.

As early as 1923, just a year after the American debut of *R.U.R.*, *robot* had acquired a figurative sense, that of a person who acts mechanically or with no evidence of emotion. Derivations of the word followed shortly after. Famed science fiction writer Isaac Asimov—who, incidentally, referred to Čapek's play as "a terribly bad one"—coined the adjective *robotic* in 1942, and by 1945 the prefix *robo-* was in use.

While Čapek—or, more accurately, his brother Josef, to whom he gave coinage credit—gave us the word, he did not invent the idea. The idea of humanlike automatons had been around for centuries. The word *android*, denoting an automaton that resembles a human, dates to the 1700s. And the Czech origin of *robot* recalls the Jewish tradition of the *golem*, a robotlike being made from mud. *Golem* is Hebrew for "shapeless mass."

Čapek's robots were humanlike androids, but the word, especially in its actual application to real-world automatons, is often used for machines that do not resemble humans at all. This shift is not recent, and such usage has been around almost as long as the word has been in use in English. In 1930, for example, automated traffic signals in London were dubbed *robots*. This particular use of the word, which never caught on in the United States (although the *Oxford English Dictionary* cites an early example from Canada), has fallen out of use in British English but survives in South Africa.

Self-driving Cars

Almost as quickly as it garnered worldwide attention over the previous decade, self-driving car technology fell far behind schedule after 2015. That year, Tesla head Elon Musk predicted a fully autonomous vehicle (AV) by 2018. Google anticipated having one on the road, too. Those were two of the countless predictions coming from the car manufacturers and tech companies across the globe that proved to be at least a few years—and perhaps more—too optimistic.

* * * *

THE REASONS FOR the optimism, however, remain plain to see. Waymo, which began in 2009 as Google's self-driving car project, has been considered the leader in AV technology over the last several years. In 2018, it invited residents of Phoenix, Arizona to become riders in the first public trial of its AVs. Putting driverless cars on the roads of suburban Phoenix was a giant leap not only for Google and Waymo, which touts that it drives more than 25,000 autonomous miles each day, but also for the industry.

Other companies continued to push the boundaries, too. GM said in 2018 that it would be sending a fully autonomous car, with no steering wheel or driver intervention, into production in 2019. Tesla says it now equips every model it produces with the hardware required for full self-driving capability at a level "substantially greater" than that of an actual human behind the wheel. So as the rubber prepares to meet the road, so to speak, it's interesting to look back and consider what the earliest AV visionaries saw as they explored the concept.

There Goes Nobody

In 1928, a *Milwaukee Sentinel* headline screamed that a "Phantom Auto" would tour the city. "Driverless," the article noted, "it will start its own motor, throw in its clutch, twist its steering wheel, toot its horn, and it may even 'sass' the policeman at the corner." The article went on to explain that the "ghost vehicle" would be steered by a remote-control device.

At the 1939 World's Fair, the General Motors Company unveiled an exhibit that pointed to a time when there would be "abundant sunshine, fresh clean air [and] fine green parkways" upon which cars would drive themselves. Almost 20 years later, in 1958, GM tested a 1958 Chevrolet with pick-up coils on the front that could sense the alternating current of a wire lining the road and adjust the steering wheel accordingly. It worked on a two-lane test track but never made it into production.

What is considered the world's first truly autonomous car, with the ability to process images in its path, was developed in 1977 by Japan's Tsukuba Mechanical Engineering Laboratory. The vehicle carried two cameras capable of tracking white markers in the road. It used analog computer technology—the best of its day—to interpret the signals, make necessary adjustments and keep moving at speeds up to 19 miles per hour. The car's operation was supported by an elevated rail.

A series of robotic car projects in the 1980s earned German aerospace engineer Ernst Dickmanns the title "pioneer of the autonomous car." His VaMoRs, in 1987, used two cameras, eight microprocessors and other sensory and software equipment to drive more than 90 kilometers per hour (56 mph) over about 20 kilometers (12.5 miles).

Dickmanns' follow-up project, in 1994, saw VaMP make considerable improvements over its predecessor. The car could process images at a range of 100 meters—more than the length of a football field—while recognizing road markings, its position in its lane, and the presence of other vehicles sharing the road. It topped 80 miles per hour in simulated traffic on a test drive in Paris, even recognizing when it could change lanes safely. The next year, Dickmanns and his team "drove" a Mercedes S-Class from Munich to Denmark at speeds reaching 112 miles an hour, with automated driving handling some 95 percent of the trip.

Such successful experiments turned the attention away from systems that required buried cables or wires in the road to vehicles that had "vision" and the ability to detect objects around them. In 1991, the United States government steered $650 million toward research on the National Automated Highway System (NAHS). The United States military also provided funding through the Defense Advanced Research Projects Agency (DARPA).

A combined effort of the NAHS and DARPA produced a milestone trek in 1995 for Navlab, a series of autonomous and semi-autonomous vehicles developed at Carnegie Mellon University. A team took an automated car 2,848 miles across America, with 98 percent of the trip handled autonomously. That record stood a whopping 20 years until Delphi went 3,400 miles with an autonomous Audi in 2015.

The Society of Automotive Engineers, now SAE International, released a scale to describe five different levels of automation as engineers and automakers race to be first in class and best in class. Their scale was adopted by the National Highway Traffic Safety Administration, which oversees auto safety in the United States. With that kind of universal acceptance, the categories have become part of industry lexicon and a great way to track progress toward full automation.

Level 1 is Driver Assistance. In this first level, the driver maintains responsibility for safety and operation of the vehicle, but the car can take over at least one aspect of the operation like speed or steering. Many new cars offer this level of automation today. For example, cars with adaptive cruise control can adjust their speed based on the movement of traffic in the road ahead.

Level 2 is Partial Automation. Some of today's cars come with this level of automation. Tesla has Autopilot, Mercedes-Benz offers Drive Pilot, BMW has Active Driving Assistant Plus and Volvo touts Pilot Assist. In each case, the driver is responsible for safe operation but the car can take over, in certain conditions, functions like braking, steering, and acceleration.

Level 3 is Conditional Automation. This is the point where a vehicle is able to drive itself. The human, however, is expected to pay full attention and can take over at any time. The car would notify its human driver when intervention is necessary. Because early studies have shown that people tend to trust the technology and stop paying attention, this level seems to be a plateau in AV development.

Level 4 is High Automation. This is the level many manufacturers predict will be hitting the roads in high volume in the coming years. The computer driver takes charge in certain situations—highway driving or pre-determined routes, for example. The human driver needs to pay attention only when the car gives that notification. The car comes equipped with the technology to remain operational even when the technology fails. The vehicle will also pull over and shut down if it determines it's not safe to drive.

Level 5 is Full Automation. As the name implies, the vehicle does it all. No human interaction is expected. There is no need for a steering wheel or gas/brake pedals. This is the gold standard to which the industry aspires. It would change vehicle travel for everyone and be especially welcomed by those who are unable to drive.

AI Winter

Given that the first autonomous car took to the road more than 40 years ago and the fact huge technology innovations have come about in just the last few turns of the calendar, why is it that the brightest engineering minds in the world have not been able to fill the streets with self-driving vehicles? Artificial intelligence (AI) experts have warned that it might be a long wait before autonomous cars can "learn" to avoid the kinds of accidents that happen—so often unexpectedly—in the real world. Some have dubbed the delay the "AI Winter."

The strides made by artificial intelligence in the technology world have reshaped many parts of our lives. From news delivery to shopping personalization to weather prediction and language translation, AI has changed the world. Driving a car, though, has always required not only human motor function, but also human thought and judgement. How does one train a computer in a car to recognize and possibly handle every situation that can arise when driving in traffic?

While autonomous vehicles have the potential to cut way down on the millions of accidents where human error is the culprit, could they also lead to more accidents in situations where a human might have used his or her instincts to navigate safely? "Driverless cars are like a scientific experiment where we don't know the answer," wrote New York University scientist and AI expert Gary Marcus. General Motors President Dan Ammann called developing AVs that can navigate their way through urban traffic "the engineering challenge of our generation."

Massive amounts of data can be programmed into vehicle computers and countless situations can be covered in virtual environments. Until it's road tested, though, all the programming in the world is not likely to convince a parent to put his or her child's fate in the hands of a robotic car. Not helping the quest to fill freeways with autonomous vehicles was a series of tragic mishaps involving some of the earliest test cars.

In March 2018, an AV operated by Uber was going 40 mph in Tempe, Arizona, when it fatally hit a 49-year-old pedestrian crossing the street. It was dark, and apparently the vehicle's perception system got confused by the bicycle she was walking across the road. In another fatal accident in 2016, a Model S Tesla rear-ended a tractor-trailer without slowing down, apparently mixed up by the height of the vehicle and the reflection of the sun on the white truck. Another mishap saw a Tesla Model X in California turn toward a barrier and speed up before impact for unknown reasons.

Uber and Waymo are among the companies that equip their vehicles with light detection and ranging sensors (lidar) that use rapid light pulses to track the surroundings of the vehicle at 360 degrees. Lidar, however, has had problems in heavy rain and snow and struggles to detect the small plastic markers sometimes used to divide lanes in California and other states. Cameras, radar and GPS systems also help self-driving cars detect objects all around them. However, how does a vehicle replicate the eye contact that frequently occurs from driver to driver or driver to pedestrian? How does it listen to the direction of a traffic cop or crossing guard? "We're not even remotely close to being able to be truly autonomous in diverse conditions," explained Austin Russell, CEO of Luminar, the manufacturer of lidar.

That self-driving cars will be showing up en masse is not a matter of "if," but of "when." Money and brain power are being poured into the endeavor, and it makes sense. According to some research, 94 percent of automobile accidents are the result of human error. One hundred percent of the time, autonomous vehicles will not be drinking, texting, on drugs, asleep, or distracted while driving.

There are also ethical questions to be considered. An autonomous vehicle in an emergency situation cannot decide—at least not like a human—whether to veer into something like a pole to avoid an even worse scenario like hitting a group of people in a quick-response, emergency situation. Some would argue that there is no substitute for experience behind the wheel.

Close to two million Americans drive heavy trucks for a living. More than 1.5 million drive buses, taxis and delivery vehicles. Talk of AVs taking over U.S. highways is sure to have many Americans concerned about their job security. Uber has tried to alleviate those concerns by emphasizing that human drivers will be needed onboard the 500,000 to 1.5 million self-driving trucks that could be hitting the U.S. roadways by 2028.

Whether that becomes the case is anyone's guess. Perhaps the last sprint to mass use of AVs will indeed happen in the next few years. Or perhaps all the questions, as they have over the last few years, will lead to as many more questions as answers.

ARPANET: Grandfather of the Internet

Rumors abound that ARPANET was designed as a communications network that would withstand nuclear attacks. That simply isn't true. The creators of ARPANET weren't seeking invulnerability, but reliability—in order to fulfill one man's vision of an "inter-galactic" computer network.

* * * *

ON OCTOBER 4, 1957, the Soviet Union launched the world's first artificial satellite, *Sputnik I*, into space. It was a clear message that Russian technology was more advanced than American technology. To amend this oversight, the Advanced Research Projects Agency (ARPA) was formed to fund technical research. The United States already had a substantial financial investment in computer tech—the initial purpose of ARPA was to figure out the best way to put that to use. Though it fell under the auspices of the U.S. Department of Defense (and was renamed DARPA), the research was never intended to be used solely for military purposes. Instead, the agency's purpose was to develop technology that would benefit civilization and the world in general.

Not Connected to Other Galaxies—Yet

The expert chosen to head ARPA's initial effort was Joseph "Lick" Licklider, a leading computer scientist. Lick had a vision of a worldwide communications network connected by computers, which he referred to as the "inter-galactic computer network." Lick departed ARPA in 1965, before his plan could be implemented, but he left a lasting impression on his successor, Bob Taylor.

Taylor selected a new leader for the system design team that would make Lick's vision a reality: Dr. Lawrence "Larry" G. Roberts, an MIT researcher. He became one of the four people most closely associated with the birth of the Internet. (The other three are Vinton Cerf, Leonard Kleinrock, and Robert Kahn.) Roberts had gained experience in computer linking while at MIT, having linked computers using the old-fashioned telephone method of circuit switching.

The concept of packet switching was at first controversial, but it proved to be one of the key factors in linking multiple computers to form the network. The other important technical achievement was the use of small computers, then known as interface message processors (IMPs), to store and handle the data packets.

By 1968, the concept for ARPANET was in place, and invitations to bid on the project were sent to 140 institutions; only 12 actually replied. The others apparently believed the concept to be impractical, even bizarre, and never bothered to bid. In the end, BBN Technologies—Licklider's former employer—got the nod.

A Hesitant Start

The first piece went to UCLA, thanks to the reputation of Professor Kleinrock, an expert in computer statistical analysis and measurement. The first IMP link was with Stanford Research Institute (SRI). The first message was sent on October 29, 1969, and was supervised by Kleinrock—it was an omen of things to come. The message was supposed to be "login," but after two letters, the system crashed, and only "lo" was sent.

About an hour later, the system was up and running again, and the full message was transmitted. By December 5, 1969, four IMPs were linked: UCLA, SRI, University of California at Santa Barbara, and the University of Utah. These IMP sites were chosen on the basis of their ability to research and implement the protocol that would allow for the continued growth of ARPANET.

ARPANET was no longer just a vision—it was a reality. The growth of ARPANET during the 1970s was phenomenal, as newer and better protocols were designed. In 1971, e-mail was born; in 1972, telnet was developed; and in 1973, file transfer protocol came into play.

By 1986, ARPANET had serious competition from the National Science Foundation Network (NSFNET), which became the true backbone of the Internet. ARPANET closed up shop in 1990. In 1991, NSFNET opened to the public, introducing the Internet we know today. Within four years, more than 50 million people had traveled the information superhighway. As of March 2008, worldwide Internet usage stood at 1.4 billion—and that's only the beginning.

Taser: From Children's Book Concept to Riot Policing Tool

Like a .357 Magnum, the Taser makes troublesome suspects less troublesome. Unlike the .357, the tased suspect generally survives to stand trial, and the police save a bundle on coroner costs.

* * * *

Tom Swift

You've probably heard of the Hardy Boys and Nancy Drew. But unless you're a baby boomer or older, you may never have heard of Tom Swift books, which belonged to the same "teen adventure" genre. Tom, the precocious protagonist, is a young inventor who resolves crises and foils wickedness. One book in the series, called *Tom Swift and His Electric Rifle* (1911), has quite a stimulating legacy.

In 1967, NASA researcher Jack Cover, who grew up on Tom Swift, realized that he could actually make some of the gee-whiz gadgetry from the series. In 1974, he finished designing an electricity weapon he named the "Thomas A. Swift Electric Rifle," or TASER. (In so doing, he departed from canon. Tom Swift never had a middle initial, but Cover inserted the "A" to make the acronym easier on the tongue.)

How It Worked

Cover's first "electric rifle," the Taser TF-76, used a small gunpowder charge to fire two barbed darts up to 15 feet. Thin wires conducted electricity from the weapon's battery to the target, causing great pain and brief paralysis with little risk of death—except in the young, elderly, or frail. That was okay, since the police rarely felt compelled to take down children or senior citizens.

The police saw potential in the Taser. The TF-76 showed great promise as a nonlethal wingnut takedown tool.

Federal Shocker

Never underestimate the creativity-squelching power of government. The Bureau of Alcohol, Tobacco, and Firearms (BATF) wondered: How do we classify this thing? It's not really a pistol or a rifle. It uses gunpowder — *Aha!* The BATF grouped the TF-76 with sawed-off shotguns: illegal for most to acquire or possess. A .44 Magnum? Carry it on your hip if you like. An electric stunner that took neither blood nor life? A felony to possess, much less use. This BATF ruling zapped Taser Systems (Cover's new company) right out of business.

Second and Third Volleys

Taser Systems became Tasertron, limping along on sales to police. In the 1990s, an idealist named Rick Smith wanted to popularize nonlethal weapons. He licensed the technology from Cover, and they began changing the weapon. To deal with the BATF's buzzkill, Smith and Cover designed a Taser dart propelled by compressed air. They also loaded each cartridge with paper and confetti with a serial number. If bad guys misused a Taser, they wouldn't be able to eradicate the evidence.

To Tase or Not to Tase: That Is the Question . . .

Modern Tasers reflect the benefits of experience. In 1991, an LAPD Taser failed to subdue a defiant motorist named Rodney King. The events that followed (including the cops beating King with clubs) put the Taser on the public's radar as something unreliable. This was not offset (on the contrary, it was compounded) by deaths from tasing. The public might justly ask: "Does this thing really work? Does it work too well?" One fact isn't in question. A nightstick blow to the head or a 9mm police bullet are both deadlier than a Taser. As a result, the debate revolves more around police officers' over-willingness to tase rather than whether or not police should carry Tasers in the first place.

In 2007, the United Nations ruled that a Taser could be considered an instrument of torture.

✳ Chapter 2

Scientists Acting Strangely

Following Freud

As it turns out, the "Father of Psychoanalysis" was a case study of neurotic behavior himself.

✳ ✳ ✳ ✳

A NEUROLOGIST AND PSYCHIATRIST, Sigmund Freud's research on human behavior left a lasting impact on the field of psychology. Freud himself was not without issues. He was a heavy smoker—smoking as many as 20 cigars a day for most of his life—and as a result, endured more than 30 operations for mouth cancer. In the 1880s, he conducted extensive research on cocaine, advocating use of the drug as a cure for a number of ills, including depression. Reports indicate that Freud was probably addicted to cocaine for several years during this time. And a friend for whom he prescribed cocaine was later diagnosed with "cocaine psychosis" and subsequently died in what is referred to by biographers as the "cocaine incident."

Freud suffered psychosomatic disorders and phobias, including agoraphobia (a fear of crowded spaces) and a fear of dying. Though his Theory of Sexuality was being widely denounced as a threat to morality, he decided that sexual activity was incompatible with accomplishing great work and stopped having sexual relations with his wife. Yet he is thought to have had a long affair with his wife's sister, Minna Bernays, who lived with the couple. Freud denied these persistent rumors, but in

2006, a German researcher uncovered a century-old guest book at a Swiss hotel in which Freud registered himself and Minna as "Dr. Freud and wife."

Freud fled his native Austria after the Nazi Anschluss in 1938 and spent his last year of life in London. Dying from mouth cancer, in September 1939, he convinced his doctor to help him commit suicide with injections of morphine.

Modern Rocketry and the Moonchild

American rocket scientist Jack Parsons wielded the power to bring about mass devastation—though not in the way he anticipated.

* * * *

A Flight of Fancy

IN 1936, 22-YEAR-OLD Jack Parsons was the epitome of "tall, dark, and handsome." Parsons's natural charisma and fierce intellect, however, did not prevent him from leaving his chemistry studies at Caltech, where he drew criticism for his abiding interest in rocketry, specifically the quest to develop a workable rocket fuel. In Depression-era America, rocketry was seen as nothing more than a flight of fancy. Exiled from campus for their explosive experiments, Parsons and his fellow enthusiasts trekked out to the isolated Arroyo Seco Canyon. On October 31, they conducted a test that, instead of resulting in an explosion, led to a successful launch. Soon thereafter, they formed the Jet Propulsion Laboratory (JPL).

So Far, So . . . Bad

With the advent of the Second World War, Parsons's talents were suddenly very much in demand. Thanks to Parsons's intuitive understanding of chemistry, the company was able to produce a working jet-assisted take-off (JATO) rocket for aircraft. The military found the JATO particularly useful on the short runways that dotted the South Pacific islands.

Parsons soon started another successful enterprise—AeroJet Corporation. The founding of AeroJet seemed like just another chapter in Parsons's successful life. But then everything seemed to take a much darker turn. Jack sold his AeroJet shares to finance his other abiding interest—the occult.

British writer, hedonist, and self-proclaimed master of the occult Aleister Crowley led a religious organization called Ordo Templi Orientis, through which he spread his mystical life philosophy of Thelema, dictating "Do what thou wilt." This pagan, power-based religion of the individual appealed to Parsons, who had regularly invoked the Greek god Pan when conducting rocket tests for JPL. Jack joined the West Coast chapter in 1941 and was quickly recognized as a likely successor to Crowley himself. A year later, Parsons was made the leader of the West Coast church and began conducting "sex magick" rituals intended to bring about the end of the world.

Then things got really weird.

Enter (and Exit) L. Ron Hubbard

Parsons's first wife, Helen Parsons Smith, left him shortly after he joined Crowley's church. And no wonder—Parsons and his new friend, fellow church member L. Ron Hubbard, were busy conducting lewd rituals intended to call forth an "elemental" partner to sire a "moonchild," who was to be the harbinger of the apocalypse. After a particularly vigorous ritual, a young woman knocked on the door. Redheaded (a prerequisite for the elemental), she was willing to carry Parsons's moonchild. Despite many attempts at conception, it was to no avail. Finally, Hubbard absconded with all of Parsons's money, as well as his girlfriend. Soon thereafter, Hubbard used the money to finance his first book, and lo, Scientology was born.

But, Getting Back to Jack

Parsons's life began to unravel. The moonchild didn't materialize, his friend had stolen his money, and he was under investigation by the FBI. For a time he worked at a gas station; later he remarried and began making special effects for movies. He was reportedly at work on a new kind of artificial fog when an explosion in his apartment took his life on June 17, 1952.

By then, the Cold War was new and fear was rampant. Soon both the Soviet Union and the United States developed intercontinental ballistic missiles, and the threat of global nuclear apocalypse grew increasingly real. Parsons may not have conceived a magical moonchild to bring about the apocalypse, but he had created the technology needed to hurl atomic weapons through space—enabling a possible apocalypse in itself.

Beatrix Potter's Scientific Side

Although the name Beatrix Potter brings to mind bunnies and briar patches more than it does algae and Agaricineae, *this iconic children's author was much more than just the creator of Flopsy, Mopsy, Cottontail, and Peter Rabbit.*

⁕ ⁕ ⁕ ⁕

A Victorian Upbringing

CALL IT A case of living in the wrong place at the wrong time. If not for the strict Victorian society of her upbringing, Beatrix Potter might have been too busy conducting breakthrough scientific research to introduce readers worldwide to the tale of Peter Rabbit. Born in 1866 to a wealthy family in London, her parents left her upbringing to a string of tutors and governesses. In fact, Potter was so sharp that most of her teachers could not keep up with her. At one point, she learned six of Shakespeare's plays by heart in less than a year.

Potter the Scientific Illustrator

The Victorian era was a time when it was not considered necessary to send girls away for a proper education. Instead, Potter's teachers were to instruct her in "womanly" subjects such as French and drawing. Certainly, the study of mushrooms and lichen was not part of Potter's formal curriculum.

But Potter had been fascinated by nature since childhood. She kept all sorts of animals as pets, and she drew beautiful and accurate pictures of them. This interest continued to manifest itself in her botanical drawings, particularly of mushrooms. In her late 20s, unmarried and increasingly at odds with her parents who expected their daughter to take on the domestic responsibilities of their household, Potter found a much-needed escape in her drawings and nature studies.

Potter began visiting the Royal Botanical Gardens at Kew to learn more about the fungi that were being researched there. In a journal (in which she wrote in code to keep her studies a secret from her mother), Potter expressed her excitement over recent scientific findings such as that of scientist Louis Pasteur. She grew her own spores, observed them under a microscope, and carefully recorded her findings. Entirely self-taught, Potter began developing theories of her own.

A Dream Denied

At the time, the British scientific community deemed the notion that lichen could actually be made up of two organisms—an algae and a fungus—absurd. However, Potter's readings of recent findings from Continental Europe contradicted this. In her studies, she had observed firsthand how the algae and fungus found in lichen did, in fact, need each other to survive. She wrote her findings in an 1897 paper, *On the Germination of the Spores of Agaricineae*. Unfortunately, because she was a woman, she was not allowed to be present when it was read at a meeting of a society of naturalists. Her work had little impact on the biologists of her time; decades later, however, her findings would be accepted as pioneering work in the understanding of symbiotic relationships in biology.

Potter also tried to gain acceptance as a student at the Kew Gardens in order to formalize her research and gain credentials as a scientist. When she went to meet the director, however, botanical drawings in hand, it was immediately clear to her that he did not take her application seriously. The director would not even look at her drawings. The visit was a great humiliation for the shy Potter; afterward, she gave up what she called "grown-up science" altogether.

The Birth of a Bunny

The same year her paper was rejected, Potter began throwing her energies into the writing and illustration of children's books. With her 25 little books, she would eventually achieve what few women of her era managed: financial and intellectual independence. She may have given up her formal studies, but she never stopped drawing with scientific accuracy. After all, even if her rabbits are wearing sweaters, the trees and flowers in the background are drawn with a biologist's eye for detail.

How to Electrocute an Elephant

A circus elephant named Topsy became the pawn in a feud between two of the most prolific inventors of the era. The result was the first public electrocution and an act of animal cruelty that came close to sullying Thomas Edison's reputation.

* * * *

IN 1903, TWO important inventors were engaged in brisk competition to sell the merits of their respective forms of electricity. Nikola Tesla was the inventor of alternating current (AC), which varies in direction and magnitude as it flows. Thomas Edison, inventor of the version that maintains a constant direction, called direct current (DC), vigorously advocated his invention as the sensible and safe choice for illumination. He went to great lengths to discredit AC as being unsafe for the public. His opportunity to publicly demonstrate AC's supposed dangers came about when a popular Coney Island elephant named Topsy was condemned by her owners for killing her third handler in as many years. The aggressive act by the 3-ton, 10-foot-tall, 20-foot-long pachyderm was in retaliation for cruelly being fed a lit cigarette by an abusive handler named J. F.

Blount. Topsy lifted Blount into the air and dashed him to the ground, killing him instantly and becoming a candidate for the gallows. The American Society for the Prevention of Cruelty to Animals strenuously objected to her hanging, a cue for Edison to proffer his services and suggest electrocution instead.

"A Rather Inglorious Affair"

The public event drew more than 1,500 onlookers to Coney Island's Luna Park. The elephant was first fed a ration of carrots containing 460 grams of potassium cyanide in preparation for the main event. She was then fitted with copper-lined wooden sandals and electrodes and jolted with 6,600 volts of electricity. A few sparks and Topsy was dead; one newspaper referred to it as "a rather inglorious affair."

The event was filmed, and Edison exhibited the film throughout the country, but his plan to discredit AC backfired. Topsy's execution, which mercifully took less than one minute, was a success due to the power of AC and was not a condemnation of the current as Edison had hoped.

The Manhattan Project's Odd Couple

In the world of creative tension, perhaps no more unlikely pair ever collaborated on a project to change world history.

* * * *

BRIGADIER GENERAL LESLIE R. Groves was the Manhattan Project's military leader. Tall and bombastic, the general tapped J. Robert Oppenheimer, a brilliant theoretician, to head the team building the bomb. Oppenheimer, a wiry, intellectual Communist supporter, was a security risk and an odd fit for the project, but he threw himself into his work and assembled a brilliant group of scientists—including Enrico Fermi, Edward Teller, Hans Bethe, and Richard Feynman—at the Los Alamos laboratory not far from his ranch in New Mexico.

Groves ran roughshod over the intellectually gifted group. Needing the atomic bomb to be completed as quickly as possible, he (and the FBI) kept a close eye on Oppenheimer. As an example of Groves's whip-cracking approach, he promised his scientists on Christmas Eve 1944, "If this weapon fizzles, each of you can look forward to a lifetime of testifying before congressional investigating committees."

Personality differences led to an intense personal feud between Groves and Oppenheimer, but their sense of duty kept their personal feelings from affecting the breakneck pace of the project. As Paul Tibbets, pilot of the *Enola Gay*, recalled his impressions of the two men years later:

"[Oppenheimer's] a young, brilliant person. And he's a chain smoker and he drinks cocktails. And he hates fat men. And General Leslie Groves, he's a fat man, and he hates people who smoke and drink. The two of them are the first, original odd couple." However, this "original odd couple" managed to get past their personal differences long enough to change history.

There's More to Know About Tycho

A golden nose, a dwarf, a pet elk, drunken revelry, and . . . astronomy? Read about the wild life of this groundbreaking astronomer.

* * * *

Look to the Stars

TYCHO BRAHE WAS a Dutch nobleman who is best remembered for blazing a trail in astronomy in an era before the invention of the telescope. Through tireless observation and study, Brahe became one of the first astronomers to fully understand the exact motions of the planets, thereby laying the groundwork for future generations of star gazers.

In 1560, Brahe, then a 13-year-old law student, witnessed a partial eclipse of the sun. He reportedly was so moved by the event that he bought a set of astronomical tools and a copy of Ptolemy's legendary astronomical treatise, *Almagest*, and began a life-long career studying the stars. Where Brahe would differ from his forbearers in this field of study was that he believed that new discoveries in the field of astronomy could be made, not by guesswork and conjecture, but rather by rigorous and repetitious studies. His work would include many publications and even the discovery of a supernova now known as SN 1572.

Hven, Sweet Hven

As his career as an astronomer blossomed, Brahe became one of the most widely renowned astronomers in all of Europe. In fact, he was so acclaimed that when King Frederick II of Denmark heard of Brahe's plans to move to the Swiss city of Basle, the King offered him his own island, Hven, located in the Danish Sound.

Once there, Brahe built his own observatory known as Uraniborg and ruled the island as if it were his own personal kingdom. This meant that his tenants were often forced to supply their ruler (in this case Brahe) with goods and services or be locked up in the island's prison. At one point Brahe imprisoned an entire family—contrary to Danish law.

Did We Mention That He Was Completely Nutty?

While he is famous for his work in astronomy, Brahe is more infamous for his colorful lifestyle. At age 20, he lost part of his nose in an alcohol-fueled duel (reportedly using rapiers while in the dark) that ensued after a Christmas party. Portraits of Brahe show him wearing a replacement nose possibly made of gold and silver and held in place by an adhesive. Upon the exhumation of his body in 1901, green rings discovered around the nasal cavity of Brahe's skull have also led some scholars to speculate that the nose may actually have been made of copper.

While there was a considerable amount of groundbreaking astronomical research done on Hven, Brahe also spent his time hosting legendarily drunken parties. Such parties often featured a colorful cast of characters including a Little Person named Jepp who dwelled under Brahe's dining table and functioned as something of a court jester; it is speculated that Brahe believed that Jepp was clairvoyant. Brahe also kept a tame pet elk, which stumbled to its death after falling down a flight of stairs—the animal had gotten drunk on beer at the home of a nobleman.

Brahe also garnered additional notoriety for marrying a woman from the lower classes. Such a union was considered shameful for a nobleman such as Brahe, and he was ostracized because of the marriage. Thusly all of his eight children were considered illegitimate.

However, the most lurid story of all is the legend that Brahe died from a complication to his bladder caused by not urinating, out of politeness, at a friend's dinner party where prodigious amounts of wine were consumed. The tale lives on, but it should be pointed out that recent research suggests this version of Brahe's demise could be apocryphal: He may have died of mercury poisoning from his own fake nose.

Inventors Killed by Their Inventions

The success of history's most famous inventors rested not just upon brilliant ideas but also upon having the dedication and confidence to pursue those ideas in the face of public doubt. Unfortunately, inventors have sometimes been too confident in their work—with disastrous consequences. Here are six inventors whose inventions got the better of them.

❋ ❋ ❋ ❋

Henry Winstanley, Lighthouse Architect

WHILE 17TH-CENTURY LIGHTHOUSE-SMITH Henry Winstanley didn't invent lighthouses, he did design a new kind of lighthouse—the Eddystone Lighthouse, an octagonal-shaped structure built to withstand treacherous conditions on tenuous ground. Despite observers' doubts that the lighthouse would stand up to serious meteorological assault, Hank believed in his design—so much so that he insisted on taking shelter in it during a terrible storm in November 1703. It was a poor decision—the lighthouse collapsed, ending Winstanley's life.

Marie Curie, Radiation Pioneer

Marie Curie is known to schoolchildren as the discoverer of the elements radium and polonium, the first woman to win a Nobel Prize, a pioneer in the field of radioactivity, and the inventor of a method for isolating radioactive isotopes. Unfortunately, she is also known as a pioneer in the field of radiation-induced cancer. Curie, who was working with radioactive isotopes well before the dangers of radiation were fully known, contracted leukemia from radiation exposure and died at age 66.

William Bullock, Inventor of the Web Rotary Printing Press

Before the 19th century, the printing press hadn't advanced much beyond Gutenberg's first effort back in the 15th century. In 1863, William Bullock changed everything by coming up with the idea of a web rotary press—a self-feeding, high-speed press that could print as many as 10,000 pages per hour. Unfortunately, Bullock forgot a basic rule of printing presses: Don't stick your foot into the rotating gears. In 1867, Bullock got tangled up in his invention, severely injuring his foot. Gangrene set in, and he died shortly afterward.

Karel Soucek, Inventor of the "Stunt Capsule"

Soucek shot to fame in 1984 by designing a special stunt capsule that he used to plunge over Niagara Falls. Seeking to capitalize on his newfound popularity, Soucek decided to repeat the stunt in 1985—only this time from an artificial waterfall running from the top of the Houston Astrodome down to a tank of water. It seemed like a bad idea, and it was: The capsule exploded upon impact, and Soucek suffered fatal injuries.

Otto Lilienthal, Inventor of the Hang Glider

Until the late 19th century, human flight was little more than a pipe dream. Otto Lilienthal changed all of that with his hang glider, and his successful glides made him famous the world over. Unfortunately, in 1896, Lilienthal plunged more than 50 feet during one of his test runs. The fall broke his spine, and he died shortly after.

Cowper Phipps Coles, Inventor of the Rotating Ship Turret

The splendidly named Cowper Phipps Coles was a captain in the British navy who invented a "rotating gun turret" for British naval vessels during the Crimean War. After the war, Coles patented his invention and set about building ships equipped with his new turret. Unfortunately, the first ship that he built, the HMS *Captain*, turned into the HMS Capsized. In order to accommodate his turret design, the shipbuilders were forced to make odd adjustments to the rest of the ship, which seriously raised the center of gravity. End result? The ship sank on one of its first voyages, killing Coles and much of his crew.

Founding Father with Odd Habits

Benjamin Franklin was an unsurpassed diplomat, writer, and inventor. He may also have been the most eccentric, and funniest, of the founding fathers.

* * * *

IN THE EARLY 1760s, early-rising Londoners passing the apartment of the noted representative from the American colonies would get quite a surprise. Each morning, stout Benjamin Franklin would step naked through his rooms, opening up the windows to let in the fresh air he found so invigorating. If the weather was mild, Franklin would step outside and peruse the morning's newspaper outside his digs, a gentle breeze lapping at his bare body.

The famous printer and scientist urged others to try his scanty approach to apparel, but he was less encouraging to strangers who crowded around his property. Interlopers pressing against the iron fence of one domicile were shocked, literally, by an electric charge sent coursing through the metal by the discoverer of electricity.

Franklin on the Human Body

As his displays of nudity might attest, Franklin was very comfortable with his body. He was evidently also proud of his bodily functions. He penned an essay dubbed "Fart Proudly," in which he proposed to:

"Discover some Drug wholesome & not disagreeable, to be mixed with our common Food, or Sauces, that shall render the natural Discharges of Wind from our Bodies, not only inoffensive, but agreeable as Perfumes."

With regard to another key physical function, Franklin urged a friend, who was having trouble landing a young wife, to take an elderly woman as his mistress. He counseled:

"Because in every animal that walks upright, the Deficiency of the Fluids that fill the Muscles appears first in the highest Part: the Face first grows lank and wrinkled; then the Neck; then the Breast and Arms; the lower Parts continuing to the last as plump as ever: So that covering all above with a Basket, and regarding only what is below the Girdle, it is impossible of two Women to know an old from a young one.

And as in the dark all Cats are grey, the Pleasure of corporal Enjoyment with an Old Woman is at least equal, and frequently superior."

Franklin on Religion

Franklin also had distinctive religious views, often skeptical, sometimes more traditional. He anonymously copublished an *Abridgment of the Book of Common Prayer,* which shortened funeral services to six minutes to better "preserve the health and lives of the living."

In the run-up to the American Revolution, Franklin's Committee of Safety—a sort of state provisional government—was deadlocked over whether Episcopal priests should pray for King George. Franklin told the committee:

"The Episcopal clergy, to my certain knowledge, have been constantly praying, these twenty years, that 'God would give the King and his Council wisdom,' and we all know that not the least notice has ever been taken of that prayer."

The prayers were canceled.

Yet, Franklin believed religion had a salutary effect on society and men's morals. In a letter to pamphleteer Thomas Paine, a decided agnostic, he wrote: "If men are so wicked with religion, what would they be if without it?" Asked to design the

Great Seal of the United States, Franklin submitted a sketch of Moses and the Israelites drowning the Pharaoh's army in the Red Sea, with the motto: "Rebellion Against Tyrants is Obedience to God."

On ultimate questions of faith—the afterlife, the nature of God—Franklin was sometimes droll:

"As to Jesus of Nazareth . . . I think the System of Morals and his Religion . . . the best the world ever saw or is likely to see; but I apprehend it has received various corrupt changes, and I have, with most of the present Dissenters in England, some Doubts as to his divinity; tho' it is a question I do not dogmatize upon . . . when I expect soon an Opportunity of knowing the Truth with less Trouble."

The End

In 1785, a Frenchman wrote a satire of Franklin's famous literary character Poor Richard in which the notoriously frugal Richard bequeathed a small amount of money, whose accumulating interest was not to be touched for 500 years. Franklin wrote the satirist, thanking him for a "great idea," and proceeded to bequeath 1,000 pounds each to his home and adopted towns, Philadelphia and Boston, to be placed in trust for 200 years. By 1990, the amount in the Philadelphia fund had reached $2 million and was dispensed as loans to townspeople and scholarships for students. The other trust tallied close to $5 million, paying for a trade school that blossomed into the Franklin Institute of Boston. Lastly, the eccentric, sharp-witted founder composed his own epitaph. It read:

"The Body of B. Franklin Printer; Like the Cover of an old Book, Its Contents torn out, And stript of its Lettering and Gilding, Lies here, Food for Worms. But the Work shall not be wholly lost: For it will, as he believ'd, appear once more, In a new & more perfect Edition, Corrected and Amended By the Author."

The Snowflake Man

His name was Wilson Bentley—and he was the person who assured us that no two snowflakes are alike.

❄ ❄ ❄ ❄

A Life Spent Studying Snowflakes

IN 1885, WILSON Bentley became the first person to photograph a single snow crystal. By cleverly adapting a microscope to a bellows camera, the 19-year-old perfected a process that allowed him to catch snowflakes on a black-painted wooden tray and then capture their images before they melted away. A self-educated farmer from the rural town of Jericho, Vermont, Bentley would go on to attract worldwide attention for his pioneering work in the field of photomicrography. In 1920, the American Meteorological Society elected him as a fellow and later awarded him its very first research grant, a whopping $25.

His Famous Words

Over 47 years, Bentley captured 5,381 pictographs of individual snowflakes. Near the end of his life, the Snowflake Man said that he had never seen two snowflakes that were alike: "Under the microscope, I found that snowflakes were miracles of beauty. Every crystal was a masterpiece of design and no one design was ever repeated."

Since Bentley's observation, physicists, snowologists, crystallographers, and meteorologists have continued to photograph and study the different patterns of ice-crystal growth and snowflake formation (with more technologically advanced equipment, of course). But guess what? Bentley's snow story sticks.

No Proof to the Contrary

Scientists still agree: It is unlikely that two snowflakes can be exactly alike. It's so unlikely, in fact, that Kenneth G. Libbrecht, a professor of physics at Caltech, says, "Even if you looked at every one ever made, you would not find any exact duplicates."

How so? Says Libbrecht, "The number of possible ways of making a complex snowflake is staggeringly large." A snowflake may start out as a speck of dust, but as it falls through the clouds, it gathers up more than 180 billion water molecules. These water molecules freeze, evaporate, and arrange themselves into endlessly inventive patterns under the influence of endless environmental conditions.

And that's just it—snow crystals are so sensitive to the tiniest fluctuations in temperature and atmosphere that they're constantly changing in shape and structure as they gently fall to the ground. Molecule for molecule, it's virtually impossible for two snow crystals to have the exact same pattern of development and design.

"It is probably safe to say that the possible number of snow crystal shapes exceeds the estimated number of atoms in the known universe," says Jon Nelson, a cloud physicist who has studied snowflakes for fifteen years. Still, we can't be 100 percent sure that no two snowflakes are exactly alike—we're just going to have to take science's word for it. Each winter, trillions upon trillions of snow crystals drop from the sky. Are you going to check them all out?

Louis Pasteur

"No, a thousand times no; there does not exist a category of science to which one can give the name applied science. There are science and the applications of science, bound together as the fruit to the tree which bears it."

* * * *

LOUIS PASTEUR'S CONTRIBUTIONS to modern biology are immense: he discovered pasteurization, vaccinations for anthrax and rabies, and microbial fermentation. His discoveries advanced the germ theory of disease, disproved the theory of spontaneous generation, and laid the groundwork for the study of bacteriology.

Pasteur was born in Dole, France, in 1822. Although he was an average student, he obtained a Bachelor of Arts in 1840, a Bachelor of Science in 1842, and a Doctorate at the École Normale in Paris in 1847. He spent the first several years of his career as a teacher and researcher in Dijon Lycée before becoming a chemistry professor at the University of Strasbourg. He married Marie Laurent, the daughter of the university's director, and together they had five children. Three of them survived to adulthood.

Chirality and Isomerism

In 1849, Pasteur was studying the chemical properties of tartaric acid, a crystal found in wine sediments, and comparing them to paratartaric acid, a synthetic compound which had the same chemical composition. The two behaved differently, however, and he determined to discover why. He passed polarized light through each crystal, and found that while tartaric acid rotated the light, paratartaric acid did not. In doing so he discovered the principles of molecular chirality and isomerism; the former describes compounds that are mirror images, and the latter the fact that compounds can have identical molecular formulas and different chemical structures.

Fermentation and Germ Theory

In 1856, a winemaker asked Pasteur his advice on preventing stored alcohol from going bad. Pasteur surmised that fermentation and spoiling are caused by microorganisms, and demonstrated that the presence of oxygen was not necessary for this to occur. He showed experimentally that wine soured when lactic acid was produced by bacterial contamination. Having established this, he then realized that heat could kill the majority of microorganisms that were present, preventing them from spoiling. He patented the process in 1865, calling it pasteurization.

His discovery that bacteria were responsible for fermentation and spoiling led him to suggest that the same microorganisms also caused human and animal diseases—called the germ theory of disease. He proposed that protection from bacteria could reduce disease, leading to the development of antiseptics.

Disproving Spontaneous Generation

For centuries, the prevailing wisdom held that living organisms arose spontaneously from nonliving matter; fleas were believed to appear from dust and maggots from dead flesh. Pasteur suspected that this was not the case when he observed that yeast did not grow on sterilized grapes. His assertion that spontaneous generation was incorrect sparked a furious debate, and the French Academy of Sciences proposed a cash prize for anyone who could prove or disprove the theory. Pasteur devised an experiment in which he boiled broth in swan-necked flasks. The necks of the flasks prevented airborne particles from reaching the broth. He also exposed a control set of boiled broth to the air. Nothing grew in the swan-necked flasks, but microorganisms did grow in the control set, demonstrating that spontaneous generation was incorrect.

Vaccination

Pasteur discovered the principle of vaccination almost by accident. While he was studying chicken cholera, his assistant inoculated a group of chickens with a culture of the disease that had spoiled. While the chickens became ill, they did not die. Pasteur attempted to re-infect the chickens, but discovered that the weakened culture had made them immune to cholera. Pasteur would go on to apply this principle to developing vaccinations for anthrax and rabies.

He founded the Pasteur Institute in 1887. In 1894, he suffered a stroke, and died the next year. He was buried at Notre Dame Cathedral.

The Way the Future Wasn't . . . Cities Under Glass

The vision of cities flourishing under glass domes dates back to the science fiction of the late 19th century. During the early 20th century, futuristic pictures of urban life frequently depicted gigantic overturned bowls protecting people and buildings from the elements. But it was little Winooski, Vermont, that almost became the world's first real domed city.

⁕ ⁕ ⁕ ⁕

IN 1979, RISING fuel oil prices had given Winooski residents a chill. One fateful night, Mark Tisan, a 32-year-old community development planner, came up with a novel idea. Why not enclose the entire city under a glass dome?

Though it seems incredible in retrospect, city officials encouraged Tisan to pursue federal funding for the newly christened Winooski Dome Project. Tisan called a press conference. Within a few days the story had gone viral, and he began to receive a flood of mail from people all over the world eager to help.

Tisan didn't even know what the dome would look like, so he hired conceptual architect John Anderson, who quickly whipped up a design for a clear vinyl structure supported by a web of metal struts. Shaped rather like a hamburger bun with a wide, flat top, the dome would measure 250 feet high and encompass an area of one square mile—most of the residential city. Of course there were still a few kinks in the plan—like how to get rid of auto exhaust and other fumes. Tisan remained undeterred. He assured skeptics that electric cars and similar innovations would solve any potential problems.

All systems were go for Winooski to become the city of the future. Engineer Buckminster Fuller, designer of the geodesic dome, endorsed the project, as did President Jimmy Carter. Carter, however, lost his bid for reelection to conservative Ronald Reagan. In 1980, the government turned down Tisan's request for funds, and the dome disappeared into the mists of history as if it had never happened.

Facing Facts

John Tyndall's discoveries— of the Greenhouse Effect, magnetism, and more— are still applicable today, and he was a tireless promoter of the cause of science.

* * * *

JOHN TYNDALL WAS born in County Carlow, Ireland, on August 2, 1820. His family was fairly poor, but he was able to have a basic education, and decided to pursue a career as a surveyor following his schooling. In 1847, after performing surveys of England and Ireland for eight years, he became interested in the sciences. Tyndall had saved enough money from his work to be able to afford to attend the University of Marburg, Germany, from 1848 to 1850, earning a Ph.D. there. After he earned his degree, however, he had difficulty finding work for the first few years, until he was able to secure a position in 1853 at the Royal Institution in London as a professor of natural philosophy.

Scientific Study

Tyndall began studying magnetism and magnetic polarity, and pursued this research until 1856. His work on magnetism attracted the attention of other physicists of the day, and in 1852 he was inducted as a Fellow of the Royal Society. While he was at the Royal Institution, he became friends with the brilliant physicist and chemist Michael Faraday. Faraday was already extremely popular because of his entertaining and enlightening lectures. Tyndall would eventually be appointed to the chair held by Faraday upon his mentor's retirement.

Faraday had encouraged Tyndall in his successful study of magnetism, but in the late 1850s, the protégé would embark on his own area of novel research: the effects of heat and light energy (called radiant energy) on vapors and gases in the air. He was able to show experimentally that the heat in the Earth's atmosphere was caused by the ability of the gases present in the air to absorb radiant energy. He invented a device called a thermopile that converted radiant energy to electricity, which allowed him to make accurate measurements of air temperature when various gases were present. He correctly measured the radiant energy absorption potential of several atmospheric gases in 1859. His work was the first to prove the existence of the Greenhouse Effect.

In order to make his measurements as accurate as possible, Tyndall needed to ensure that there were no microscopic dust particles present in the air sample he was measuring. In the 1860s he hit upon the idea of shining intense light at the air sample; if dust particles were present, the light would be scattered. This effect, which he first described, is now called Tyndall Scattering. His work led to many other discoveries that are still relevant: in 1862 he determined a means of measuring the amount of carbon dioxide in a patient's breath, and this is still used to monitor anesthetized patients in hospitals today.

Science Education

Tyndall was also a dedicated promoter of scientific inquiry. He regularly gave public lectures on scientific topics to lay audiences at the Royal Society. On a tour of the United States, during which he delivered dozens of lectures, he was praised for his ability to not only teach science accurately, but in a manner that made it entertaining and captivated his audiences. He wrote multiple books on science, and eventually grew to be one of the most famous physicists of his day. He often wrote and spoke about the importance of science education and how teaching was the noblest profession he could think of. He was a vocal supporter of Darwin's theory of Natural Selection, and worked against the influence of religion on science.

Tyndall died on December 4, 1893.

Visualizing a Better Life

In advocating for animal welfare and individuals with autism, Temple Grandin pictures a better world.

❋ ❋ ❋ ❋

IN THE OPENING chapter of her book, *Thinking in Pictures*, Dr. Temple Grandin writes "Words are like a second language to me. I translate both spoken and written words into full-color movies, complete with sound, which run like a VCR tape in my head." Throughout her life, this visual thinker has sought to explain what it is like to live with autism. Born in 1947, she was diagnosed with the developmental disorder in 1950. By sharing her perspective and experiences, she hopes that she will enlighten and empower others.

Grandin is also a Professor of Animal Science, a world-renowned advocate for animal welfare, and a prolific author. She also has had considerable influence in the livestock industry, where she has helped design more humane facilities, served as a consultant for firms such as McDonald's and Burger King, and educated people about proper animal handling. These accomplishments earned her such nicknames as "The Woman Who Thinks Like a Cow." She has said that "using animals for food is an ethical thing to do," but that it requires respect: "We've got to do it right. We've got to give those animals a decent life and we've got to give them a painless death."

The Squeeze Machine

As is common with autistic children, Grandin did not speak until age three and a half. Instead, she would communicate via screaming or humming. She was also highly sensitive to touch and sound. Doctors told her parents that she should be placed in an institution, but Grandin remained in school. Although she often endured ridicule from classmates, she was an imaginative thinker and developed her own strategies for coping with stress and anxiety. In one of her more profound instances of "thinking like a cow," in 1992, Grandin developed something called the "squeeze machine" or "hug box" for those with autism. Modeled after the squeeze chutes used to restrain cattle while they're being given veterinary treatment, this machine applies deep pressure stimulation (similar to a firm and long-lasting hug) to the person using it. "As a little kid, I wanted to experience the nice feeling of being held, but it was just too much overwhelming stimulation," Grandin said in a BBC documentary about her life, *The Woman Who Thinks Like a Cow*. Using the machine gives her more control of the situation, which allows her to relax and enjoy the feeling.

Other Innovations

Another one of Grandin's groundbreaking inventions is a curved chute or race system for corralling cattle. Designed to lower stress and fear in the animals, the curved chutes are more efficient than straight chutes, because, Grandin explains, "they take advantage of the natural behavior of cattle." (Cows have a natural tendency to return to where they came from.) Now processing plants throughout the world—businesses that slaughter millions of cattle and pigs for human consumption—

use this type of corralling method. Grandin also developed an objective scoring system to assess how well cattle and pigs are handled at these plants. In addition, she has studied bull fertility, stunning methods for cattle and pigs, and cattle temperament. Grandin credits her strong visual thinking skills for her sensitivity to animals' experiences and ability to come up with humane treatment solutions.

Telling Her Story

In addition to working as a consultant to the livestock industry and teaching courses on livestock behavior and facility design, Grandin is the author of a number of bestselling books, including, *Emergence: Labeled Autistic, Animals in Translation: Using the Mysteries of Autism to Decode Animal Behavior,* and *Animals Make Us Human: Creating the Best Life for Animals.* She has also written articles for numerous magazines and is a go-to expert on most things cow-related. The neurologist Oliver Sacks wrote about her in his book, *An Anthropologist on Mars.* In sharing her experiences, Grandin aims to provide hope and insight to individuals on the autism spectrum.

Newton's Apple

Could a falling apple have triggered one of the greatest scientific discoveries of all time? Probably not—but it's a cute story.

* * * *

THE TALE OF the apple landing on Isaac Newton's head during an afternoon nap has been told for hundreds of years as the explanation for his discovery of the law of gravity. If only it were so simple. Newton enjoyed taking walks in his orchard and probably even indulged in a nap or two under an apple tree. However, his understanding of gravity did not suddenly come to him as a flash of insight. Rather, it was the result of years of painstaking study.

The Plague of 1665 probably had more to do with Newton's intellectual feat than a round, red fruit. Newton was a 23-year-old student at Cambridge when the plague gripped England. As a result, the university closed and students were sent back to their homes in the countryside. Newton used this time to devote himself to his private studies, and in later years, he would refer to this period as the most productive of his life. He spent days working nonstop on computations and nights observing and measuring the skies. These calculations provided the seeds for an idea that would take years of covert and obsessive work to formulate—his Theory of Universal Gravitation.

Accounts of the apple story began appearing after Newton's death in 1727, probably written by the French philosopher Voltaire, who was famous for his wit but not his accuracy. He reported having heard the story about the apple from one of Newton's relatives, but there is no sound evidence to support that claim.

The falling apple will always be associated with Newton's great discovery. Many universities claim to own trees grown from grafts of trees from Newton's orchard, perhaps to remind overworked students that the theory of gravity was no piece of pie but, rather, the fruit of hard labor.

Heads Up: The Study of Phrenology

Sure, someone may look like a nice enough guy, but a phrenologist might just diagnose the same fella as a potential axe murderer.

❋ ❋ ❋ ❋

He Had the Gall

THERE ARE BUMPS in the road and bumps in life. Then there are the bumps on our heads. In the last half of the 19th century, the bumps and lumps and shapes of the human skull became an area of scientific study known as *phrenology*.

Early in the century, an Austrian physicist named Franz Joseph Gall theorized that the shape of the head followed the shape of the brain. Moreover, he wrote, the skull's shape was determined by the development of the brain's various parts. He described 27 separate parts of the brain and attributed to each one specific personality traits.

Gall's theories hit the public at a time of widespread optimism in Europe and North America. New and startling inventions seemed to appear every week. No problem was insurmountable, no hope unattainable. Physical science prevailed.

By mid-century, Gall's theories had spread favorably throughout industrialized society. What was particularly attractive about phrenology was its value as both an indicator and predictor of psychological traits. If these traits could be identified—and phrenology presumably could do this—they could be re-engineered through "moral counseling" before they became entrenched as bad habits, which could result in socially unacceptable behavior. On the other hand, latent goodness, intellect, and rectitude could also be identified and nurtured.

As it grew in popularity, phrenology found its way into literature as diverse as the Brontë family's writings and those of Edgar Allen Poe. It also influenced the work of philosopher William James. Famed poet Walt Whitman was so proud of his phrenological chart that he published it five times. Thomas Edison was also a vocal supporter. "I never knew I had an inventive talent until phrenology told me so," he said. "I was a stranger to myself until then."

Criminal Minds

Early criminologists such as Cesare Lombroso and Èmile Durkheim (the latter considered to be the founder of the academic discipline of sociology) saw remarkable possibilities for phrenology's use in the study of criminal behavior. Indeed, according to one tale, the legendary Old West figure Bat Masterson invited a phrenologist to Dodge City to identify horse thieves and cattle rustlers. A lecture before an audience of gun-toting citizenry ended with the audience shooting out the lights and the lecturer hastily departing through the back exit.

In 1847, Orson Fowler, a leading American phrenologist, conducted an analysis of a Massachusetts wool trader and found him "to go the whole figure or nothing," a man who would "often find (his) motives are not understood." Sure enough, years later Fowler was proven to be on the money. The man was noted slavery abolitionist John Brown, and he definitely went the "whole figure."

Bumpology Booms

By the turn of the century, the famous and not so famous were having their skulls checked. Phrenology had become a fad and, as such, it attracted a few charlatans. By the 1920s, the science had degenerated into a parlor game. Disrepute and discredit followed, but not before new expressions slipped into the language. Among these: "low brow" and "high brow" describe varying intellectual capacity, as well as the offhand remark, "You should have your head examined."

Nevertheless, phrenology did figure in the early development of American psychiatry, and it helped point medical scientists in new directions: neurology for one and, more recently, genomics—the study of the human genome.

The Drink of Doctors?

There were no postgraduate degrees involved in the creation of Dr Pepper (the company dropped the period from "Dr." in the 1950s), and it was never considered a health drink. But soda lore does tell of a real doctor who inspired the name.

⁕ ⁕ ⁕ ⁕

CHARLES ALDERTON—A PHARMACIST at Morrison's Old Corner Drug Store in Waco, Texas—invented the drink in 1885. (In those days, a drugstore often featured well-stocked soda fountains.) Alderton loved the smell of various fruit syrups mixed together and experimented to create a drink that captured that aroma. Customers eagerly gulped down the result, which was initially called a "Waco." Alderton's boss, Wade Morrison, renamed the beverage "Dr. Pepper" and started selling it to other soda fountains.

A long-standing legend holds that Morrison named the drink after Dr. Charles T. Pepper, a physician and druggist who had been Morrison's boss back in his home state of Virginia. One version of the story claims that Morrison was simply honoring the man who had given him his start in the business.

However, the more popular variation contends that Morrison was in love with Pepper's daughter, but that Pepper didn't approve. Heartbroken, Morrison moved to Texas. He eventually called his popular beverage Dr. Pepper—either to flatter Pepper and perhaps get another shot at his daughter, or just as a joke.

This was the official story for years, but researchers eventually uncovered evidence that largely debunked it. Census records show a Dr. Charles Pepper living in Virginia at the time, but his daughter would have been eight years old when Morrison left the state, and it's not clear whether Morrison actually worked for Pepper. However, census records also show that when Morrison was a teenager, he lived near another Pepper family, which included a girl who was a year younger than him. The star-crossed-lovers story might be true—just with a different Pepper. Another possibility is that Morrison simply came up with a good name. "Doctor" could have suggested that the drink was endorsed by a physician for its health benefits, while "Pepper" may have indicated that it was a good pick-me-up.

So the original good doctor was either an MD or a figment of a pharmacist's imagination. In any case, the name worked—Dr Pepper is the oldest soda brand in the world. It just goes to show that people like a drink with a good education.

Bunsen's Burner: Scientific Error

A staple in science classes for generations, this device's hot blue flame has heated the experiments of countless scientists. But the so-called "Bunsen burner" is actually a misnomer.

* * * *

Among the achievements of 19th-century scientist Robert Wilhelm Bunsen are the co-discovery of chemical spectroscopy—the use of an electromagnetic light spectrum to analyze the chemical composition of materials—and the discovery of two new elements, cesium and rubidium. Bunsen is best known, however, for his invention of several pieces of laboratory equipment, including the grease-spot thermometer, the ice calorimeter, and a gas burner that became the standard for chemical laboratories the world over.

In 1852, Bunsen joined the University of Heidelberg and asked for a new laboratory with built-in gas piping. Although already in use, gas burners at the time were smoky and produced flickering flames of low heat intensity. Bunsen had the idea of improving an existing burner by pre-mixing gas with air before combustion, giving the device a hotter-burning and non-luminous flame. He took his concept to mechanic Peter Desaga, who then designed and built the burner according to Bunsen's specifications, adding a valve that regulated the amount of oxygen mixed with gas. Bunsen gave Desaga the right to manufacture and sell the burner, and Desaga's son, Carl, started a company to fill the orders. Bunsen and Desaga, however, did not apply for a patent, and soon other manufacturers were selling their versions. Competitors applied for their own patents, and Bunsen and Desaga spent decades refuting these claims.

99 Percent Perspiration

Thomas Alva Edison was one of the most prolific inventors in American history. Called "The Wizard of Menlo Park," Edison held over a thousand US Patents in his name, and had a profound impact on industry and life in the twentieth century.

* * * *

EDISON BEGAN HIS career as an inventor in 1877, and continued to be active in developing new commercial products right up until his death in 1931.

Phonograph

One of Edison's personal favorites among his inventions, the phonograph had a groundbreaking design. It used a needle that would vibrate when the user spoke into the receiver, causing it to mark a rotating drum wrapped in foil. Eventually, Edison would develop a phonograph that used discs and cylinders to allow music to be recorded—the forerunner of the record player. Edison's first message was a recitation of the poem "Mary Had a Little Lamb."

Edison Light Bulb

Edison is probably best known for having invented the light bulb. This belief isn't precisely accurate, however: light bulbs had existed for several years before Edison's invention, but they were expensive, unreliable, and only lasted a few hours before they burned out. There was an ongoing race to perfect the electric light bulb when Edison took on the challenge.

His innovation was to create a vacuum inside the bulb, use a carbon filament, and reduce the voltage to make the bulb stable and long-lasting. His first success created a bulb that lasted for thirteen and a half hours. He later switched to using a carbonized bamboo filament. This bulb lasted over 1,200 hours.

Motion Picture

Edison's first motion picture device was similar to the design of his phonograph. Tiny pictures were arranged on a cylinder and viewed under a microscope as it rotated. Thanks to George Eastman's invention of 35mm celluloid film, Edison was able to develop the Strip Kinetograph. The film, cut into long strips that were perforated along the edges, moved past a shutter, and the viewer observed twenty-second films through a peephole. Later, Edison manufactured and marketed Thomas Arnat's Vitascope, which was the first movie projector.

Electric Power Grid

With the success of his light bulb, Edison realized that an electric distribution system was necessary to make it commercially viable. He patented a system for distributing electricity. In January 1882, Edison's first steam-powered electric power station was turned on in London. It powered street lamps and a few houses close by. In September of that year his first generating station in the United States was switched on, providing 110 volts to 59 customers in lower Manhattan. And on January 19, 1883, the first incandescent electric streetlight system that used overhead wires was installed in Roselle, New Jersey.

✳ Chapter 3

Only in the Military

Beware, Balloon Bombs!

In a last-ditch effort to attack America, Japan relied on the wind.

✳ ✳ ✳ ✳

BY LATE 1944, the United States had cut off much of Japan's supply of food, fuel, and other war materials. Despite a shortage of resources, the Japanese Ninth Research Division laboratory developed a new weapon.

The Japanese knew that a strong wind current swept across the Pacific from Japan to North America (later this current would be called the jet stream). Researchers supposed they could float a large number of missiles on the current to explode over America. They thought Japan could achieve indefensible terror and destruction similar to Germany's buzz bombs and V-1s in Britain. Military brass called in their engineers and laid out the requirements—the prime purpose of the missiles would be to burn America's food crops and forests, and the weapons would need to carry antipersonnel explosives to prevent anyone from interfering with the devices. Originally, targeting major cities was an objective, but it was soon realized the guidance would rely on the whimsy of wind currents.

So began project Fu-go. Planners set about the task. They decided a hydrogen-filled balloon would be the best method of transportation. With rubber in short supply, the engineers created a balloon skin using thick, impermeable paper called washi, made from mulberry trees. Meteorologists agreed the plan was feasible. The Japanese government evacuated large island warehouses to provide assembly sites for this priority project. In all, about 30,000 soldiers and an equal number of civilians were put to work on the weapon. The product was then sent to northern Honshu, where technicians attached explosives and incendiary devices. The weapons were called *fusen bakudan*, which means "balloon bomb" but has also been translated as "fire balloon."

Engineers refined the mechanisms to ensure the balloons would be carried along the jet stream at an altitude near 30,000 feet. If they slipped below 30,000 feet, a mechanism would release a pair of sandbags, and the balloons would rise. If they got as high as 38,000 feet, a vent was activated to release some hydrogen from the balloons.

Released from northern Honshu, the balloons would take three days to cross the Pacific. With its sandbags spent, a mechanism would drop the bombs and light a fuse that would burn for 84 minutes before detonating a flash bomb that would destroy the balloon. By causing the balloons to self-destruct in midair, the Japanese hoped to add mystery to the source of the fires.

The first balloons were launched November 3, 1944, and one was spotted two days later off the coast of San Pedro, California. They continued to turn up throughout the northwest United States and western Canada, reaching as far east as Farmington, Michigan and south to northern Mexico. Rather than incite widespread panic, the balloons were largely ineffective and rarely discussed. *Newsweek* ran a report on the weapons in January 1945, and the Office of Censorship issued a notice to the media not to report further incidents.

During January to February 1945, debris showed up as far inland as Arizona and Texas. One bomb exploded near the Boeing plant in Seattle that produced B-29s. Another shorted a high-tension wire, temporarily blacking out one of the Manhattan Project's reactors in Hanford, Washington. A balloon killed a woman in Helena, Montana, and one in Oregon claimed six lives. On March 5, 1945, a minister's wife and children from the Sunday school were on a fishing trip. They discovered a grounded balloon and tried to move it, but it exploded. After this incident, the media ban was lifted so that people could be warned of the potential danger.

When the balloons arrived, they were indeed a mystery—no one knew where they had come from. Some feared they could carry biological weapons. Researchers examined balloon bombs that were found unexploded and analyzed the sand. Finding that it was not from America or the mid-Pacific, they eventually determined it came from beaches in northeast Japan. Troops flew photo reconnaissance missions over the area, and photo interpreters identified two of the three hydrogen plants near Ichinomiya. B-29s were sent to destroy the plants, grinding the balloon-bomb production to a halt.

The Japanese government suspended funding for project Fu-go in April 1945. While Japanese propaganda had declared casualties as high as 10,000, they had no evidence the bombs were actually reaching or exploding in America. Of more than 9,000 bombs launched, only about 300 reached our shores.

A Perfect Container, but How Do You Open It?

The can opener was certainly a sharp invention, but it was also long overdue.

* * * *

BEFORE THE CAN opener, there was a revolutionary (albeit somewhat half-baked) invention called the can. The process for canning foodstuffs was patented by Peter Durand of Britain in 1810, and the first commercial canning factory opened three years later. The British Army quickly became a leading customer for the innovative product. After all, the can greatly simplified the logistics of keeping the nation's soldiers fed. In 1846, a new machine that could produce 60 cans per hour increased the rate of production for tinned food tenfold. Life made easier through the wonders of technology, right?

Well, not exactly. In all that time, no one came up with a way to address the most serious drawback of this perfectly sealed, airtight, solid-iron container: It was incredibly difficult to open! In fact, some cans actually came with instructions that read, "Cut 'round the top near the outer edge with a chisel and hammer."

It wasn't until the middle of the century—when manufacturers devised methods for producing thinner cans made of steel—that there was any hope of creating a simple and safe way to open them. In 1858, Ezra Warner of Connecticut patented the first functional can opener, a bulky thing resembling a bent bayonet that you shoved through the top of the can and then carefully forced around the lid. It was an improvement, no doubt, but still a tiresome and potentially dangerous way to get at your potted meat.

Then, in 1870, a full 60 years after the canning process was perfected, William Lyman designed an easy-to-use wheeled blade that could cut a can open as it rolled around the edge of the lid—essentially the same function as that of the can openers we use today. The next two big inventions in can-opening technology were a long time coming: The electric can opener appeared in the 1930s, and pull-open cans arrived in 1966.

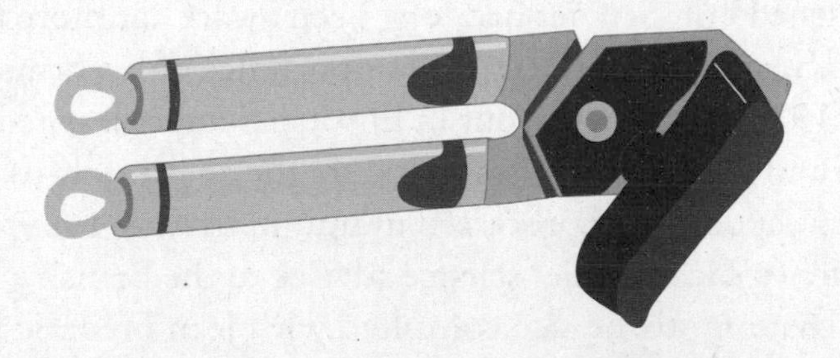

The Secret Stash of the Bomb Project's Foreign Agent

U.S. officials working on the atomic bomb project frantically searched the world for sources of uranium. However, they were unaware that enough nuclear material to make several bombs lay for the taking at a New York City warehouse.

* * * *

U.S. ARMY COLONEL Kenneth Nichols was engaged in what he thought was a nearly impossible task. With the war at its height in September 1942, U.S. researchers were beginning work on the Manhattan Project. Yet the United States possessed no uranium, the essential building block for the atomic bomb. The project's chief, Army General Leslie Groves, had ordered Nichols to obtain some of the precious material.

Nichols paid a visit to the New York City office of Edgar Sengier, a Belgian tycoon. Sengier was director of the Union Minière du Haut Katanga, a company that owned the world's most precious uranium mine, located in the Belgian Congo's Katanga province.

Nichols asked Sengier if his company could procure uranium for the U.S. government. He added that he was well aware this was an unusual request that could not be swiftly fulfilled. Sengier carefully checked Nichols's credentials. Then he replied, "You can have the ore now. It is here in New York, a thousand tons of it. I was waiting for your visit."

As it turned out, Sengier had long been aware, far more than most U.S. officials, of the critical importance of his firm's product. In 1939 as war broke out in Europe, Sengier learned from British and French scientists that it might be possible to devise a superweapon out of processed uranium. In May 1939, he met with Henry Tizard, chief science adviser to the British government. That month, he also consulted with Jean Frédéric Joliot, the co-winner with wife Irène Joliot-Curie—daughter of Marie Curie—of the Nobel Prize in Chemistry. Joliot had taken out patents on path-breaking work to induce chain reactions in atomic piles using uranium and heavy water (deuterium oxide).

In fact, Sengier negotiated with the French to supply them with 55 tons of uranium ore. When the Germans invaded France in May 1940, Union Minière shipped some eight tons of the promised uranium oxide to Joliot's research team. The material, hidden in French Morocco during five years of German occupation, became the basis for France's postwar nuclear program.

Sengier also sought to aid other nations in the fight against the Nazis. He reestablished his firm's headquarters in the safe harbor of New York and had 1,250 tons of uranium ore shipped there. The consignment cleared customs and was unloaded at Staten Island. It was stored—unguarded and unnoticed for nearly two years—at a warehouse there.

Jesse Johnson, the Atomic Energy Commission's director of raw materials, later remarked, "M. Sengier told members of the State Department and other government officials about the shipment, but the secret of the atomic bomb was so closely guarded that no one with whom he talked recognized the importance of the information."

The shipment was a true mother lode. The only other source of uranium at the time, with a uranium content of only .02 percent, was at distant Great Bear Lake in northern Canada. The Katanga ore, in contrast, was up to 65 percent uranium.

After Colonel Nichols's visit, the army bought and transferred Sengier's stockpile. The ore, said Johnson, supplied "the bulk of the uranium for the [Manhattan Project's] development work and early production of fissionable material for bombs." The army, in the meantime, in one of the war's little-known and most far-reaching operations, acted to exploit the Union Minière's other valuable holdings.

The Katanga mine, Shinkolobwe, had been closed since 1939 because its mineshafts had flooded. The Army Corps of Engineers sent units to restore the mine, to construct a port on the Congo River, and to renovate an airport in Léopoldville. The United States, with British backing, negotiated a ten-year deal for exclusive rights to the Shinkolobwe uranium with the Belgian government-in-exile. Sengier stayed on to manage the mine, and before the war's end, the army had bought up 12 times Sengier's original hoard—some 30,000 tons.

Sengier was also awarded the Belgian Ordre de la Couronne, the French Légion d'Honneur, and the title of Knight Commander in the Order of the British Empire. Sengierite, a radioactive crystal, is named after him. In 1946, General Groves, in President Truman's presence, presented Sengier with the prestigious Medal for Merit—he was the only non-U.S. citizen to receive the accolade up to that time. Details of the Manhattan Project were still hush-hush, so the citation was cloaked in vague language that lauded the Belgian's "wartime services in the realm of raw materials."

Quirky Inventor's Explosive Career Leads to Noble Testament

He created the Nobel Peace Prize. He also invented dynamite. When Alfred Nobel wrote his legendary will, was it to atone for his devastating invention? Or was it simply the final good deed of a man dedicated to peace?

* * * *

Protecting a Secret Prize

IN DECEMBER 1896, Ragnar Sohlman rode nervously through Paris, sitting on a box containing one of the world's great fortunes. He was the main executor for the will of Alfred Nobel, the Swedish industrialist—and inventor of dynamite—who had just died at an Italian villa, after years of residing in Paris.

Sohlman had his hands full. Nobel's will consisted of one long, vague paragraph, directing that the then-huge amount of $4.2 million be awarded to those "who during the preceding year shall have conferred the greatest benefit to mankind" in the fields of medicine, chemistry, physics, literature, and peace. Nobel had named the organizations he wanted to dispense the awards but hadn't bothered to tell them. Nor had he set up a foundation. Nobel had lived in Italy, Russia, Germany, the United States, and other lands. Thus, many different nations could claim his wealth.

Setting Up a Famous Award

After secreting the treasure with Swedish authorities, Sohlman undertook drawn-out negotiations with the awarding bodies Nobel had chosen. These were the Swedish Academy of Sciences (for chemistry and physics), the Stockholm-based Caroline Institute (medicine), the Swedish Academy (literature), and the Norwegian Parliament (peace). The first Nobel Prizes were finally granted on December 10, 1901—the fifth anniversary of Nobel's death.

A Polymath's Polyglot Interests

Born in 1833, Alfred Nobel had science in his blood. He was a descendant of medical professor Olof Rudbeck the Elder, who discovered the lymphatic system. Alfred's father Immanuel, an architect and engineer, was the inventor of plywood. His brothers founded Russia's oil industry. Apart from ancestry, Nobel's interest in medicine derived from his bad health. He endured awful migraines and black depression. He said of himself: "Alfred Nobel—a pitiful half-life which ought to have been extinguished by some compassionate doctor as the infant yelled its way into the world." He suffered from angina. The popularizer of the explosive as well as the medicine known as nitroglycerin wrote: "Isn't it the irony of fate that I have been prescribed nitroglycerin to be taken internally?" He was an amateur authority on blood transfusions. He even underwrote early experiments in physiology, funding the work of Ivan Pavlov.

As for literature, Nobel was a poet and authored the play *Nemesis*. It concerned a terrified family who bludgeoned a violent father to death. His family had 97 of the 100 copies destroyed because the play was considered blasphemous.

One subject Nobel wasn't interested in was law. He wrote: "Lawyers have to make a living, and can only do so by inducing people to believe that a straight line is crooked." It's little wonder he penned his own will.

It's often thought that Nobel's interest in peace was due to guilt over his invention of dynamite. An 1888 French obituary that mistakenly reported his death called him the "merchant of death." It read, "Dr. Alfred Nobel, who became rich by finding ways to kill more people faster than ever before, died yesterday."

However, evidence regarding his feelings of guilt (or lack thereof) is mixed. Recently revealed correspondence with his mistress Sophie, an Austrian flower clerk half his age, indicates he did harbor doubts about his work. However, since he lived in a time of relative peace, almost none of his inventions were used in battle. Indeed, he long believed the destructive power of creations such as dynamite would deter war. Anticipating atomic weapons, he once told a peace activist, "When two armies of equal strength can annihilate each other in an instant, then all civilized nations will retreat and disband their troops."

Devising a Double-Edged Sword

Nobel's family was among the leading explosives and armaments manufacturers of the time. His father established a factory in Russia to build naval mines. During the Crimean War, the family business prospered but went belly-up when peace arrived in 1856.

This spurred Alfred into a rabid search for new products. He and his brothers experimented with volatile nitroglycerin, invented by Ascanio Sobrero decades before. In 1864, his youngest brother Emil died when the Nobel nitro factory near

Stockholm blew up. Two years later, another accident destroyed a Nobel nitro plant outside Hamburg, Germany. Undeterred, that same year Alfred tested the substance in a safer place—a raft on the German river Elbe. Finally, at a demonstration in Manhattan, he proved nitro could be used safely.

His famous breakthrough came the next year. He added an inert substance, silicon-laden soil, to nitro, which yielded dynamite, a material that could be handled and transported safely.

Next came blasting caps, patents for 355 inventions, and multiple profitable businesses in many lands. Most of Nobel's explosives were employed for constructing mine shafts, dams, canals, and buildings.

If taken at his word, then Nobel's quirky idealism was the reason for his eventual interest in peace: "I've got a mass of screws loose and am a superidealist who can digest philosophy better than food."

The Super Weapon That Fired Sound

One of the most important weapons of World War II never fired a shot and was helpless if attacked—but it brought down hundreds of enemy aircraft and saved thousands of lives.

⁕ ⁕ ⁕ ⁕

THE RADIO DETECTION and Ranging device, or radar for short, was developed independently by researchers in eight nations, beginning nearly four decades before World War II. The technology used sound waves to create an echo that would bounce off metal objects such as ships. When the echo returned to the sending unit, it could be analyzed to determine the distance—and to a lesser extent, the size—of the target.

Radar Developments

By the late 1930s, the major belligerents began crash programs to develop the technology for naval- and air-defense purposes. The U.S. Naval Research Laboratory, the British Meteorological Office, and the German Kriegsmarine had developed a series of workable sets by the time hostilities broke out in Europe.

As an island, Britain was protected by the sea and an impressive navy, but its vulnerability to air attacks meant it had much to gain from developing radar. In the late 1930s, it developed a rudimentary radar network called Chain Home. While the technology was merely adequate, the devices could be rushed into production in time to help defend the Home Isles against the Luftwaffe in the summer of 1940. That same year, two British researchers at the University of Birmingham developed the cavity magnetron, a device that allowed radar operators to use higher-frequency sound waves that could be focused more tightly. Britain shipped the prototype in secrecy to the United States, where researchers at Massachusetts Institute of Technology (MIT) developed production models for an improved radar system.

The Soviet Union had one ship-based radar device, the Redut-K system, in place by 1940. For its other radar needs, the USSR relied heavily upon Lend-Lease sets from the United States and Britain.

Neither Germany nor Japan elected to keep pace with the Allies' radar development. Although Germany's Freya system was more sophisticated than the early Chain Home units fielded by Great Britain, Germany had only eight operational units in the field at the outbreak of the war. Further, Freya technology did not accurately determine altitude. The Kriegsmarine received several Freya systems in 1937, and in 1942, after the conquest of France and the Low Countries, the Third Reich established the Kammhuber Line, a chain of radar installations running from Denmark to central France. These stations helped the Germans defend against RAF attacks, but by 1942 British air planners could overwhelm the flak and air interception potential of the Kammhuber Line by concentrating bomber formations.

Japanese radar lagged well behind advances in the United States and Europe. Early in the war, the Japanese also created a small number of reasonably effective naval-radar sets. They relied on some captured devices, including a British model taken after the fall of Singapore and two American devices found when the army overran the Philippines.

Radar Evolves

Ships were equipped with radar to spot enemy craft (and periscopes) at night. De Havilland Mosquito fighter-bombers and Bristol Beaufighter fighters, among others, were fitted with miniature radar sets: These allowed the fighters to locate Luftwaffe bombers at night or in bad weather. Even artillery shells were equipped with tiny radar systems known as proximity fuses, which were especially useful in antiaircraft roles.

As radar became more sophisticated, air forces began using countermeasures, such as chaff (metal strips that reflected radar waves). On radar devices, chaff looked the same as a formation of enemy planes. Raiders would drop the metal strips to divert interceptor resources from their planned targets and protect their planes.

Outstanding and Interesting Tanks

The 20th century saw the armored fighting vehicle (AFV) supplant heavy cavalry as a swift, hard-hitting arm of war. Some tanks have been a cut above their contemporaries, while others were funny looking, or failures, or both.

❋ ❋ ❋ ❋

The Greatest

FT-17 (France, 1917; 37mm or machine gun)

Unlike most elephantine World War I armor, the small FT-17 had a 360-degree rotating gun turret, and it crossed trenches with the help of extended rear rockers rather than oversized, diamond-shaped treads. It was the first World War I tank that looked like World War II tanks and the only one to see significant World War II service.

Somua S-35 (France, 1935; 47mm)

The S-35's tough, cast hull was a big leap from riveted and bolted designs, which tended to spall (meaning "rivets, bolts, coffee cups, and such flying around inside") when hit. Captured S-35s ended up in German service.

Matilda II (U.K., 1937; 40mm)

If you want to stop the enemy in its tracks, bring a tank that can barely be harmed. Mattie's crews could laugh at German tank cannons until 1942, when most Soviet Matildas were phased out in favor of faster, more powerful tanks. Although slow, its bull-dog presence in the battle area commanded respect.

KV-1 (U.S.S.R., 1939; 76mm)

This hulking brute owes its fame partly to timing. When Germany invaded the Soviet Union, the KV was a monster only the Stuka dive-bomber or 88mm flak gun could slay. Outside Leningrad, one KV-1 withstood 135 German cannon hits.

T-34 (U.S.S.R., 1941; 76mm, then 85mm)

Looking for credibility around World War II zealots? Call this the best tank of the war. Its amazing speed, sloped armor, and strong gunnery enabled aggressive tank tactics, perfectly suited to Eastern Front warfare and deadly to Germany. The 1944 T-34 (85) model was still serving in some armies in 2000.

Panzerkampfwagen V Panther (Germany, 1943; 75mm)

Remember the guys who sneered when you heaped praise on the T-34? They have nothing on Panther advocates. Speedy and well armored up front, the Panther's hard-hitting long gun could engage at ranges that allowed few enemies to harm it.

Sherman Firefly (U.K., 1944; 77mm)

Early Shermans burned too easily and hit too gently. Then the British fitted their outstanding 17-pound cannon to Yankee-made Shermans. The result combined the Sherman's reliability with gunnery that could cook even a German Tiger or Panther.

M-60 Patton (U.S., 1960; 105mm)

For two decades, this reliable tank was NATO's mainstay—and perhaps the finest tank of the 1960s. Its variants formed the backbone of Israeli armor during the Yom Kippur War, and the U.S. Marines drove some into battle during the 1991 Gulf War. M-60s were routinely updated until the end of the Cold War.

T-72 (U.S.S.R., 1971; 125mm)

Any tank this prolific rates mention. When introduced, the T-72's 125mm cannon raised the gunnery bar above the common 105mm. Fast, reliable, and with a lot of bang for the ruble, it was once one of the world's most widely deployed tanks.

Merkava (Israel, 1977; 105mm, then 120mm)

After three tank wars with the Arabs, Israel had learned the following: Tank crews experience horrifying deaths; tanks can be replaced, but crewmembers can't; and a crew that feels safe will fight boldly. The innovative Merkava ("Chariot") broke all convention by putting the engine in front, just to help shield the crew. In its first full-dressed engagement (1982), seven Merkavas became battlefield kills—but all of their crewmembers lived.

The Oddest

A7V (Germany, 1917; 57mm)

This clodhopping clunker looked like a railroad caboose covered with a big steel drop cloth, and it fought about as well. It couldn't cross trenches, which was disappointing, because the goal of tanks was to break up trench warfare stalemates. If you romanticize German tanks, try not to look at pictures of this armored banana slug.

Fiat 2000 (Italy, 1917; 65mm)

This looked like a German A7V (see above) with a little observatory on top. While heavily armored and bristling with seven machine guns, the slow F-2000 could barely outrun a briskly marching infantryman. Fortunately for its crews, it never had to try; it only served in peacetime.

M13/40 (Italy, 1940; 47mm)

Lousy armor, weak gun, underpowered and unreliable engine, prone to stalling or catching fire when hit—what a combination. This mantrap was the mainstay of Italian armor in North Africa. With equipment like this, who can blame Italian crews for bailing out of their tanks and surrendering?

Elefant tank destroyer (Germany, 1943; 88mm)

A perennial candidate for Dumbest World War II Armor Design, the glacially slow Elefant mounted the famous 88mm cannon and looked like today's self-propelled howitzers. It had no machine guns of its own, so enemy infantry were welcome to spray tags on it or use its massive front housing as a latrine.

Churchill Crocodile (U.K., 1944; 75mm, flamethrower)

Most World War II powers developed flamethrowing tanks; this was one of the best designs. The Crocodile pulled a trailer of modified gasoline "ammo" yet still mounted a standard tank cannon. Hosing flaming gas the length of a football field, the Croc was hell for dug-in defenders.

Sherman DD (U.S./U.K., 1944; 75mm)

Tanks don't swim well, which is an issue when you're planning a D-Day. Allied engineers invented a watertight skirting for the Sherman so that it could float, and then added a little propeller for instant buoyancy—at least enough for the tank to reach the beach. Ten Sherman DD battalions, the equivalent of a full division, waded ashore this way on June 6, 1944. A number swamped and sank, but on that day any Allied tank on shore was welcome.

How Does a Flak Jacket Stop a Bullet?

⁕ ⁕ ⁕ ⁕

"Flak" is an abbreviation of Fliegerabwehrkanone, *a German word that looks rather silly (as many German words do). There's nothing silly, however, about its meaning: anti-aircraft cannon.*

SERIOUS DEVELOPMENT OF flak jackets began during World War II, when Air Force gunners wore nylon vests with steel plates sewn into them as protection against shrapnel. After the war, manufacturers discovered that they could remove the steel plates and instead make the vests out of multiple layers of dense, heavily woven nylon.

Without the steel plates, the vests became a viable option for ground troops to wear during combat. Anywhere from sixteen to twenty-four layers of this nylon fabric were stitched together into a thick quilt. In the 1960s, DuPont developed Kevlar, a lightweight fiber that is five times stronger than a piece of steel of the same weight. Kevlar was added to flak jackets in 1975.

It seems inconceivable that any cloth could withstand the force of a bullet. The key, however, is in the construction of the fabric. In a flak jacket, the fibers are interlaced to form a super strong net. The fibers are twisted as they are woven, which adds to their density. Modern flak jackets also incorporate a coating of resin on the fibers and layers of plastic film between the layers of fabric. The result is a series of nets that are designed to bend but not break.

A bullet that hits the outer layers of the vest's material is flattened into a mushroom-like shape. The remaining layers of the vest can then dissipate the misshapen bullet's energy and prevent it from penetrating. The impact of the offending bullet usually leaves a bruise or blunt trauma to internal organs, which is a minor injury compared to the type of devastation a bullet is meant to inflict.

While no body armor is 100 percent impenetrable, flak jackets offer different levels of protection depending on the construction and materials involved. At the higher levels of protection, plates of lightweight steel or special ceramic are still used. But all flak jackets incorporate this netlike fabric as a first line of defense. *Fliegerabwehrkanone*, indeed.

The Rocket Scientist

Through an unlikely turn of events, the scientist in charge of developing the Nazi's V-2 rocket would help the Americans reach the moon.

* * * *

BORN IN 1912, Wernher Magnus Maximilian Freiherr von Braun had a pedigree of greatness. His father Baron Magnus von Braun was the minister of agriculture under the Weimar Republic, and his mother was descended from

Swedish and German aristocrats. At age 12, inspired by Fritz von Opel's land speed records, he took a "rocket-propelled" trip after lighting six large skyrockets fitted onto a wagon. "The wagon careened crazily about," Braun recalled, "trailing a tail of fire like a comet." The propellants made a thunderous noise; alarmed police briefly put Braun under custody.

From Skyrockets to the Vengeance Weapon

Inspired by scientist Hermann Oberth's landmark 1923 work *The Rocket Into Interplanetary Space*, in 1930 Braun attended Berlin's Charlottenburg Institute, where he worked with Oberth on liquid-fueled rockets. As Nazi Germany rearmed in the 1930s and banned research on civilian rockets, Braun began conducting missile tests for the Wehrmacht's Ordnance Corps. He formed a long-term friendship with an artillery captain, Walter Dornberger, who arranged funding for Braun's doctorate. Braun and Dornberger went on to work at Peenemünde, as technical and military directors. Along the way Braun joined the Nazi Party. He later stated, "My refusal [to join] . . . would have meant abandon[ing] the work of my life."

Braun's team designed the A-4 ballistic missile, renamed by Josef Goebbels as the Vergeltungswaffe 2, the "Vengeance Weapon 2," or V-2. (The shorter-range V-1 "buzz bomb" was designed by engineer Robert Lusser under Luftwaffe supervision.) The designer of the V-2 engine was Walter Thiel, who was killed during a 1943 British air raid on Peenemünde.

The 46-foot-long, ethanol-and-water-fueled missile could hurl its 2,800-pound warhead some 200 miles. Its highest recorded altitude was 117 miles, making it the first craft to reach outer space. But its guidance system was inaccurate: Chances were about even that it would come within 10 miles of a target. Ironically, Braun regarded the V-2 primarily as a device for space travel, and was briefly imprisoned by the Gestapo in 1944 for his presumed disinterest in weaponry.

Death from the Skies

Hurtling down from an altitude of 60 miles at supersonic speeds no fighter could catch, the V-2 struck without warning. "The V-2 was a truly remarkable machine for its time," recalled Braun. From September 1944 to March 1945, 3,172 V-2s were successfully launched: 1,610 hit the vital port of Antwerp; 1,358 hit London. In one incident, 567 people died after a V-2 landed in a Belgian movie house. The attacks on the British capital killed 2,754 and wounded 6,523. The Nazis relied on the weapons even when impractical: 11 V-2s were fired toward the Remagen bridge while Patton's troops poured over the Rhine into Germany.

The British constantly bombed fixed launching sites, so the V-2s were fired instead from mobile launchers, the equipment and fuel borne by 30 trucks. It took two hours to set up the rocket, launch the V-2, and repack equipment and crew.

The German missile program also successfully tested a "rocket U-boat," which fired V-2s from a submarine-towed platform. Had the method been developed sooner, Germany may have used the subs to launch rockets from off the U.S. coastline. German scientists worked on chemical weapons agents for the V-2. Braun's technicians also adapted the V-2 into the Wasserfall, perhaps the first antiaircraft missile.

Created in Deplorable Conditions

Much of this ordnance was built with slave labor. The V-2's chief assembly plants, called Mittelwerk, were in the Harz Mountains near the town of Nordhausen. There, Russian, Polish, and French inmates from the nearby Dora concentration camp dug out an underground factory. The V-2's top engineer, Arthur Rudolph, helped arrange for the transfer of the inmates, after SS General Hans Kammler, an engineer who had built Auschwitz, came up with the idea of using slave labor for rockets. (Decades later Rudolph fled the United States after the Justice Department accused him of war crimes.)

Braun visited the Mittelwerk plant enough to know, he later admitted, that many laborers died from brutal treatment and wretched conditions. Perhaps 15,000 perished. One eyewitness noted, "You could see piles of prisoners every day who had not survived the workload and had been tortured to death by the vindictive guards . . . But Professor Wernher von Braun just walked past them, so close that he almost touched the bodies." At the time Braun remarked, "It is hellish. My spontaneous reaction was to talk to one of the SS guards, only to be told with unmistakable harshness that I should mind my own business, or find myself in the same striped fatigues! . . . I realized that any attempt of reasoning on humane grounds would be utterly futile."

A New Life for the Nazi Scientist

As the war ground to a close, Braun again faced the SS. In spring 1945, Soviet troops closed to 100 miles of Peenemünde. Most of the V-2 staff decided to surrender to the Western Allies, but in the meantime the SS was ordered to liquidate the rocket engineers and burn their records. With forged documents, Braun and 500 of his staff put together dozens of train cars, as well as about 1,000 automobiles and trucks, and headed toward advancing American troops. At the end of the journey, Braun's brother Magnus buttonholed a GI: "My name is Magnus von Braun. My brother invented the V-2. We want to surrender." The Americans took Braun and his staff into custody, recovered their hidden records, and seized hundreds of freightloads of V-2 components.

Braun, his team, their families, caches of scientific records, and enough V-2 components for 100 missiles, were brought to America. Since their Nazi associations would have barred many from visas, the scheme was hush-hush. Stationed at Fort Bliss, Texas, they called themselves "Prisoners of Peace." At the White Sands Proving Grounds in New Mexico, they continued their work; progress was rapid. By October 24, 1945, one of their reconstituted V-2s snapped photos from space.

Thereafter, Braun was transformed into an honored, naturalized American citizen, and fulfilled his boyhood dreams. He married his German sweetheart and had three children. In 1950 his group moved to Huntsville, Alabama, and designed the army's Jupiter ballistic missile, which later launched the first U.S. satellite. He made television programs with Walt Disney that argued for manned space flight. In 1960, in the wake of Sputnik, he was made head of the NASA team that built the Saturn V rocket, which ferried Americans to the moon. He retired in 1972 when NASA opted for the earthbound space shuttle instead of a piloted mission to Mars.

Cabin Pressurization

When mankind first found ways to thrust itself into the skies, not enough was known about human physiology or atmospheric pressure to ensure the safety of doing so. Decades of experimentation were necessary to discover the parameters of safety at high altitudes.

⁕ ⁕ ⁕ ⁕

IN 1875, THREE French balloonists attempted to break the world record for human ascent. It was 36,000 feet at the time. They brought up with them several bags of "breathing air," thinking this would be sufficient to get them to their goal. When *Zenith*, their balloon, reached about 26,000 feet or so, all three lost consciousness due to a lack of oxygen. Gaston Tissandier, the sole survivor, awoke to find colleagues Joseph Croce-Spinelli and Théodore Sivel dead and the vessel plummeting back to the earth. "One becomes indifferent," Tissandier offered. "One thinks neither of the perilous situation nor of any danger; one rises and is happy to rise."

With education came change. Humans continued to rise. As the altitude increased, however, so did the quest to reach new heights safely. The Wright Brothers put the first airplane into the sky near Kitty Hawk, North Carolina, in 1903, but their altitude of 10 feet was, of course, not a problem. The climb to great heights, though, came faster than you can say "fasten your seatbelts." By the 1920s, Lt. John A. Macready took a biplane past 37,000 feet over Dayton, Ohio. To make that possible despite the lack of oxygen at that altitude, stored oxygen was released into an enclosed cabin in the cockpit. The oxygen mask had not been invented yet. Pilots managed to reach 40,000 feet in this manner, but the feats came at a cost. The lack of atmospheric pressure at those altitudes caused at least one pilot's heart to enlarge, leading to serious health problems.

A completely airtight cabin was built into a Wright-Dayton USD-9A biplane in 1921. Small turbines pushed air into it and a valve could then release it. Macready tried twice to operate the plane at high altitude. The first time, he discovered that the valve was not able to release air as quickly as the turbines were pumping it in, resulting in too much pressure. He abandoned his second attempt after reaching 31,000 feet upon realizing he was too tall to close the cabin's hatch.

Passengers May Now Board

After experimentation in many parts of the world, the U.S. Army Air Corps sent its XC-35 into the skies as the first airplane featuring room in its pressurized cabin for passengers in addition to the crew of three. The Army Air Corps won the 1937 Collier Trophy for most significant development of the year for the feat. The following year, the Boeing 307 Stratoliner became the first commercial airliner to feature a pressurized cabin. By 1940, Boeing was flying passengers in pressurized planes at 20,000 feet.

Anyone who has had their ears pop or plug has noticed the impact that altitude has when it comes to oxygen and pressurized flight. As we climb through the atmosphere, there is less oxygen in the air. Most of the oxygen in the earth's 300-mile-thick atmosphere is contained in the bottom layer. As we climb, it becomes more difficult to breathe. In Denver, the "Mile High

City," the atmospheric air pressure is 12 pounds per square inch versus the 14.7 mark at sea level. At about 18,000 feet, the pressure is 7.3 psi, about half of what it is at sea level. Because most airliners fly between 30,000 and 43,000 feet, where the atmosphere provides less than 4 pounds of pressure, a pressurized cabin is a matter of life and death.

How this works is relatively simple. Because no airplane is 100% airtight, sealing it with ground-level air is not the solution. Instead, outside air is pumped into the fuselage, typically around 11 or 12 psi, and an outflow valve then allows the air to exit. It's usually located on the side or bottom of the fuselage and it opens and closes according to how much pressure is needed. To increase the pressure, the door closes. To decrease pressure, it opens. In addition to providing crew and passengers the ability to breathe, this pressurization method also ensures fresh, clean air inside the plane.

Another important number to note is differential pressure: the difference between the pressure outside the aircraft and that in the cabin. Surpassing that limit is what causes an overblown balloon to pop. Needless to say, that's not a situation airplanes want to experience. That's why the cabin pressure is set a few psi lower than the 14.7 sea-level mark, rather than risking a scenario where the interior pressure is too high.

Hollywood has given some false impressions of what happens if an airplane loses cabin pressure. While rapid decompression *could* take place in a catastrophic situation, the FAA Advisory Circular has a less frightening chart showing how long crew members can perform duties with an insufficient oxygen supply. Even in a decompression situation, the FAA reports that a crew would have five minutes of "useful consciousness" at 22,000 feet. But that time goes down as altitude goes up. Fortunately, these situations are amazingly rare. Pressurization technology has advanced to a point where human beings can enjoy the comfort of the friendly skies, even at high altitudes.

Inside the M3 Tank

Despite several design flaws, M3 tanks did their best to rival Germany's panzers before being replaced by M4s in the latter part of the war.

* * * *

WHEN WORLD WAR II began in September 1939, German tank technology and armored doctrine were far superior to anything the Allies had developed. When Germany's panzers, mechanized infantry, and dive-bombing planes rolled the French and British armies back to the shores of the English Channel in a matter of mere weeks, American military planners were put on notice that any future wars would be fought in a manner and with weapons not previously experienced.

While the Germans adhered to the doctrine of unified tank formations with a mix of armament, American strategists of the day expected tanks would be used only in conjunction with mobile infantry. Based on their limited World War I experiences, U.S. Army commanders believed tanks would take a secondary, supporting role to infantry units, which they incorrectly assumed would again dominate the fields of battle.

Germany's so-called Blitzkrieg tactics served as a wake-up call to American military planners. In July 1940 the War Department authorized the design and development of a new medium-class tank. The resulting M3 was eventually sold to Great Britain and Russia through the Lend-Lease program.

The M3 was designed as a stopgap. The new tank was built to counter the superior armor found on German tanks, and to that end was well armed. It packed a 75-mm gun, which was on par with anything the Germans produced at the time, and could fire armor piercing and high-explosive (HE) rounds.

A small turret with a high-velocity 37-mm gun and coaxial .30-caliber machine gun were mounted atop the M3's hull.

A small cupola on top of this turret held a second .30-caliber machine gun, while two additional .30-caliber machine guns were mounted in a fixed position on the left side of the hull.

Design Flaws

Because of time and production constraints, the major drawback of the M3's main gun was that it was not turret mounted. The desire to get the tank into production in the shortest possible amount of time precluded the design of a rotating turret big enough to house the new gun. As the U.S. Army had not yet developed a turret capable of holding a large caliber gun, the M3's 75-mm gun was instead mounted in an ungainly fashion in a sponson in the right front side of the tank's hull. This archaic design feature was a throwback to the first tank designs of World War I and restricted the gun's range of movement to a mere 15 degrees to each side.

Another shortcoming was the tank's ungainly height. It stood a whopping 10-feet 4-inches tall, making the vehicle a prime target on any battlefield. Having the 75-mm gun located midpoint on the tank's vertical frame meant the top half of the M3 would be exposed to enemy gunners before the M3's crew could fire on their target.

Instead of being welded, the M3's steel components were riveted together. A hit on an M3 by an enemy shell could pop rivets out of place, which might then ricochet around the inside of the tank, wounding or killing crewmen. Later variants were produced with welded components.

The M3 also suffered from poor off-road performance. Tanks sold to Russia were often referred to by their crews as the "grave for six brothers," and Russian tankers complained of continual track failure. The combination rubber-metal tracks had a tendency to burn out during the heat of battle, causing the tracks to collapse and the tank to become immobilized.

M3 In Battle

M3s were first used by the British in North Africa. On May 26–27, 1942, 167 of these tanks took part in the Battle of the Gazala Line, which pitted elements of the British 8th Army Division against an offensive launched by General Erwin Rommel's Afrika Korps. The M3 performed admirably and was more than a match for the German's Panzer III and IV tanks. By the end of the battle, the British, sporting their new American-made tanks, had cost Rommel almost one-third of his panzers. In North Africa, British crews used the M3's 75-mm HE rounds with great effect against the enemy's 88-mm antiaircraft gun, which the Germans had ingeniously converted into an antitank gun. The use of HE rounds enabled the British to shell German antitank positions while remaining outside the effective range of the German guns. The war progressed, however, and the M3 was quickly outclassed. The introduction of the German Panther and Tiger tanks rendered the M3 obsolete, and the tank was withdrawn from service in every theater of operation, save one—the China-Burma-India frontier, where the British employed the tank until the end of the war. While M3 variants continued to operate in Europe as tank recovery vehicles or howitzer gun platforms, the M4 (the famed "Sherman") effectively replaced the M3 as the Allies' main battle tank by mid-1943.

The Ancient Pedigree of Biological and Chemical Warfare

Considered the pinnacle of military know-how, biological and chemical warfare has actually been around for millennia.

* * * *

China's Deadly Fog

INVENTORS OF GUNPOWDER and rockets, the Chinese were also among the first to use biological and chemical agents.

Fumigation to purge homes of vermin in the 7th century BC likely inspired the employment of poisonous smoke during war. Ancient Chinese military writings contain hundreds of recipes for such things as "soul-hunting fog," containing arsenic, and "five-league fog," which was laced with wolf dung. When a besieging army burrowed under a city's walls, defenders struck back. They burned piles of mustard in ovens, then operated bellows to blow the noxious gas at the subterranean attackers. In the 2nd century AD, authorities dispersed hordes of rebellious peasants with a kind of tear gas made from chopped bits of lime.

Ancient Greek Poisons

The ancient Greeks were also experienced with biological and chemical weapons. Herodotus wrote in the 5th century BC about the Scythian archers, who were barbarian warriors dwelling near Greek colonies along the Black Sea. By his account, Scythian bowmen could accurately fire an arrow 500 yards every three seconds. Their arrows were dipped in a mixture of dung, human blood, and the venom of adders. These ingredients were mixed and buried in jars until they reached the desired state of putrefaction. These poison arrows paralyzed the lungs, inducing asphyxiation.

A bioweapon figured prominently in the First Sacred War. Around 590 BC, fighters from the city of Kirrha attacked travelers on their way to the Oracle of Delphi and seized Delphic territories. Enraged at the sacrilege, several Greek city-states formed the League of Delphi and attacked Kirrha. For a time, the town's defenses stymied the attackers. However, according to the ancient writer Thessalos, a horse stepped through a piece of a buried pipe that brought water into the city. A medicine man named Nebros convinced the Greeks to ply the water with the plant hellebore, a strong purgative. The defenders, devastated by diarrhea, were rendered too weak to fight, and the Greeks captured the town and killed every inhabitant.

Flying Corpses Spread the Black Plague

In 1340, during the siege of a French town during the Hundred Years War, it was reported that catapults "cast in deed horses, and beestes stynking . . . the ayre was hote as in the myddes of somer: the stynke and ayre was so abominable." Vlad the Impaler, the 15th-century Romanian warlord and real-life model for Dracula, used a similar method against his foes.

Scholars believe that this ghastly biological warfare tactic played a big role in spreading the worst plague in human history, the bubonic plague, better known as the Black Death. In 1346, merchants from Genoa set up a trading outpost in Crimea, which was attacked by Tartars, a warlike horde of Muslim Turks. However, during the siege, the attacking forces were decimated by the plague. To even the score, the Tartars catapulted the corpses of plague victims over the walls of the Genoan fortress.

Horrified, the Genoan merchants set sail for home. In October 1347, their galleys, carrying rats and fleas infested with the Black Death, pulled into Genoa's harbor. Within several years, the plague would spread from Italy to the rest of Europe, felling more than a third of its population.

A Pox on All Their Houses

In America, biological warfare darkened the French and Indian War. In 1763, during the vast rebellion of Native Americans under Chief Pontiac, the Delaware tribe allied with the French and attacked the British at Fort Pitt. Following the deaths of 400 soldiers and 2,000 settlers, the fort's defenders turned to desperate means. William Trent, the commander of Fort Pitt's militia, knew that a smallpox epidemic had been ravaging the area, and he concocted a plan. He then made a sinister "peace offering" to the attackers. Trent wrote in his journal, "We gave them two Blankets and an Handkerchief out of the Small Pox Hospital. I hope it will have the desired effect." It did. Afflicted with the disease, the Delaware died in droves, and the fort held.

Trent's idea caught on. Soon after the Fort Pitt incident, Lord Jeffrey Amherst, the British military commander in North America, wrote to Colonel Henry Bouquet, "Could it not be contrived to send the Small Pox among those disaffected tribes of Indians? We must on this occasion use every stratagem in our power to reduce them." Amherst, for whom Amherst, Massachusetts, is named, added, "Try every other method that can serve to Extirpate this Execrable Race."

The Da Vinci Formula

Even one of history's best and brightest minds, Leonardo da Vinci, dabbled with chemical weapons. The artist, and sometime inventor of war machines, proposed to "throw poison in the form of powder upon galleys." He stated, "Chalk, fine sulfide of arsenic, and powdered verdigris [toxic copper acetate] may be thrown among enemy ships by means of small mangonels [single-arm catapults], and all those who, as they breathe, inhale the powder into their lungs will become asphyxiated." Ever ahead of his time, the inveterate inventor even sketched out a diagram for a simple gas mask.

Lightning Strikes from on High

The most successful WWII American fighter was the P-38 Lightning. The single-seat, twin-engine craft and its skilled pilots compiled an astonishing ten-to-one kill ratio against Japanese opponents.

* * * *

THE P-38 WAS DESIGNED by Clarence "Kelly" Johnson and his team of Lockheed engineers, the nucleus of the legendary "Skunk Works." The prototype cost $134,284. NASA's predecessor, the National Advisory Committee for Aeronautics, performed the wind-tunnel tests. In 1939, the P-38 broke the cross-continental flight record with a time of seven hours, two minutes. Although the pilot crashed on landing, the Army Air Force had already contracted Lockheed for 66 P-38s.

High Marks for the Lightning

The plane set many other performance marks. Its two turbocharged V-1710 engines propelled the craft at 413 mph, a speed unmatched until the jet-powered P-80 Shooting Star entered service in 1944. Its range—1,100 miles—was tops for any U.S. fighter. Moreover, the P-38 could climb to 40,000 feet, and it zoomed up far faster than hostile planes. The twin engines made it more durable than the single-engine P-51 Mustang fighter.

Advanced features led to unmatched versatility. The P-38 could double as a bomber and carry 4,000 pounds of munitions, almost as much as a B-17. Equipped with two torpedoes, it had the potential to attack large ships, although it never did so in combat. Outfitted with radar, it served as a nighttime reconnaissance plane. But most impressive was its firepower. You may have seen archival footage of U.S. fighters strafing and blowing up Japanese ships; in all likelihood the footage was shot by camera-equipped P-38s, firing four .50-caliber machine guns and a 20-mm cannon.

A Top Secret Task

The P-38's strengths played to its most famous mission, the killing of Yamamoto Isoroku, Commander in Chief of the Japanese Combined Fleet and mastermind of the Pearl Harbor attack. In April 1943, naval intelligence intercepted plans for Yamamoto's airborne arrival at Ballale Airfield off Bourgainville in the Solomon Islands.

A U.S. ambush force arrived at its destination at 9:34 a.m., exactly as the punctual Yamamoto's entourage appeared, with its two Mitsubishi "Betty" bombers and six Zeroes flying cover. Four attacking P-38s sent the admiral's bomber crashing into the jungle. All but one P-38 returned safely to base. U.S. newspapers published a cover story stating that an Allied coast watcher had spotted Yamamoto, leading the Japanese to believe their communications were still secure.

Minor Setbacks

Still, the P-38 had its problems. Earlier models had a terrifying tendency to freeze up during dives. Because the engines were off to the side, they provided no warmth to the cockpit, which was freezing cold during bomber support runs over Germany. Because the P-38 was so fast, pilots could not open the canopy in flight, so they kept the canopy shut and sweltered in the tropical Pacific.

Lightning Aces

The Lightning's best pilots overcame the obstacles. One was Major Thomas "Tommy" McGuire, credited with 38 kills, 7 in one battle alone. He scored the second most kills ever among American pilots. The top American ace of all time was Major Richard "Dick" Bong, with 40 kills, and like McGuire, a winner of the Congressional Medal of Honor. Bong once buzzed the San Francisco area on a P-38 training flight, knocking the clothes off a woman's laundry line. His enraged commander ordered, "Monday morning you check this address out in Oakland, and if the woman has any washing to be hung out on the line, you do it for her!"

But fun and games were few for such aces. The physical forces of an all-out air fight would literally warp the metal of their planes. But the exertions were worth it. Said Sakai, the Japanese ace, "The P-38 . . . destroyed the morale of the Zero fighter pilot."

Bomb in a Bottle

You don't toast the bride with this cocktail.

* * * *

THE MOLOTOV COCKTAIL is a crude but effective hand-held weapon that can be put together as simply as a grade-school science experiment: Fill a slim, easily gripped bottle to about the three-quarter mark with gasoline or other flammable liquid (kerosene and even wood alcohol will do); push a wick-like shred of oily rag snugly into the neck of the bottle; and wait for the perfect moment. Then light the rag, fling the bottle, and watch for the red-orange explosion of fire. The ideal outcome is a panicked scramble of enemy troops, some of them ablaze and many out in the open, where they can be picked off by gunfire.

Originally known as the petrol or gasoline bomb, this simple but intimidating weapon took its most enduring name from Soviet Foreign Minister Vyacheslav Molotov, who claimed on radio broadcasts during his nation's brutal 1939–40 Winter War against Finland that the USSR wasn't dropping bombs on the Finns, despite all evidence to the contrary. The Finnish Army jokingly responded to the foreign minister's lie by calling the Soviet air bombs "Molotov breadbaskets." Finnish troops had already been using petrol bombs against the invading Soviets (the devices had been widely used in the Spanish Civil War of 1936–39) and soon dubbed them "Molotov cocktails."

During the Winter War, Finland's national alcohol retailing monopoly, Alko, manufactured 450,000 bottle bombs made from a mix of ethanol, tar, and gasoline—perhaps the only time the weapon has been professionally produced on a mass scale.

That winter, the Finns bloodied the Soviets far more seriously than the world anticipated, and although they ultimately lost the brutish little war, "Molotov cocktail" would shortly enter the popular lexicon.

Cheers.

The Norden Bombsight

Civilian populations in some areas may have been spared during the war due to the tinkering of a humble engineer.

* * * *

IN 1920, DUTCH immigrant Carl Norden began working on an advanced bombsight for the U.S. Navy. Two years later, he partnered with Theodore Barth, and during the next four years the team designed and built a bombsight from locations in both the United States and Switzerland. The firm incorporated in 1928 and finished a revolutionary bombsight the same year. Its key feature was a timing mechanism that indicated when the bombardier should release his payload.

Norden demonstrated his bombsight's capabilities in 1931 to senior Navy personnel, who were duly impressed and placed a contract for 40 sights. The Army Air Corps also placed an order. In 1935, the Air Corps tested the new bombsights, which had been installed on B-10 bombers: By the end of testing, bombardiers were able to drop 50 percent of their bombs within just 75 feet of their targets.

How Did It Work?

The Norden bombsight was a complex piece of equipment consisting of an analog computer, gyros, levels, mirrors, electric motors, and a small telescope. The final version, the Mark XV, comprised more than 2,000 individual components.

Approximately 30 to 45 minutes before reaching the target, the bombardier would run through a series of precise, carefully predetermined functions to prepare the bombsight for use. He would input altitude, airspeed, wind speed, and the angle of drift into the bombsight's computer, which then calculated both the trajectory of the bombs and the exact time the payload should be released.

Antiaircraft fire and weather conditions severely decreased the accuracy of the bombsight. To help counter these adverse conditions, Norden had a hand (Minneapolis-Honeywell was also involved) in the invention of an automatic pilot, which worked in conjunction with the sight. The autopilot helped reduce turbulence and overcontrol by anxious pilots, which increased the accuracy of the sight.

The actual bombing run began when the airplane reached a point specified at the preflight briefing. The bombardier then ran through a series of final steps that began with aligning the vertical and horizontal cross hairs on the target and ended with the bomber's auto-pilot engaging and the bombs dropping.

The Bombsight Becomes a Must

Norden's factory had prewar production capabilities of about 800 bombsights per month. After the Japanese attack on Pearl Harbor, Norden factories began producing some 2,000 units per month. By the fall of 1945, more than 43,000 sights had been manufactured.

Norden was paid just $250 for his bombsight. In the spirit of patriotism typical of the times, he sold the rights to his invention to the U.S. government for just one dollar more. Norden was satisfied with the fact that his invention enabled the military to more accurately strike targets, thus minimizing the collateral damage done to civilian populations.

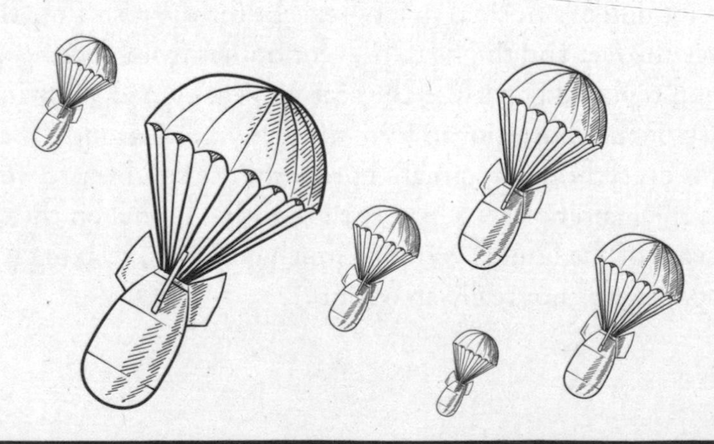

Chapter 4

Spaced Out

Taking Care of Business in Outer Space

Weightlessness sure seems fun. You see those astronauts all effortlessly floating around, mugging for the camera, and magically spinning their pens in midair. But what you don't get to see is what happens when nature calls.

* * * *

The Final Frontier

YOU CAN BE sure that as much as astronauts enjoy swimming through the air like waterless fish, there's one place on Earth where all astronauts thank their lucky stars for gravity: the bathroom.

During shuttle flights, astronauts sat on a commode with a hole in it, not unlike a normal toilet—except for the restraints that fit over the feet and thighs to prevent bodies from floating away. Suction replaced gravity, so the seat was cushioned, allowing the astronaut's posterior to form an airtight seal around the hole. If everything was situated properly, the solid waste went down the main hole: A separate tube with a funnel on the end took care of the liquids. With so much going on, relaxing with a newspaper was not really an option.

But it was worse in the early days of the Apollo missions (1961–1975). When an Apollo astronaut had to go number two, he attached a specially designed plastic bag to his rear end. The bag had an adhesive flange at its opening to ensure a proper seal.

But if you think that this procedure couldn't have been any more undignified, consider this: There was no privacy. The astronauts would usually carry on with their duties while they were, you know, doing their duty. In the words of Apollo astronaut Rusty Schweickart, "You just float around for a while doing things with a bag on your butt." With no gravity and no suction, getting the feces to separate from the body was, generally, an hour-long process. It began with removing the bag—very carefully—and ended with lots and lots of wiping.

Waste Management

Where does all this stuff go? In the past, fecal material was dried, compressed, and stored until the ship returned to Earth. The ISS periodically shoots collections of poop at Earth, to burn up in the atmosphere. And spaceships have been leaving trails of urine in the sky for rather a long time. The memory of this event caused Schweickart to wax darn-near poetic, calling a urine dump at sunset, "one of the most beautiful sights" he saw in space.

"As the stuff comes out and hits the exit nozzle," Schweickart went on, "it instantly flashes into ten million little ice crystals, which go out almost in a hemisphere. The stuff goes in every direction, all radially out from the spacecraft at relatively high velocity. It's surprising, and it's an incredible stream of . . . just a spray of sparklers almost. It's really a spectacular sight."

And you thought stars looked cool.

Lunar Legacy: The Big Whack?

Here's one thing we know for certain about the moon: It isn't made of cheese. Most everything else, including its origins, is a matter of scientific reasoning and speculation.

⁕ ⁕ ⁕ ⁕

Mooning on about the Moon

OUR PLANET'S MOON, our only true natural satellite, has stimulated romance, mystery, and scientific curiosity. And no wonder—besides the sun, the moon is the most noticeable member of our solar system, measuring about one-quarter the size of Earth. Only one side faces our planet, and every month, because of its orbit around us, we watch the moon change phases, from full to quarter to gibbous to new and back again. The moon is also the subject of various origin theories, which alternately laud it as a deity or discount it as a flying chunk of rock, depending on the culture.

Blinded by Science

The list of scientific theories concerning the moon's origin is a bit smaller. One theory suggests that the moon was "captured" by Earth's gravity as it traveled by our planet; another theory posits that our planet and its satellite formed side by side as the solar system developed some 4.56 billion years ago. The moon has simply tagged along with us ever since.

The most recently accepted theory has its origins in the 19th century. In 1879, the British astronomer George Darwin (son of Charles Darwin) suggested that a rapidly spinning Earth threw off material from the Pacific Ocean, creating the moon. The idea drew criticism on and off for decades. But thanks to the advent of modern computers, scientists have created a similar theoretical scenario that makes parts of Darwin's suggestion more reasonable. The data suggests that while Earth was still in a semimolten state, it was hit by a space body—a protoplanet, or planetesimal—almost the size of Mars, or about half the size of Earth. The massive collision would have sent a huge chunk of broken material into orbit around Earth; over time, those larger pieces could have gathered together—thanks to gravity—creating our moon.

It's All Relative?

Why do scientists now agree with the "Moon, daughter of Earth" theory? One of the main reasons is the Apollo program, the U.S. moon missions. Astronauts gathered and delivered more than 800 pounds of lunar material back to Earth. The dates of those rocks—ranging from 3.2 to 4.2 billion years old for material gathered from the flat, dark maria (lava seas) and 4.3 to 4.5 billion years old for rocks from the highlands—along with their composition, have led scientists to believe that the moon is definitely related to Earth.

The evidence is in the fact that the rocks are similar to Earth's mantle material—the moving, molten layer of our planet just under the crust. If a huge planetary body struck our planet, it would make sense that the resulting material would be similar to rock deep below Earth's surface. In addition, moon rocks have exactly the same oxygen isotope composition as Earth's rocks. Materials from other parts of the solar system have different oxygen isotope compositions, which means that the moon probably formed around Earth's neighborhood.

Is the moon our only satellite? Scientists know there are other space bodies circling our planet, but none of the objects can be considered a moon. They are more likely asteroids caught in the Earth's and moon's gravitation. For example, the asteroid 3753 Cruithne looks like it's following Earth in the orbit around the sun; the asteroid 2002 AA29 follows a horseshoe path near Earth. Neither is a moon, and so far neither rock has been in danger of striking our planet. Another object once caught scientists' eyes: Nearby J002E3 was considered a possible new moon of Earth until it was determined to be the third stage of the *Apollo 12* Saturn V rocket.

Like Money from Heaven

Money may not grow on trees, but it does occasionally fall from the sky. Welcome to the strange but lucrative market in space rocks (and the objects they impact).

❊ ❊ ❊ ❊

Heads Up!

EVERY DAY THE Earth is bombarded by meteorites. Most originate in our solar system's asteroid belt, some are pieces of passing comets, and others originate from the Moon or Mars. On a clear dark night it's possible to observe as many as a few per hour; during a meteor shower the count may rise to as many as 100 per hour.

Despite the large numbers of meteorites, most of them burn up upon hitting the Earth's atmosphere. Of those that survive, most fall into an ocean; the meteorites that fall on land are typically the size of a pea or smaller.

Aside from the few kilograms of moon rocks retrieved from various Apollo and Luna space missions, meteorites are the only extraterrestrial artifacts on the planet. Not only does this make them valuable to scientists seeking knowledge about other planets, but it also makes them valuable to collectors seeking rarities and oddities. As such, finding a bona fide meteorite is a rare and potentially lucrative event.

Watch the Skies . . . for Cash!

In 2007, a small meteorite from Siberia sold for $122,750 at an auction in New York City. In 1972, a cow in Valera, Venezuela, was hit by a falling meteorite. Tiny fragments of the stone that hit the beast are worth more than $1,000. (Unfortunately, the cow was eaten—perhaps its carcass could have fetched a good deal more. "Space steaks," anyone?)

Looking to get married? As of 2009, you can purchase a custom wedding ring made from fragments of the Gibeon

Meteorite that hit Africa more than 30,000 years ago (after spending 4 billion years hurtling through space) for as little as $195. If you'd rather save some money, a dime-size sliver of Mars goes for about $100 online.

Lucrative Destruction

Interestingly, the most sought-after meteorite fragments are those that hit other objects. These meteorites are called "hammers," and the smaller the object they hit the better. Even more valuable than hammer meteorites, however, are the objects themselves.

While many people would order their meteorite off the Internet, one woman had hers delivered in a more direct manner. On the evening of December 10, 1984, Carutha Barnard of Claxon, Georgia (who was not a meteorite collector at the time), received a surprise package when a small stone slammed into the back of her mailbox and thudded into the ground below. Though the box itself was knocked from its post, the outgoing mail flag intrepidly remained in the upright position. In October 2007, the mailbox, without the accompanying meteorite, sold for a whopping $83,000 at auction.

The Most Valuable Chevy on Earth

On October 9, 1992, thousands of people across the Mid-Atlantic States witnessed one of the most spectacular meteor showers on record. Glowing brighter than a full moon, and breaking into more than 70 fragments as it arched through the heavens, the meteorite was recorded by more than a dozen video cameras.

In Peekskill, New York, Michelle Knapp was startled by a loud noise outside her home and emerged to find the rear end of her red 1980 Chevy Malibu badly damaged. The police were soon on the scene and filed a criminal report, not realizing that the perpetrator was still on the premises. The smell of leaking gasoline, however, brought the fire department, which discovered the fallen meteorite fragment in the Malibu's gas tank. Knapp sold the car and rock to collector R. A. Langheinrich for an estimated $30,000. Since then, the car has traveled the world as a display piece to places such as New York, Paris, Tokyo, and Munich.

Tracking Tektite Truths

The origin of strangely shaped bits of glass called tektites has been debated for decades—do they come from the moon? From somewhere else in outer space? It seems the answer is more down-to-earth.

* * * *

THE FIRST TEKTITES were found in 1787 in the Moldau River in the Czech Republic, giving them their original name, "Moldavites." They come in many shapes (button, teardrop, dumbbell, and blobs), have little or no water content, and range from dark green to black to colorless.

Originally, many geologists believed tektites were extraterrestrial in origin, specifically from the moon. They theorized that impacts from comets and asteroids—or even volcanic eruptions—on the moon ejected huge amounts of material. As the moon circled in its orbit around our planet, the material eventually worked its way to Earth, through the atmosphere, and onto the surface.

One of the first scientists to debate the tektite-lunar origin idea was Texas geologist Virgil E. Barnes, who contended that tektites were actually created from Earth-bound soil and rock. Many scientists now agree with Barnes, theorizing that when a comet or asteroid collided with the earth, it sent massive amounts of material high into the atmosphere at hypervelocities. The energy from such a strike easily melted the terrestrial rock and burned off much of the material's water. And because of the earth's gravitational pull, what goes up must come down—causing the melted material to rain down on the planet in specific locations. Most of the resulting tektites have been exposed to the elements for millions of years, causing many to be etched and/or eroded over time.

Unlike most extraterrestrial rocks—such as meteorites and micrometeorites, which are found everywhere on Earth—tektites are generally found in four major regions of the world called *strewn* (or splash) fields. The almost 15-million-year-old Moldavites are mainly found in the Czech Republic, but the strewn field extends into Austria; these tektites are derived from the Nordlinger Ries impact crater in southern Germany. The australites, indochinites, and chinites of the huge Australasian strewn field extend around Australia, Indochina, and the Philippines; so far, no one has agreed on its source crater. The georgiaites (Georgia) and bediasites (Texas) are North American tektites formed by the asteroid impact that created the Chesapeake Crater around 35 million years ago. And finally, the 1.3-million-year-old ivorites of the Ivory Coast strewn field originate from the Bosumtwi crater in neighboring Ghana. Other tektites have been discovered in various places around the world but in very limited quantities compared to the major strewn fields.

Weird Things Happen to an Astronaut's Body in Outer Space

They sure do. Playing zero-gravity paddleball and cruising around on the surface of the moon may look like good, clean fun, but space travel is no picnic.

* * * *

ASTRONAUTS' BODIES ENDURE some crazy changes in the celestial firmament, and it takes awhile for them to recover once they're back on solid ground. What's the problem? In a word, weightlessness.

The most immediate consequence of the zero-G lifestyle is something that astronauts call space adaptation syndrome. It occurs because the structure of the inner ear that gives you your sense of balance is acclimated to the constant force of gravity; when that force disappears, your inner ear tells you that you're perpetually falling forward. This typically causes nausea, vomiting, dizziness, and disorientation.

There are other, more serious consequences to leaving Earth, and they have to do with gravity's effect on the rest of the body. Here on Earth, we spend our entire lives within the planet's gravitational field. Even when we're not trying to leap over a puddle or pull off a Superman dunk, the force of gravity is constantly compressing our bodies—and, importantly, our bodies are fighting back.

To understand how your body is forever fighting gravity, just think of your circulatory system. The force of gravity tries to pull your blood supply down into your lower extremities; your heart meanwhile works hard against gravity to keep the blood flowing into your upper body, too. If you were to leave Earth's gravitational field, your heart would be working too hard at forcing the blood upwards, causing your face to swell and leading to nasal congestion and bulging eyes. But eventually, your body would adjust to its new environment; your heart would pump less intensely and your blood pressure would lower.

And without the force of gravity, your muscles atrophy—they get smaller and weaker. Hardest hit are postural muscles like your hamstrings and back muscles, the ones that fight against gravity to keep you standing tall. But all of your muscles begin to wither, as it requires less effort to make any movement. Even your heart is affected, in part because of your lowered blood pressure. As a perk, your intervertebral discs—essentially shock absorbers for the spine—expand, making you two to three inches taller. But even this height enhancement can be painful.

The biggest problem is the effect of weightlessness on the bones. In order to stay strong enough to meet the demands of daily living, your bones constantly regenerate themselves in accordance with the level of strain that they experience. For example, if you lift weights regularly, your bones will grow stronger and more calcium-fortified.

In space, however, the reduced level of stress causes bones to weaken and lose their mass. Studies on astronauts working in Skylab in the 1970s showed a 0.3 percent loss of bone mass during each month of weightlessness.

Due to all of these physical changes, astronauts are in pretty bad shape when they get back home. Their lowered blood pressure can lead to fainting. (To alleviate this problem, they sometimes wear special suits that compress the legs and feet, forcing more blood to the torso and head.) Their sense of balance is out of whack for about a week, making it hard for them to remain steady on their feet. And they're very weak because of the muscle degeneration; it can take months to fully regain lost muscle mass. Bone regeneration can take years, and extended missions—like a three-year Mars trip—would cause permanent bone damage.

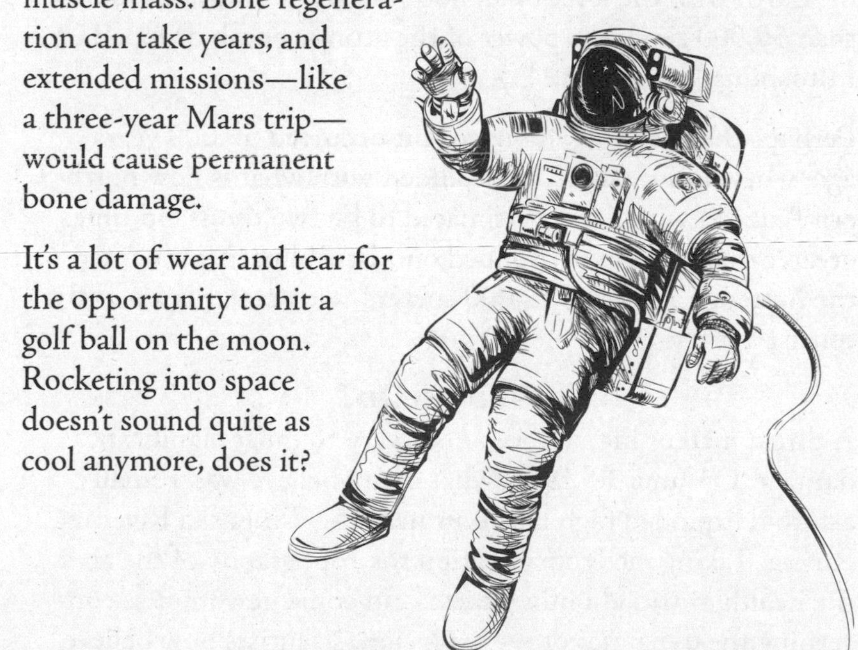

It's a lot of wear and tear for the opportunity to hit a golf ball on the moon. Rocketing into space doesn't sound quite as cool anymore, does it?

Watch Out for That Meteor

NASA estimates that about once every 100 years, a meteorite substantial enough in size to cause tidal waves wallops Earth's surface. About once every few hundred thousand years, an object strikes that is large enough to cause a global catastrophe. So future hits—both larger and smaller—are inevitable.

* * * *

Scanning the Skies

NASA'S NEAR EARTH Objects program surveys the heavens and observes comets and asteroids that could potentially enter Earth's neighborhood. It has been keeping close tabs on an asteroid called Apophis, a.k.a. MN2004. According to NASA, on April 13, 2029, Apophis will be close enough to Earth that it will be visible to the naked eye. At one time, the odds were estimated to be as great as 1-in-300 that Apophis would hit Earth. However, NASA has now ruled out a collision, which is a good thing because the asteroid would have hit Earth with the force of an 880-megaton explosion (more than 50,000 times the power of the atomic bomb dropped on Hiroshima, Japan, in 1945).

Perhaps the best-known meteor hit occurred 50,000 years ago, when an iron meteorite collided with what is now northern Arizona with a force estimated to be two thousand times greater than the bomb dropped on Hiroshima. Now named the Meteor Crater, the 12,000-meters-wide crater is a popular tourist attraction.

What to Do?

A direct meteor hit isn't even necessary to cause significant damage. On June 30, 1908, what many believe was a small asteroid exploded high in the air near the Tunguska River in Russia. Taking into consideration the topography of the area, the health of the adjoining forest, and some new models concerning the dynamics of the explosion, scientists now believe that the force of the explosion was about three to five megatons. Trees were knocked down for hundreds of square miles.

NASA hopes to provide a few years' warning if there is a meteor approaching that could cause a global catastrophe. The organization anticipates that our existing technology would allow us to, among other things, set off nuclear fusion weapons near an object in order to deflect its trajectory. Or we can simply hope that Bruce Willis will save us, just like he did in the 1998 movie *Armageddon*.

You Say Uranus, I Say George

Its name has been the butt of countless bad jokes, but was the planet Uranus—the dimmest bulb in our solar system and nothing more than a celestial conglomeration of hydrogen, helium, and ice—first known as George?

* * * *

THERE'S ACTUALLY MORE truth than rumor in this story, but the lines of historical fact and fiction are blurred just enough to make the discovery and naming of the seventh planet fascinating. The heavenly globe that eventually was saddled with the name Uranus had been seen for years before it was given its just rewards. For decades, it was thought to be simply another star and was even cataloged as such under the name 34 Tauri (it was initially detected in the constellation Taurus). Astronomer William Herschel first determined that the circulating specimen was actually a planet. On the evening of March 13, 1791, while scanning the sky for the odd and unusual, Herschel spotted what he first assumed was a comet.

After months of scrutiny, Herschel announced his discovery to a higher power, in this case the Royal Society of London for the Improvement of Natural Knowledge, which agreed that the scientist had indeed plucked a planet out of the night sky. King George III was duly impressed and rewarded Herschel with a tidy bursary to continue with his research. To honor his monarch, Herschel named his discovery Georgium Sidus, or George's Star, referred to simply as George. This caused some consternation among Herschel's contemporaries, who felt the planet should be given a more appropriate—and scientific—appellation. It was therefore decided to name the new planet for Uranus, the Greek god of the sky. Let the mispronunciations begin!

The Ups and Downs of the Vomit Comet

This aptly named airplane has been used by NASA for more than 50 years to train astronauts and conduct experiments in a zero-gravity environment.

* * * *

THE VOMIT COMET simulates the absence of gravity by flying in a series of parabolas—arcs that resemble the paths of especially gut-wrenching roller coasters. When the Vomit Comet descends toward the earth, its passengers experience weightlessness for the 20 to 25 seconds it takes to reach the bottom of the parabola. Then the plane flies back up to repeat the maneuver, beginning a new dive from an altitude of over thirty thousand feet.

Being weightless and buoyant might bring on nausea all by itself, but when the plane arcs, dips, and ascends again, the occupants feel about twice as heavy as usual. The wild ride induces many of its otherwise steely stomached passengers to vomit—hence, the name. (The plane is also called, by those with a greater sense of propriety, the Weightless Wonder.)

NASA has used the Vomit Comet to train astronauts for the Mercury, Gemini, Apollo, Skylab, Space Shuttle, and Space Station programs. The first Vomit Comets, which were unveiled in 1959 as part of the Mercury program, were

C-131 Samaritans. A series of KC-135A Stratotankers came next. The most famous of these, the NASA 930, was retired in 1995 after 22 years of service as NASA's primary reduced-gravity research plane. This is the aircraft that was used to film the scenes of space weightlessness in the 1995 movie *Apollo 13*. It is now on public display at Ellington Field, near Johnson Space Center in Houston. After the 930 was put out to pasture, another KC-135A—the NASA 931, which was retired in 2004—took over. The 931 flew 34,757 parabolas, generating some 285 gallons of vomit. Yes, the engineers at NASA measured the barf.

What Is the Speed of Dark?

Most of us believe that nothing is faster than the speed of light. In high school physics, we learned that something traveling faster could theoretically go back in time. This would allow for the possibility that you could go back in time and kill your grandfather and, thus, negate your existence—a scenario known as the Grandfather Paradox.

* * * *

YET THERE IS something that *may* be faster than the speed of light: the speed of dark. The speed of dark may not even exist. When you're talking about astrophysics and quantum mechanics, nothing is certain (indeed, uncertainty might be said to be the defining principle of modern physics). Observations and experiments in recent years have helped astrophysicists shape a more comprehensive understanding of how the universe operates, but even the most brilliant scientists are operating largely on guesswork. To understand how the speed of dark theoretically might—or might not—exceed the speed of light, we'll have to get into some pretty wild concepts.

As with much of astronomy, our explanation is rooted in the Big Bang. For those of you who slept through science class or were raised in the Bible Belt, the Big Bang is the prevailing scientific explanation for the creation of the universe. According to the Big Bang theory, the universe started as a pinpoint of dense, hot matter. About fourteen billion years ago, this infinitely dense point exploded, sending the foundations of the universe into the outer reaches of space.

The momentum from this initial explosion caused the universe to expand its boundaries outward. For most of the twentieth century, the prevailing thought was that the rate of expansion was slowing down and would eventually grind to a halt. Seemed logical enough, right?

In 1998, however, astronomers who were participating in two top-secret-sounding projects—the Supernova Cosmology Project and the High-Z Supernova Search—made a surprising discovery while observing supernovae events (exploding stars) in the distant reaches of space. Supernovae are handy for astronomers because just prior to exploding, these stars reach a uniform brightness. Why is this important? The stars provide a standard variable, allowing scientists to infer other statistics, such as how far the stars are from Earth. Once scientists know a star's distance from Earth, they can use another phenomenon known as a redshift (a visual analogue to the Doppler effect in which light appears differently to the observer because an object is moving away from him or her) to determine how much the universe has expanded since the explosion.

Still with us? Now, based on what scientists had previously believed, certain supernovae should have appeared brighter than what the redshift indicated. But to the scientists' amazement, the supernovae appeared dimmer, indicating that the expansion of the universe is speeding up, not slowing down. How could this be? And if the expansion is quickening, what is it that's driving it forward and filling up that empty space?

Initially, nobody had any real idea. But after much discussion, theorists came up with the idea of dark energy. What is dark energy? Ultimately, it's a made-up term for the inexplicable and incomprehensible emptiness of deep space. For the purposes of our question, however, dark energy is theoretically far faster than the speed of light—it's so fast, in fact, that it is moving too quickly for new stars to form in the empty space. No, it doesn't make a whole heck of a lot of sense to us either, but rest assured, a lot of very nerdy people have spent a long time studying it.

Of course, there may be a far simpler answer, one posited by science-fiction writer Terry Pratchett: The speed of dark must be faster than the speed of light—otherwise, how would dark be able to get out of the way?

Mercury 13: The Mission That Never Went

In 1961, a group of highly qualified women was selected for astronaut flight training. They passed every test and endured every poke, prod, and simulation. In some cases, they actually fared better than their male counterparts. But was America really ready to send women into space? Apparently not.

* * * *

THE SOVIETS FIRED the starter's gun in the space race by launching the Sputnik satellite on October 4, 1957. Threatened by the Soviets' ability to beat them to the punch, the U.S. accelerated its own space initiatives, including the formation of the National Aeronautics and Space Administration (NASA). Next time, the United States would be first.

There was much yet to learn about space. What could humans tolerate? Jet pilots required pressure suits; what of weightless space? What of the confinement? Military test pilots—fit, brave, and calm during flight crises—seemed logical candidates. Of course, since women weren't allowed to be military test pilots, they weren't considered for astronaut training. At least, not at first.

Secret Experiments

Freethinking researcher Dr. Randy Lovelace II helped screen the first seven male astronauts as part of the Mercury 7 program. Then Lovelace had a flash of inspiration, thinking: A space rocket needs every joule of energy. Every gram of weight counts. Women are lighter; they use less oxygen and food. We know for sure they can fly; heck, Jackie Cochran helped me start my research foundation. Maybe they're actually better suited! Let's explore this!

Cochran herself was well over the age limit of 35, but Oklahoman Geraldyn "Jerrie" Cobb wasn't. A record-setting aviator, Cobb had earned her private pilot's license when she was just 17 years old. Between 1957 and 1960, she set four aviation world records for speed, distance, and absolute altitude.

When Jerrie received an invitation from Dr. Lovelace to train for space flight, she dropped everything for what seemed like the opportunity of a lifetime. She arrived in Albuquerque in 1960 and began the torture tests. She underwent barium enemas and had all her body fluids sampled. Supercooled water was squirted into her ear canal to test her reaction to vertigo. She endured the infamous "Vomit Comet" spin simulator—and many more tests besides.

Cobb blew the trials away. When Lovelace announced this to the media, they gushed over the "astronette." Cobb was the first of 25 women tested for astronaut potential. Only some of the women met one another in person, but Cobb was involved in their recruitment and knew them all. Thirteen passed all the tests to become FLATs: Fellow Lady Astronaut Trainees, an acronym taken from Cobb's written salutation to them.

During this phase, Soviet cosmonaut Yuri Gagarin orbited the planet, lapping NASA once again in the space race.

The women's next planned step was testing at navy facilities in Pensacola. Each went home to wait. But when Lovelace asked to use Pensacola, the navy called NASA. The organization was less than enthused about the female astronauts, so the navy pushed the training overboard.

In September 1961, each FLAT got a telegram: Sorry, program cancelled. You may now resume your normal lives.

"Let's Stop this Now!"

The women couldn't have been more dismayed. All that work—for nothing! They didn't give up, but they also didn't coordinate their lobbying. Cobb, the FLATs' self-appointed spokesperson, didn't get along well with Cochran—who in turn had her own ideas. Cobb's appeals up the national chain of command were honest, impassioned, and naive. Cochran, with personal contacts ranging from Chuck Yeager to VP Lyndon Johnson, preferred to work gradually within the sexist system rather than have an open challenge slapped down. One FLAT, Jane Hart, was the wife of a U.S. senator and was arguably the savviest political spokesperson available. Hart fumed as Cochran testified to Congress that she was against a "special program for women."

Was It Ever Possible?

What if all the Mercury 13 women and Cochran had spoken to Congress and to the country in a unified voice? We can guess the outcome based on LBJ's reaction to the memo across his desk concerning the female astronauts. He scrawled: "LET'S STOP THIS NOW!" If President Kennedy's space tsar had that attitude, there had never been any real hope. The men leading the nation weren't ready to send women into space.

Was Lovelace deluded? Give him credit for trying, but he also didn't clue NASA in until news of the Pensacola plans blind-sided them, resulting in a reflex "no way." On the other hand, had he sought advance permission, odds of NASA giving it were — astronomical.

On June 16, 1963, about a year after Cobb, Hart, and Cochran spoke before Congress, cosmonaut Valentina Tereshkova of the Soviet Union flew in space. She was not a test pilot but a parachute hobbyist and textile worker. It would be 20 more years before the first American woman, Sally Ride, made it as far.

A Traffic Jam in Space

No one gives much thought to all the stuff we launch into space and don't bring back, but it creates a major hazard.

* * * *

Steer Clear of That Satellite

IF YOU THINK it's nerve-wracking when you have to swerve around a huge pothole as you cruise down the highway, just imagine how it would feel if you were hundreds of miles above the surface of Earth, where the stakes couldn't be higher. That's what the crew of the International Space Station (ISS) faced in 2008, when it had to perform evasive maneuvers to avoid debris from a Russian satellite.

And that was just one piece of orbital trash—all in all, there are tens of millions of junky objects that are larger than a millimeter and are in orbit. If you don't find this worrisome, imagine the little buggers zipping along at up to 17,000 miles per hour. Worse, these bits of flotsam and jetsam constantly crash into each other and shatter into even more pieces.

The junk largely comes from satellites that explode or disintegrate; it also includes the upper stages of launch vehicles, burnt-out rocket casings, old payloads and experiments, bolts, wire clusters, slag and dust from solid rocket motors, batteries, droplets of leftover fuel and high-pressure fluids, and even a spacesuit. (No, there wasn't an astronaut who came home naked—the suit was packed with batteries and sensors and was set adrift in 2006 so that scientists could find out how quickly a spacesuit deteriorates in the intense conditions of space.)

The U.S. and Russia: Space's Big Polluters

So who's responsible for all this orbiting garbage? The two biggest offenders are Russia—including the former Soviet Union—and the United States. Other litterers include China, France, Japan, India, Portugal, Egypt, and Chile. Each of the last three countries has launched one satellite during the past twenty years.

Most of the junk orbits Earth at between 525 and 930 miles from the surface. The ISS operates a little closer to Earth—at an altitude of about 250 miles—so it doesn't see the worst of it. Still, the ISS's emergency maneuver in 2008 was a sign that the situation is getting worse.

NASA and other agencies use radar to track the junk and are studying ways to get rid of it for good. Ideas such as shooting at objects with lasers or attaching tethers to some pieces to force them back to Earth have been discarded because of cost considerations and the potential danger to people on the ground. Until an answer is found, NASA practices constant vigilance, monitoring the junk and watching for collisions with working satellites and vehicles as they careen through space. Hazardous driving conditions, it seems, extend well beyond Earth's atmosphere.

Faster, and Slower, Than the Speed of Light

Albert Einstein taught us that the speed of light is constant at 186,282 miles per second. An unbending, iron law of nature, right? Wrong.

* * * *

FOR YEARS, DILIGENT laboratory scientists have sped up, and greatly slowed down, the components of light. They have actually made light travel many times faster than the "speed of light." And they've decelerated light to a plodding pace that wouldn't merit a speeding ticket.

Speed It Up!

At New York's University of Rochester in 2006, scientists led by optics professor Robert W. Boyd fired a laser into an optical glass fiber. The fiber had been laced with the rare metal erbium, which amplified the signal it produced—by a lot. Before the entire pulse even entered the fiber, part of it appeared at the fiber's end and then raced backward faster than the speed of light. The process was attributed to the erbium, which gave extra energy to the light.

Slow It Down!

The University of Rochester team has also taken the opposite tack and slowed down light. In 2003, Boyd's crew shone a green laser through a tiny ruby in an attempt to saturate the chromium atoms that give the gem its reddish tint. When a second green laser zapped the jewel, its light slowed to 127 miles per hour, which is 5.3 million times slower than the light of the first laser. In 1999, scientists at Harvard University slowed laser light to just 38 mph. They did this by shooting a laser through matter that was supercooled to 459 degrees below zero—a temperature at which atoms, or particles of light, practically freeze in their tracks.

Pulsars

Named for their distinctive pulses of electromagnetic radiation, pulsars are sometimes called the "lighthouses" of the universe. And like a lighthouse, these neutron stars rotate, so their beams of radiation are only detectible when pointing directly at Earth.

* * * *

A PULSAR IS FORMED when a massive star dies and detonates as a supernova. The outer layers of the star are blasted off into space, but the inner core is compacted by gravity. Now a smaller size, this tiny, dense object begins to rotate quickly, emitting radiation along its magnetic field lines. On Earth, these beams of radiation can be detected each time the pulsar rotates, which can be several times a second.

Despite the fact that humans have been observing the heavens for centuries, the existence of these interesting celestial objects was unknown until 1967. And when they were discovered, they were so unexpected that scientists initially thought they might have discovered some kind of extraterrestrial communication.

It all began with the study of quasars, the bright nuclei at the center of galaxies. A professor of radio astronomy at the University of Cambridge, Antony Hewish, designed a radio telescope to observe the objects. But building the new telescope, which would cover about four and a half acres, was more than a one-man job. So Hewish enlisted the help of graduate student Jocelyn Bell, who was pursuing her PhD in astronomy. Bell and several other students spent two years constructing the huge telescope, which consisted of 2,000 dipole antennas and 120 miles of wire and cable.

What's This Scruff?

The telescope became operational in July 1967, with Bell in charge of analyzing the data. This required the student to pore through 100 feet of paper readouts every day, which taught her what was typically expected in the data and how it related to the quasars she was studying. But one day, a few weeks into her research, Bell noticed what she called "a bit of scruff" in the data. It didn't appear to be manmade interference, and it wasn't the usual data she was used to seeing. Instead, it was a consistent, regular signal, coming from the same place in the sky.

Clockwork Aliens

Bell shared her finding with Hewish, who was just as confused as the graduate student. The data indicated a series of sharp pulses that occurred every 1.3 seconds, something that had never been observed in the skies before. The two began searching for what could possibly be causing such a signal, ruling out various sources of interference such as television signals, orbiting satellites, or radar reflected off the moon. They even used another telescope to confirm the signal, ruling out a defect with their equipment. Bell and Hewish half-jokingly considered the possibility of an extraterrestrial message, with Bell wondering, "if one thinks one may have detected life elsewhere in the universe, how does one announce the results responsibly?" They nicknamed the signal "LGM-1," for "Little Green Men."

But even the extraterrestrial theory would soon be ruled out, when Bell discovered another signal emanating from a different part of the sky. Since it would be quite unlikely that two groups of aliens from completely different parts of the galaxy would attempt to contact Earth at the same time, Bell and Hewish realized they were dealing with something natural, yet unknown. By the end of 1967, Bell had found a total of four similar signals, and the unknown sources were dubbed "pulsars," combining "pulse" and "quasar."

In February 1968, Hewish held a seminar to announce the unusual discovery, even though neither he nor Bell was yet certain what they'd found. The press, however, ran with the "extraterrestrial" story, finding it too exciting to resist. Astronomers took the discovery more seriously, searching the skies for more unusual signals, and by the end of 1968, dozens of pulsars had been detected. One theory for the source of the signals came from astrophysicist Thomas Gold, a professor of astronomy at Cornell University. Gold suggested that the signals might be from rapidly rotating neutron stars, which, due to this rotation and their strong magnetic fields, would emit consistent pulses of radiation like a rotating beacon—or a lighthouse.

Later in 1968, the discovery of the Crab Pulsar seemed to confirm Gold's theory. This pulsar is located in the center of the Crab Nebula, which was already known to be the remnant of a supernova. The detection of the Crab Pulsar confirmed that these signals were, indeed, from neutron stars.

In 1974, Hewish was awarded the Nobel Prize in Physics for his role in discovering pulsars. While Bell, who originally found the "bit of scruff" that turned out to be an amazing new find, was overlooked, she held no bitterness over the decision. Today, around 1,600 pulsars are known to be shining their beams throughout the universe, with the fastest emitting 716 pulses per second, and the slowest emitting only one pulse every 23.5 seconds. Pulsars are so incredibly regular in their rotations that they have been known to rival even atomic clocks in keeping precise time. Some scientists have even suggested that these "lighthouses" could be used to help spacecraft navigate around the universe. They may not be evidence of intelligent life, but pulsars are fascinating, nonetheless!

The Tunguska Event

What created an explosion 1,000 times greater than the atomic bomb at Hiroshima, destroyed 80 million trees, but left no hole in the ground?

* * * *

The Event

On the morning of June 30, 1908, a powerful explosion ripped through the remote Siberian wilderness near the Tunguska River. Witnesses, from nomadic herdsmen and passengers on a train to a group of people at the nearest trading post, reported seeing a bright object streak through the sky and explode into an enormous fireball. The resulting shockwave flattened approximately 830 square miles of forest. Seismographs in England recorded the event twice, once as the initial shockwave passed and then again after it had circled the planet. A huge cloud of ash reflected sunlight from over the horizon across Asia and Europe. People reported there being enough light in the night sky to facilitate reading.

A Wrathful God

Incredibly, nearly 20 years passed before anyone visited the site. Everyone had a theory of what happened, and none of it good. Outside Russia, however, the event itself was largely unknown. The English scientists who recorded the tremor, for instance, thought that it was simply an earthquake. Inside Russia, the unstable political climate of the time was not conducive to mounting an expedition. Subsequently, the economic and social upheaval created by World War I and the Russian Revolution made scientific expeditions impossible.

Looking for a Hole in the Ground

In 1921, mineralogist Leonid A. Kulik was charged by the Mineralogical Museum of St. Petersburg with locating meteorites that had fallen inside the Soviet Union. Having read old newspapers and eyewitness testimony from the Tunguska region, Kulik convinced the Academy of Sciences in 1927 to fund an expedition to locate the crater and meteorite he was certain existed.

The expedition was not going to be easy, as spring thaws turned the region into a morass. And when the team finally reached the area of destruction, their superstitious guides refused to go any further. Kulik, however, was encouraged by the sight of millions of trees splayed to the ground in a radial pattern pointing outward from an apparent impact point. Returning again, the team finally reached the epicenter where, to their surprise, they found neither a meteor nor a crater. Instead, they found a forest of what looked like telephone poles—trees stripped of their branches and reduced to vertical shafts. Scientists would not witness a similar sight until 1945 in the area below the Hiroshima blast.

Theories Abound

Here are some of the many theories of what happened at Tunguska.

Stony Asteroid: Hitting the atmosphere at a speed of about 33,500 miles per hour, a large space rock heated the air around it to 44,500 degrees Fahrenheit and exploded at an altitude of about 28,000 feet. This produced a fireball that utterly annihilated the asteroid.

Kimberlite Eruption: Formed nearly 2,000 miles below the Earth's surface, a shaft of heavy kimberlite rock carried a huge quantity of methane gas to the Earth's surface where it exploded with great force.

Black Holes or Antimatter: As early as 1941, some scientists believed that a small antimatter asteroid exploded when it encountered the upper atmosphere. In 1973, several theorists proposed that the Tunguska event was the result of a tiny black hole passing through the Earth's surface.

Alien Shipwreck: Noting the similarities between the Hiroshima atomic bomb blast and the Tunguska event, Russian novelist Alexander Kazantsev was the first to suggest that an atomic-powered UFO exploded over Siberia in 1908.

Tesla's Death Ray: Scientist Nikola Tesla is rumored to have test-fired a "death ray" on June 30, 1908, but he believed the experiment to be unsuccessful—until he learned of the Tunguska Event.

Okay, What Really Happened?

In June 2008, scientists from around the world marked the 100-year anniversary of the Tunguska event with conferences in Moscow. Yet scientists still cannot reach a consensus as to what caused the event. In fact, the anniversary gathering was split into two opposing factions—extraterrestrial versus terrestrial—who met at different sites in the city.

Cosmic Microwave Background Radiation

In 1964, Robert Wilson and Arno Penzias, two radio astronomers working for Bell Telephone Laboratories in New Jersey, made a discovery. As Penzias would later say, "we had stumbled upon something big." But at the time, neither Penzias nor Wilson realized just how big. It was, in fact, a revelation that led all the way back to the biggest event in the universe: The Big Bang.

* * * *

But before delving into the story of their discovery, let's think about the universe: First, this giant place that houses our planet and billions of other planets is indescribably vast. If we were to attempt to describe it, we'd have to think about a sphere with a 15-billion light-year radius and picture our tiny Earth somewhere inside it; and that's only the universe we know. Scientists estimate that we've only begun to scratch the surface of the mysteries of space.

And second, we know that the light we observe emanating throughout the universe is traveling at a fixed speed. The light from the sun, for instance, takes about eight minutes to reach Earth, whereas light from Pluto takes 5.3 hours. So if, theoretically, the sun were to suddenly disappear from the sky, it would take eight minutes before humans on Earth even became aware of that fact! It's also interesting to note that the sun is only about 4.5 billion years old, so its light has only traveled 4.5 billion light-years across the universe; this means that the light from our sun has yet to reach the farthest borders of our known universe.

Back to Wilson and Penzias. The two radio astronomers were tasked with creating a radio receiver for Bell Labs at the company's Crawford Hill location, to be used for radio astronomy and satellite communications experiments. They constructed a microwave radiometer, which is a very sensitive receiver that can detect energy emitted at millimeter or centimeter wavelengths, also known as microwaves. On May 20, 1964, they noticed that their new receiver was picking up a strange noise. It seemed to be a buzzing sound that was coming from everywhere, encompassing all parts of the sky at the same time.

Confused, Wilson and Penzias began looking for a source of the sound. They considered whether it could be interference from nearby New York City, or perhaps even an echo from a nuclear bomb that had been test-detonated over the Pacific years earlier. Or maybe their newly built equipment was faulty; to test this theory, they went so far as to replace parts of the receiver, but the sound remained. But then they more closely inspected the radio receiver, and thought they'd finally found their culprit: pigeons. Two birds had built nests inside the antenna, and pigeon dung was building up on the equipment. Wilson and Penzias removed the birds, cleaned the receiver, and tested their equipment again. To their surprise, the noise remained. So what could cause noise to come from all points of the sky at the same time?

A Noisy Place, Regardless

By the 1960s, there were two prevailing theories concerning the origins of the universe. The first was called the "Steady State theory," which hypothesizes that matter is continuously created as the universe expands. This theory also states that the overall density of the universe remains constant, and that the universe has existed forever. The second theory is the "Big Bang theory," which states that the universe began at a point of infinite density and then began to expand outward. Wilson and Penzias theorized that if the second theory was correct, then the universe should be filled with cosmic microwave background radiation left over from the very beginnings of its formation.

After eliminating every other cause of the strange noise (including feathered friends), the two scientists came to a startling conclusion: They had discovered the predicted cosmic microwave background radiation from the Big Bang. Their receiver was picking up a cosmic echo from the very beginnings of the universe's explosive creation, and it was literally coming from everywhere.

This radiation is the oldest light in the universe, dating from about 380,000 years after the Big Bang, which is thought to have occurred around 13.8 billion years ago. After its initial creation, the cosmos was extraordinarily hot, topping out around 273 million degrees above absolute zero. Atoms were quickly broken apart into protons and electrons, and photons of light scattered off into the hot, soupy mix. As the universe began to cool off, the photons, at first as bright and hot as the surface of a star, expanded outward. And, as with all other light in the universe, this light just kept traveling, waiting for someone to discover its presence. Now, this cosmic microwave background radiation has cooled off to a temperature of just 2.73 degrees above absolute zero, or an astoundingly cold negative 456.94 degrees Fahrenheit, and is found quite uniformly throughout the universe.

Wilson and Penzias' discovery immediately lent credence to the Big Bang theory, and the pair was awarded the 1978 Nobel Prize in physics for their work. In 1993, NASA's Cosmic Background Explorer (COBE) mission created a full-sky map of the radiation, which NASA dubbed a "baby picture" of the universe, and other detailed images have been made since then.

Studying the radiation has helped scientists unlock more mysteries of the universe, helping them to pinpoint its age, understand when the first stars were formed, and learn more about the origin of galaxies. There is even some evidence that dark matter and dark energy, mysterious theoretical forces that may affect the laws of gravity throughout the universe, do exist.

It's probably safe to say that when Wilson and Penzias were cleaning pigeon droppings out of their radio receiver, they had no idea that they were on their way to finding the "baby picture" of the universe. But no doubt their discovery inspired new generations of scientists to keep searching for answers to the most perplexing questions in the cosmos.

Is There Such a Thing as a Blue Moon?

Blue moon, You saw me standing alone, Without a dream in my heart, Without a love of my own.

—From the song "Blue Moon"

* * * *

ACCORDING TO THE performers who have recorded this Rodgers and Hart classic—including Elvis Presley, Frank Sinatra, and Bob Dylan—there really is such a thing as a blue moon, and it acts as a celestial matchmaker for the lovelorn. Not everyone is as sappy as this. When most people mention a blue moon, they are referring to an event that is highly unusual. As our crooners might say, "I have a date once in a blue moon."

The phrase "blue moon" dates back to at least 1528. It first appeared in a work by William Barlow, an English bishop, the wonderfully titled *Treatyse of the Buryall of the Masse.* "Yf they saye the mone is belewe," Barlow wrote, "we must beleve that it is true." (Trust us; he's saying something about a blue moon here.) After Barlow's usage, which no doubt confused as many readers as it edified, the term came to represent anything absurd or impossible.

It was only later that "blue moon" connoted something unusual. Most etymologists trace this usage to the wildly popular 1819 edition of the *Maine Farmer's Almanac,* which suggested that when a season experiences four full moons (instead of the usual three), the fourth was to be referred to as "blue."

As is often the case with these things, somehow the *Maine Farmer's Almanac*'s suggestion was misinterpreted—researchers blame the incompetent editors of a 1946 issue of *Sky & Telescope* magazine—to mean a second new full moon in a single month. Consequently, in present-day astronomy, that second new full moon is referred to as a "blue" moon. This frequency, ironically, isn't all that unusual, at least as astronomical events go: once every two and a half years.

As for whether the moon really can appear blue, the answer is yes. After massive forest fires swept through western Canada in 1950, for example, much of eastern North America was treated to a bluish moon in the night sky. However, events such as this occur, well, once in a blue moon.

Space Ghosts

Shortly after the Soviet Union successfully launched Sputnik 1 on October 4, 1957, rumors swirled that some cosmonauts had died during missions gone wrong, and their spacecraft had drifted out of Earth's orbit and into the vast reaches of the universe.

* * * *

IT WAS EASY to believe such stories at the time. After all, the United States was facing off against the Soviet Union in the Cold War, and the thought that the ruthless Russians would do anything to win the space race—including sending cosmonauts to their doom—seemed plausible.

However, numerous researchers have investigated the stories and concluded that, though the Soviet space program was far from perfect and some cosmonauts had in fact died, there are no dead cosmonauts floating in space. According to authors Hal Morgan and Kerry Tucker, the earliest rumors of deceased cosmonauts even mentioned their names and the dates of their doomed missions: Aleksei Ledovsky in 1957, Serenti Shiborin in 1958, and Mirya Gromova in 1959. In fact, by the time Yuri Gagarin became the first human in space in April 1961, the alleged body count exceeded a dozen.

Space Spies

So prevalent were these stories that no less an "authority" than *Reader's Digest* reported on them in its April 1965 issue. Key to the mystery were two brothers in Italy, Achille and Giovanni Battista Judica-Cordiglia, who operated a homemade listening post with a huge dish antenna. Over a seven-month period, the brothers claimed to have overheard radio signals from three troubled Soviet spacecraft:

* On November 28, 1960, a Soviet spacecraft supposedly radioed three times, in Morse code and in English, "SOS to the entire world."

* In early February 1961, the brothers are alleged to have picked up the sound of a rapidly beating heart and labored breathing, which they interpreted to be the final throes of a dying cosmonaut.
* On May 17, 1961, two men and a woman were allegedly overheard saying, in Russian, "Conditions growing worse. Why don't you answer? We are going slower . . . the world will never know about us."

The Black Hole of Soviet PR

One reason rumors of dead cosmonauts were so believable was the extremely secretive nature of the early Soviet space program. Whereas the United States touted its program as a major advance in science and its astronauts as public heroes, the Soviet Union revealed little about its program or even the people involved.

It's not surprising, then, that the Soviet Union did not report to the world the death of Valentin Bondarenko, a cosmonaut who died tragically in a fire after he tossed an alcohol-soaked cotton ball on a hot plate and ignited the oxygen-rich chamber in which he was training. He died in 1961, but it wasn't revealed publicly until 1986.

Adding to the rumors was the fact that other cosmonauts had been mysteriously airbrushed out of official government photographs. However, most had been removed because they had been dropped from the space program for academic, disciplinary, or medical reasons—not because they had died during a mission. One cosmonaut, Grigoriy Nelyubov, was booted from the program in 1961 for engaging in a drunken brawl at a rail station (he died five years later when he stepped in front of a train). Nelyubov's story, like so many others, was not made public until the mid-1980s.

Only one Soviet cosmonaut is known to have died during an actual space mission. In 1967, Vladimir Komarov was killed when the parachute on his *Soyuz 1* spacecraft failed to open properly during reentry. A Russian engineer later acknowledged that Komarov's mission had been ordered before the spacecraft had been fully debugged, likely for political reasons.

The Truth on Space Travel

Like nature, humans abhor a vacuum, and we've been filling the void of knowledge with near-truths and outright falsehoods ever since we broke the grip of Earth's gravity. Here are a few.

✻ ✻ ✻ ✻

There is no gravity in space. There is a difference between "weightlessness" and "zero-g" force. Astronauts may float inside a space shuttle, but they are still under the grasp of approximately 10 percent of Earth's gravity. Essentially, gravity will decrease as the distance from its source increases—but it never just vanishes.

Gravitational forces are powerful enough to distort a person's features. This popular notion can be traced to the fertile minds of Hollywood filmmakers, who quickly learned the value of "artistic license" when dealing with the subject of outer space. In 1955's *Conquest of Space,* director Byron Haskin portrayed space travelers stunned and frozen by the forces of liftoff, pressed deep into their seats with their faces grotesquely distorted. When humankind actually reached space in 1961, the truth became known: Although gravitational forces press against the astronauts, they are perfectly capable of performing routine tasks, and their faces do not resemble Halloween masks.

An ill-suited astronaut will explode. Filmmakers would have you believe that an astronaut who is exposed to the vacuum of space without the protection of a spacesuit would expand like a parade float. With eyes bulging and the body swelling like a big balloon, the poor soul would blow up. It would be a gruesome sight, indeed, but that's not the way it would happen. The human body is too tough to distort in a complete vacuum. The astronaut would simply double over in pain and eventually suffocate.

Stranded space travelers will be asphyxiated. The film world's take on space dangers has occasionally spilled into reality. In movies such as *Marooned,* astronauts are stuck in space as their oxygen supply runs out. Although the danger of being stranded in space is very real (*Apollo 13* comes to mind), astronauts in such a situation would not die from lack of oxygen. Carbon dioxide in a disabled spacecraft could build up to life-threatening levels long before the oxygen ran out.

The world watched as the *Challenger* "exploded." Myth even lies in one of the most tragic spaceflights in U.S. history—the *Challenger* disaster of January 1986. Stories tell of the millions of horrified viewers who watched as the spacecraft broke apart on live television. Except for cable network CNN, however, major networks had ceased their coverage. Because Christa McAuliffe was to be the first teacher in space, NASA had arranged for public schools to show the launch on live TV. Consequently, many of those who actually saw it happen were schoolchildren. It was only when videotaped replays filled the breaking newscasts that "millions" of people were able to view the catastrophe. Another misconception about the *Challenger* is that it actually "exploded." It didn't, at least not in the way most people assume. The shuttle's fuel tank ripped apart, but there was no blast or detonation.

We even have the quote wrong. History tells us that Neil Armstrong said: "That's one small step for man, one giant leap for mankind." But Armstrong was misquoted. He never intended to speak on behalf of thousands of years of human development by declaring it "one small step for man." An innocent "a" got lost in the clipped electronic transmission of nearly 250,000 miles. According to Armstrong himself, he said, "That's one small step for a man, one giant leap for mankind," giving a much more humble tone to his statement.

✳ Chapter 5

High Strangeness: UFOs

Strange Lights in Marfa

If anyone is near Marfa at night, they should watch for odd, vivid lights over nearby Mitchell Flat. The location has a reputation for extremely unusual aerial sightings.

✳ ✳ ✳ ✳

MANY PEOPLE BELIEVE that the lights from UFOs or even alien entities can be seen. The famous Marfa Lights are about the size of basketballs and are usually white, orange, red, or yellow. These unexplained orbs only appear at night and usually hover above the ground at about shoulder height. Some lights—alone or in pairs—drift and fly around the landscape.

From cowboys to truck drivers, people traveling in Texas near the intersection of U.S. Route 90 and U.S. Route 67 in southwest Texas have reported the Marfa Lights. And these lights don't just appear on the ground. Pilots and airline passengers claim to have seen the Marfa Lights from the skies. So far, no one has proved a natural explanation for the floating orbs.

Eyewitness Information

Two 1988 reports were quite specific. Pilot R. Weidig was about 8,000 feet above Marfa when he saw the lights and placed them at several hundred feet above ground. Passenger E. Halsell described them as larger than the plane and pulsating. In 2002, pilot B. Eubanks provided a similar report.

In addition to what can be seen, the Marfa Lights may also trigger low-frequency electromagnetic (radio) waves—which can be heard on special receivers—similar to the "whistlers" caused by lightning. However, unlike such waves from power lines and electrical storms, the Marfa whistlers are extremely loud. They can be heard as the orbs appear, and then they fade when the lights do.

A Little Bit on Marfa

Marfa is about 60 miles north of the Mexican border and about 190 miles southeast of El Paso. This small, friendly Texas town is 4,800 feet above sea level and covers 1.6 square miles.

In 1883, Marfa was a railroad water stop. It received its name from the wife of the president of the Texas and New Orleans Railroad, who chose the name from a Russian novel that she was reading. A strong argument can be made that this was Dostoyevsky's *The Brothers Karamazov.* The town grew slowly, reaching its peak during World War II when the U.S. government located a prisoner of war camp, the Marfa Army Airfield, and a chemical warfare brigade nearby. (Some skeptics suggest that discarded chemicals may be causing the Marfa Lights, but searchers have found no evidence of such.)

Today, Marfa is home to about 2,500 people. The small town is an emerging arts center with more than a dozen artists' studios and art galleries. However, Marfa remains most famous for its light display. The annual Marfa Lights Festival is one of the town's biggest events, but the mysterious lights attract visitors year-round.

The Marfa Lights are seen almost every clear night, but they never manifest during the daytime. The lights appear between Marfa and nearby Paisano Pass, with the Chinati Mountains as a backdrop.

Widespread Sightings

The first documented sighting was by 16-year-old cowhand Robert Reed Ellison during an 1883 cattle drive. Seeing an odd light in the area, Ellison thought he'd seen an Apache campfire. When he told his story in town, however, settlers told him that they'd seen lights in the area, too, and they'd never found evidence of campfires.

Two years later, 38-year-old Joe Humphreys and his wife, Sally, also reported unexplained lights at Marfa. In 1919, cowboys on a cattle drive paused to search the area for the origin of the lights. Like the others, they found no explanation.

In 1943, the Marfa Lights came to national attention when Fritz Kahl, an airman at the Marfa Army Base, reported that airmen were seeing lights that they couldn't explain. Four years later, he attempted to fly after them in a plane but came up empty again.

Explanations?

Some skeptics claim that the lights are headlights from U.S. 67, dismissing the many reports from before cars—or U.S. 67—were in the Marfa area. Others insist that the lights are swamp gas, ball lightning, reflections off mica deposits, or an odd nightly mirage.

At the other extreme, a contingent of people believe that the floating orbs are friendly observers of life on Earth. For example, Mrs. W. T. Giddings described her father's early 20th-century encounter with the Marfa Lights. He'd become lost during a blizzard, and according to his daughter, the lights "spoke" to him and led him to a cave where he found shelter.

Most studies of the phenomenon, however, conclude that the lights are indeed real but cannot be explained. The 1989 TV show *Unsolved Mysteries* set up equipment to find an explanation. Scientists on the scene could only say that the lights were not made by people.

Share the Wealth

Marfa is the most famous location for "ghost lights" and "mystery lights," but it's not the only place to see them. Here are just a few of the legendary unexplained lights that attract visitors to dark roads in Texas on murky nights.

* In southeast Texas, a single orb appears regularly near Saratoga on Bragg Road.
* The Anson Light appears near Mt. Hope Cemetery in Anson, by U.S. Highway 180.
* Since 1850, "Brit Bailey's Light" glows five miles west of Angleton near Highway 35 in Brazoria County.
* In January 2008, Stephenville attracted international attention when unexplained lights—and perhaps a metallic spaceship—flew fast and low over the town.

The Marfa Lights appear over Mitchell Flat, which is entirely private property. However, the curious can view the lights from a Texas Highway Department roadside parking area about nine miles east of Marfa on U.S. Highway 90. Seekers should arrive before dusk for the best location, especially during bluebonnet season (mid-April through late May), because this is a popular tourist stop.

The Marfa Lights Festival takes place during Labor Day weekend each year. This annual celebration of Marfa's mystery includes a parade, arts and crafts booths, great food, and a street dance.

The Great Texas Airship Mystery

Roswell, New Mexico, may be the most famous of potential UFO crash sites, but did a Texas town experience a similar event in the 19th century?

* * * *

ONE SUNNY APRIL morning in 1897, a UFO crashed in Aurora, Texas.

Six years before the Wright Brothers' first flight and 50 years before Roswell, a huge, cigar-shape UFO was seen in the skies. It was first noted on November 17, 1896, about a thousand feet above rooftops in Sacramento, California. From there, the spaceship traveled to San Francisco, where it was seen by hundreds of people.

A National Tour

Next, the craft crossed the United States, where it was observed by thousands. Near Omaha, Nebraska, a farmer reported the ship on the ground, making repairs. When it returned to the skies, it headed toward Chicago, where it was supposedly photographed on April 11, 1897, the first UFO photo on record (though only sketches of the photo were ever published). On April 15, near Kalamazoo, Michigan, residents reported loud noises "like that of heavy ordnance" coming from the spaceship.

Two days later, the UFO attempted a landing in Aurora, Texas, which should have been a good place. The town was almost deserted, and its broad, empty fields could have been an ideal landing strip.

No Smooth Sailing

However, at about 6 a.m. on April 17, the huge, cigar-shaped airship "sailed over the public square and, when it reached the north part of town, collided with the tower of Judge Proctor's windmill and went to pieces with a terrific explosion, scattering debris over several acres of ground, wrecking the windmill and water tank and destroying the judge's flower garden."

That's how Aurora resident and cotton buyer S. E. Haydon described the events for *The Dallas Morning News*. The remains of the ship seemed to be strips and shards of a silver-colored metal. Just one body was recovered. The newspaper reported, "while his remains are badly disfigured, enough of the original has been picked up to show that he was not an inhabitant of this world."

On April 18, reportedly, that body was given a good, Christian burial in the Aurora cemetery, where it may remain to this day. A 1973 effort to exhume the body and examine it was successfully blocked by the Aurora Cemetery Association.

A Firsthand Account

Although many people have claimed the Aurora incident was a hoax, an elderly woman was interviewed in 1973 and clearly recalled the crash from her childhood. She said that her parents wouldn't let her near the debris from the spacecraft, in case it contained something dangerous. However, she described the alien as "a small man."

Aurora continues to attract people interested in UFOs. They wonder why modern Aurora appears to be laid out like a military base. Nearby, Fort Worth seems to be home to the U.S. government's experts in alien technology. Immediately after the Roswell UFO crash in 1947, debris from that spaceship was sent to Fort Worth for analysis.

Is There Any Trace Left?

The Aurora Encounter, a 1986 movie, documents the events that began when people saw the spacecraft crash at Judge Proctor's farm. Metal debris was collected from the site in the 1970s and studied at North Texas State University. That study called one fragment "most intriguing": It was ironlike but wasn't magnetic; it was shiny and malleable rather than brittle, as iron should be.

As recently as 2008, UFOs have appeared in the north central Texas skies. In Stephenville, a freight company owner and pilot described a low-flying object in the sky, "a mile long and half a mile wide." Others who saw the ship several times during January 2008 said that its lights changed configuration, so it wasn't an airplane. The government declined to comment.

Today, a plaque at the Aurora cemetery mentions the spaceship, but the alien's tombstone—which, if it actually existed, is said to have featured a carved image of a spaceship—was stolen many years ago.

The Mysterious Orb

If Texas were a dartboard, the city of Brownwood would be at the center of the bull's-eye. Maybe that's how aliens saw it, too.

* * * *

BROWNWOOD IS A peaceful little city with about 20,000 residents and a popular train museum. A frontier town at one time, it became the trade center of Texas when the railroad arrived in 1885. Since then, the city has maintained a peaceful lifestyle. Even the massive tornado that struck Brownwood in 1976 left no fatalities. The place just has that "small town" kind of feeling.

An Invader from the Sky

In July 2002, however, the city's peace was broken. Brownwood made international headlines when a strange metal orb fell from space, landed in the Colorado River, and washed up just south of town. The orb looked like a battered metal soccer ball—it was about a foot across, and it weighed just under ten pounds. Experts described it as a titanium sphere. When it was X-rayed, it revealed a second, inner sphere with tubes and wires wrapped inside. That's all that anybody knows (or claims to know). No one is sure what the object is, and no one has claimed responsibility for it. The leading theory is that it's a cryogenic tank from some kind of spacecraft from Earth, used to store a small amount of liquid hydrogen or helium for cooling purposes. Others have speculated that it's a bomb, a spying device, or even a weapon used to combat UFOs.

It's Not Alone

The Brownwood sphere isn't unique. A similar object landed in Kingsbury, Texas, in 1997, and was quickly confiscated by the Air Force for "tests and analysis." So far, no further announcements have been made.

Of course, the Air Force probably has a lot to keep it busy. About 200 UFOs are reported each month, and Texas is among the top three states where UFOs are seen. But until anything is known for sure, those in Texas at night should keep an eye on the skies.

Ohio's Mysterious Hangar 18

An otherwordly legend makes its way from New Mexico to Ohio when the wreckage from Roswell ends up in the Midwest.

* * * *

EVEN THOSE WHO aren't UFO buffs have probably heard about the infamous Roswell Incident, where an alien spaceship supposedly crash-landed in the New Mexico desert, and the U.S. government covered the whole thing up. But what most people don't know is that according to legend, the mysterious aircraft was recovered (along with some alien bodies), secreted out of Roswell, and came to rest just outside of Dayton, Ohio.

Something Crashed in the Desert

While the exact date is unclear, sometime during the first week of July 1947, a local Roswell rancher by the name of Mac Brazel decided to go out and check his property for fallen trees and other damage after a night of heavy storms and lightning. Brazel allegedly came across an area of his property littered with strange debris unlike anything he had ever seen before. Some of the debris even had strange writing on it.

Brazel showed some of the debris to a few neighbors and then took it to the office of Roswell sheriff George Wilcox, who called authorities at Roswell Army Air Field. After speaking with Wilcox, intelligence officer Major Jesse Marcel drove out to the Brazel ranch and collected as much debris as he could. He then returned to the airfield and showed the debris to his commanding officer, Colonel William Blanchard, commander of the 509th Bomb Group that was stationed at the Roswell Air Field. Upon seeing the debris, Blanchard dispatched military vehicles and personnel back out to the Brazel ranch to see if they could recover anything else.

"Flying Saucer Captured!"

On July 8, 1947, Colonel Blanchard issued a press release stating that the wreckage of a "crashed disk" had been recovered. The bold headline of the July 8 edition of the *Roswell Daily Record* read: "RAAF Captures Flying Saucer on Ranch in Roswell Region." Newspapers around the world ran similar headlines. But then, within hours of the Blanchard release, General Roger M. Ramey, commander of the Eighth Air Force in Fort Worth, Texas, retracted Blanchard's release for him and issued another statement saying there was no UFO. Blanchard's men had simply recovered a fallen weather balloon.

Soon the headlines that had earlier touted the capture of a UFO read: "It's a Weather Balloon" and "'Flying Disc' Turns Up as Just Hot Air." Later editions even ran a staged photograph of Major Jesse Marcel, who was first sent to investigate the incident, kneeling in front of balloon debris. Most of the public seemed content with the explanation, but there were skeptics.

Whisked Away to Hangar 18?

Those who believe that aliens crash-landed near Roswell claim that, under cover of darkness, large portions of the alien spacecraft were brought out to the Roswell Air Field and loaded onto B-29 and C-54 aircrafts. Those planes were then supposedly flown to Wright-Patterson Air Force Base, outside of Dayton. Once the planes landed, they were taxied to Hangar 18 and unloaded. And according to legend, it's all still there.

There are some problems with the story, though. For one, none of the hangars on Wright-Patterson Air Force Base are officially known as "Hangar 18," and there are no buildings designated with the number 18. Rather, the hangars are labeled 1A, 1B, 1C, and so on. There's also the fact that none of the hangars seem large enough to house and conceal an alien craft. But just because there's nothing listed as Hangar 18 on a Wright-Patterson map doesn't mean it's not there. Conspiracy theorists believe that hangars 4A, 4B, and 4C might be the infamous Hangar 18. As for the overall size of the hangars, it's believed that most of the wreckage has been stored in giant underground chambers deep under the hangar, to protect the debris and keep it safe from prying eyes. It is said that Wright-Patterson is currently conducting experiments on the wreckage to see if scientists can reverse-engineer the technology.

So What's the Deal?

The story of Hangar 18 only got stranger as the years went on, starting with the government's Project Blue Book, a program designed to investigate reported UFO sightings across the United States. Between 1947 and 1969, Project Blue Book investigated more than 12,000 UFO sightings before being disbanded. And where was Project Blue Book headquartered? Wright-Patterson Air Force Base.

Then in the early 1960s, Arizona senator Barry Goldwater, himself a retired major general in the U.S. Army Air Corps (and a friend of Colonel Blanchard), became interested in what, if anything, had crashed in Roswell. When Goldwater discovered Hangar 18, he first wrote directly to Wright-Patterson and asked for permission to tour the facility but was quickly denied. He then approached another friend, General Curtis

LeMay, and asked if he could see the "Green Room" where the UFO secret was being held. Goldwater claimed that LeMay gave him "holy hell" and screamed at Goldwater, "Not only can't you get into it, but don't you ever mention it to me again."

Most recently, in 1982, retired pilot Oliver "Pappy" Henderson attended a reunion and announced that he was one of the men who had flown alien bodies out of Roswell in a C-54 cargo plane. His destination? Hangar 18 at Wright-Patterson. Although no one is closer to a definitive answer, it seems that the legend of Hangar 18 will never die.

"I certainly believe in aliens in space. They may not look like us, but I have very strong feelings that they have advanced beyond our mental capabilities . . . I think some highly secret government UFO investigations are going on that we don't know about—and probably never will unless the Air Force discloses them."

—Barry Goldwater

Mystery Spot

Area 51 is infamous for being the mystery spot to end all mystery spots. Speculation about its purpose runs the gamut from a top-secret test range to an alien research center. One thing is certain: The truth is out there somewhere.

✻ ✻ ✻ ✻

LOCATED NEAR THE southern shore of the dry lakebed known as Groom Lake is a large military airfield—one of the most secretive places in the country. It is fairly isolated from the outside world, and little official information has ever been published on it. The area is not included on any maps, yet nearby Nevada state route 375 is listed as "The Extraterrestrial Highway." Although referred to by a variety of names, including Dreamland, Paradise Ranch, Watertown Strip, and Homey Airport, this tract of mysterious land in southern Nevada is most commonly known as "Area 51."

Conspiracy theorists and UFO aficionados speculate that Area 51 is everything from the storage location of the rumored crashed Roswell, New Mexico, spacecraft to a lab where experiments are conducted on matter transportation and time travel.

The truth is probably far less fantastic and probably far more scientific. Used as a bomb range during World War II, the site was abandoned as a military base at the end of the war. The land wasn't used again until 1955, when the site became a test range for the Lockheed U-2 spy plane and, later, the USAF SR-71 Blackbird.

Whether Area 51 was ever used to house UFOs isn't known for certain, but experts believe that the site at Groom Lake was probably a test and study center for captured Soviet aircraft during the Cold War. In 2003, the federal government actually admitted the facility exists as an Air Force "operating location," but no further information was released. Today, the area, including the various runways, is officially designated as "Homey Airport."

Three Sides to Every Story

Few geographical locations on Earth have been discussed and debated more than the three-sided chunk of ocean between the Atlantic coast of Florida and the regions of San Juan, Puerto Rico, and Bermuda known as the Bermuda Triangle.

* * * *

OVER THE CENTURIES, hundreds of ships and dozens of airplanes have mysteriously disappeared while floating in or flying through the region commonly called the Bermuda Triangle. Myth mongers propose that alien forces are responsible for these dissipations. Because little or no wreckage from the vanished vessels has ever been recovered, paranormal pirating has also been cited as the culprit. Other theorists suggest that leftover technology from the lost continent of Atlantis—

mainly an underwater rock formation known as the Bimini Road (situated just off the island of Bimini in the Bahamas)—exerts a supernatural power that grabs unsuspecting intruders and drags them to the depths.

A Deadly Adjective

Although the theory of the Triangle had been mentioned in publications as early as 1950, it wasn't until the '60s that the region was anointed with its three-sided appellation. Columnist Vincent Gaddis wrote an article in the February 1964 edition of *Argosy* magazine that discussed the various mysterious disappearances that had occurred over the years and designated the area where myth and mystery mixed as the "Deadly Bermuda Triangle." The use of the adjective *deadly* perpetrated the possibility that UFOs, alien anarchists, supernatural beings, and metaphysical monsters reigned over the region. The mystery of Flight 19, which involved the disappearance of five planes in 1945, was first noted in newspaper articles that appeared in 1950, but its fame was secured when the flight and its fate were fictitiously featured in Steven Spielberg's 1977 alien opus, *Close Encounters of the Third Kind*. In Hollywood's view, the pilots and their planes were plucked from the sky by friendly aliens and later returned safely to terra firma by their abductors.

In 1975, historian, pilot, and researcher Lawrence David Kusche published one of the first definitive studies that dismissed many Triangle theories. In his book *The Bermuda Triangle Mystery—Solved*, he noted that the Triangle was a "manufactured mystery," the result of bad research and reporting and, sometimes, deliberate falsifications. Before weighing in on Kusche's conclusions, however, consider that one of his next major publications was a tome about exotic popcorn recipes.

Explaining Odd Occurrences

Other pragmatists have insisted that a combination of natural forces—a double whammy of waves and rain that create the perfect storm—is most likely the cause for these maritime misfortunes. Other possible "answers" to the mysteries include rogue waves (such as the one that capsized the *Ocean Ranger* oil rig off the coast of Newfoundland in 1982), hurricanes, underwater earthquakes, and human error. The Coast Guard has sometimes averaged 20 distress calls a day from amateur sailors attempting to navigate the slippery sides of the Triangle. Modern-day piracy—usually among those involved in drug smuggling—has been mentioned as a probable cause for odd occurrences, as have unusual magnetic anomalies that screw up compass readings. Other possible explanations include the Gulf Stream's uncertain current, the high volume of sea and air traffic in the region, and even methane hydrates (gas bubbles) that produce "mud volcanoes" capable of sucking a ship into the depths.

Other dramatic and disastrous disappearances amid the Bermuda Triangle include the USS *Cyclops*, which descended to its watery repository without a whisper in March 1918 with 309 people aboard. Myth suggests supernatural subterfuge, but the reality is that violent storms or enemy action were the likely culprits. The same deductions had been discussed and similar conclusions reached in 1812 when the sea schooner *Patriot*, a commercial vessel, was swept up by the sea with the daughter of former vice president Aaron Burr onboard.

The Kecksburg Incident

Did an acorn-shaped alien spaceship once land in a western Pennsylvania thicket?

* * * *

Dropping in for a Visit

On December 9, 1965, an unidentified flying object (UFO) streaked through the late-afternoon sky and landed in Keckburg—a rural Pennsylvania community about 40 miles southeast of Pittsburgh. This much is not disputed. However, specific accounts vary widely from person to person. Even after closely examining the facts, many people remain undecided about exactly what happened. "Roswell" type incidents—ultra-mysterious in nature and reeking of a governmental cover-up—have an uncanny way of causing confusion.

Trajectory-Interruptus

A meteor on a collision course with Earth will generally "bounce" as it enters the atmosphere. This occurs due to friction, which forcefully slows the average space rock from 6 to 45 miles per second to a few hundred miles per hour, the speed at which it strikes Earth and officially becomes a meteorite. According to the official explanation offered by the U.S. Air Force, it was a meteorite that landed in Kecksburg. However, witnesses reported that the object completed back and forth maneuvers before landing at a very low speed—moves that an un-powered chunk of earthbound rock simply cannot perform. Strike one against the meteor theory.

An Acorn-Shaped Meteorite?

When a meteor manages to pierce Earth's atmosphere, it has the physical properties of exactly what it is: a space rock. That is to say, it will generally be unevenly shaped, rough, and darkish in color, much like rocks found on Earth. But at Kecksburg, eyewitnesses reported seeing something far, far different. The unusual object they described was bronze to golden in color, acorn-shape, and as large as a Volkswagen Beetle automobile. Unless the universe has started to produce uniformly shaped and colored meteorites, the official explanation seems highly unlikely. Strike two for the meteor theory.

Markedly Different

Then there's the baffling issue of markings. A meteorite can be chock-full of holes, cracks, and other such surface imperfections. It can also vary somewhat in color. But it should never, ever have markings that seem intelligently designed. Witnesses at Kecksburg describe intricate writings similar to Egyptian hieroglyphics located near the base of the object. A cursory examination of space rocks at any natural history museum reveals that such a thing doesn't occur naturally. Strike three for the meteor theory. Logically following such a trail, could an unnatural force have been responsible for the item witnessed at Kecksburg? At least one man thought so.

Reportis Rigor Mortis

Just after the Kecksburg UFO landed, reporter John Murphy arrived at the scene. Like any seasoned pro, the newsman immediately snapped photos and gathered eyewitness accounts of the event. Strangely, FBI agents arrived, cordoned off the area, and confiscated all but one roll of his film. Undaunted, Murphy assembled a radio documentary entitled *Object in the Woods* to describe his experience. Just before the special was to air, the reporter received an unexpected visit by two men. According to a fellow employee, a dark-suited pair identified themselves as government agents and subsequently confiscated a portion of Murphy's audiotapes. A week later, a clearly perturbed Murphy aired a watered-down version of his documentary. In it, he claimed that certain interviewees requested their accounts be removed for fear of retribution at the hands of police, military, and government officials. In 1969, John Murphy was struck dead by an unidentified car while crossing the street.

Resurrected by Robert Stack

In all likelihood the Kecksburg incident would have remained dormant and under-explored had it not been for the television show *Unsolved Mysteries*. In a 1990 segment, narrator Robert Stack took an in-depth look at what occurred in Kecksburg, feeding a firestorm of interest that eventually brought forth two new witnesses. The first, a U.S. Air Force officer stationed at Lockbourne AFB (near Columbus, Ohio), claimed to have seen a flatbed truck carrying a mysterious object as it arrived on base on December 10, 1965. The military man told of a tarpaulin-covered conical object that he couldn't identify and a "shoot to kill" order given to him for anyone who ventured too close. He was told that the truck was bound for Wright–Patterson AFB in Dayton, Ohio, an installation that's alleged to contain downed flying saucers. The other witness was a building contractor who claimed to have delivered 6,500 special bricks to a hanger inside Wright–Patterson AFB on December 12, 1965. Curious, he peeked inside the hanger and saw a "bell-shaped" device, 12-feet high, surrounded by several men wearing anti-radiation style suits. Upon leaving, he was told that he had just witnessed an object that would become "common knowledge" in the next 20 years.

Will We Ever Know the Truth?

Like Roswell before it, we will probably never know for certain what occurred in western Pennsylvania back in 1965. The more that's learned about the case, the more confusing and contradictory it becomes. For instance, the official 1965 meteorite explanation contains more holes than Bonnie and Clyde's death car, and other explanations, such as orbiting space debris (from past U.S. and Russian missions) reentering Earth's atmosphere, seem equally preposterous. In 2005, as the result of a new investigation launched by the Sci-Fi Television Network, NASA asserted that the object was a Russian satellite. According to a NASA spokesperson, documents of this investigation were somehow misplaced in the 1990s. Mysteriously, this finding directly contradicts the official air force version that nothing at all was found at the site. It also runs counter to a 2003 report made by NASA's own Nicholas L. Johnson, Chief Scientist for Orbital Debris. That document shows no missing satellites at the time of the incident. This includes a missing Russian Venus Probe (since accounted for)—the very item that was once considered a prime crash candidate.

Brave New World

These days, visitors to Kecksburg will be hard-pressed to find any trace of the encounter—perhaps that's how it should be. Since speculation comes to an abrupt halt whenever a concrete answer is provided, Kecksburg's reputation as "Roswell of the East" looks secure, at least for the foreseeable future. But if one longs for proof that something mysterious occurred there, they need look no further than the backyard of the Kecksburg Volunteer Fire Department. There, in all of its acorn-shaped glory, stands an full-scale mock-up of the spacecraft reportedly found in this peaceful town on December 9, 1965. There too rests the mystery, intrigue, and romance that have accompanied this alleged space traveler for more than 40 years.

Unidentified Submerged Objects

Much like their flying brethren, unidentified submerged objects captivate and mystify. But instead of vanishing into the skies, USOs, such as the following, plunge underwater.

* * * *

Sighting at Puerto Rico Trench

In 1963, while conducting exercises off the coast of Puerto Rico, U.S. Navy submarines encountered something extraordinary. The incident began when a sonar operator aboard an accompanying destroyer reported a strange occurrence. According to the seaman, one of the subs traveling with the armada broke free from the pack to chase a USO. This quarry would be unlike anything the submariners had ever pursued.

Underwater technology in the early 1960s was advancing rapidly. Still, vessels had their limitations. The U.S.S. *Nautilus*, though faster than any submarine that preceded it, was still limited to about 20 knots (23 miles per hour). The bathyscaphe *Trieste*, a deep-sea submersible, could exceed 30,000 feet in depth, but the descent took as long as five hours. Once there, the vessel could not be maneuvered side to side.

Knowing this, the submariners were bewildered by what they witnessed. This particular USO was moving at 150 knots (170 miles per hour) and hitting depths greater than

20,000 feet! No underwater vehicles on Earth were capable of such fantastic numbers. Even today, modern nuclear subs have top speeds of about 25 knots (29 miles per hour) and can operate at around 800-plus feet below the surface.

Thirteen separate crafts witnessed the USO as it criss-crossed the Atlantic Ocean over a four-day period. At its deepest, the mystery vehicle reached 27,000 feet. To this day, there's been no earthly explanation offered for the occurrence.

USO with a Bus Pass

In 1964, London bus driver Bob Fall witnessed one of the strangest USO sightings. While transporting a full contingent of passengers, the driver and his fares reported seeing a silver, cigar-shape object dive into the nearby waters of the River Lea. The police attributed the phenomenon to a flight of ducks, despite the obvious incongruence. Severed telephone lines and a large gouge on the river's embankment suggested something far different.

Shag Harbour Incident

The fishing village of Shag Harbour lies on Canada's East Coast. This unassuming hamlet is to USOs what Roswell, New Mexico, is to UFOs. Simply put, it played host to the most famous occurrence of a USO ever recorded.

On the evening of October 4, 1967, the Royal Canadian Mounted Police (RCMP) were barraged by reports of a UFO that had crashed into the bay at Shag Harbour. Laurie Wickens and four friends witnessed a large object (approximately 60 feet in diameter) falling into the water just after 11:00 p.m. Floating approximately 1,000 feet off the coast they could clearly detect a yellow light on top of the object.

The RCMP promptly contacted the Rescue Coordination Center in Halifax to ask if any aircraft were missing. None were. Shortly thereafter, the object sank into the depths of the water and disappeared from view.

When local fishing boats went to the USO crash site, they encountered yellow foam on the water's surface and detected an odd sulfuric smell. No survivors or bodies were ever found. The Royal Canadian Air Force officially labeled the occurrence a UFO, but because the object was last seen under water, such events are now described as USOs.

Pascagoula Incident

On November 6, 1973, at approximately 8:00 p.m., a USO was sighted by at least nine fishermen anchored off the coast of Pascagoula, Mississippi. They witnessed an underwater object five feet in diameter that emitted a strange amber light.

First to spot the USO was Rayme Ryan. He repeatedly poked at the light-emitting object with an oar. Each time he made contact with the strange object, its light would dim and it would move a few feet away, then brighten once again.

Fascinated by the ethereal quality of this submerged question mark, Ryan summoned the others. For the next half hour, the cat-and-mouse game played out in front of the fishermen until Ryan struck the object with a particularly forceful blow. With this action, the USO disappeared from view.

The anglers moved about a half mile away and continued fishing. After about 30 minutes, they returned to their earlier location and were astounded to find that the USO had returned. At this point, they decided to alert the Coast Guard.

After interviewing the witnesses, investigators from the Naval Ship Research and Development Laboratory in Panama City, Florida, submitted their findings: At least nine persons had witnessed an undetermined light source whose characteristics and actions were inconsistent with those of known marine organisms or with an uncontrolled human-made object. Their final report was inconclusive, stating that the object could not be positively identified.

The Nazca Lines—Pictures Aimed at an Eye in the Sky?

Ancient works of art etched into a desert floor in South America have inspired wild theories about who created them and why. Did space aliens leave them on long-ago visits? Decades of scientific research reject the popular notion, showing that the lines were the work of mere Earthlings.

⁕ ⁕ ⁕ ⁕

FLYING ABOVE THE rocky plains northwest of Nazca, Peru, in 1927, aviator Toribio Mejía Xesspe was surprised to see gigantic eyes looking up at him. Then the pilot noticed that the orbs stared out of a bulbous head upon a cartoonish line drawing of a man, etched over hundreds of square feet of the landscape below.

The huge drawing—later called "owl man" for its staring eyes—turned out to be just one of scores of huge, 2,000-year-old images scratched into the earth over almost 200 square miles of the parched Peruvian landscape.

There is a 360-foot-long monkey with a whimsically spiraled tail, along with a 150-foot-long spider, and a 935-foot pelican. Other figures range from hummingbird to killer whale. Unless the viewer knows what to look for, they're almost invisible from ground level. There are also geometric shapes and straight lines that stretch for miles across the stony ground.

The Theory of Ancient Astronauts

The drawings have been dated to a period between 200 BC and AD 600. Obviously, there were no airplanes from which to view them then. So why were they made? And for who? In his 1968 book *Chariots of the Gods?*, Swiss author Erich Von Däniken popularized the idea that the drawings and lines were landing signals and runways for starships that visited southern Peru long before the modern era. In his interpretation, the owl man is instead an astronaut in a helmet. Von Däniken's theory caught on among UFO enthusiasts. Many science-fiction novels and films make reference to this desert in Peru's Pampa Colorado region as a unique site with special significance to space travelers.

Coming Down to Earth

Examined up close, the drawings consist of cleared paths—areas where someone removed reddish surface rocks to expose the soft soil beneath. In the stable desert climate—averaging less than an inch of rain per year—the paths have survived through many centuries largely intact.

Scientists believe the Nazca culture—a civilization that came before the Incas—drew the lines. The style of the artwork is similar to that featured on Nazca pottery. German-born researcher Maria Reiche (1903–1998) showed how the Nazca could have laid out the figures using simple surveying tools such as ropes and posts. In the 1980s, American researcher Joe Nickell duplicated one of the drawings, a condor, showing that the Nazca could have rendered parts of the figures "freehand"—that is, without special tools or even scale models. Nickell also demonstrated that despite their great size, the figures can be identified as drawings even from ground level. No alien technology would have been required to make them.

Still Mysterious

As for why the Nazca drew giant doodles across the desert, no one is sure. Reiche noted that some of the lines could have astronomical relevance. For example, one points to where the sun sets at the winter solstice. Some lines may also have pointed toward underground water sources—crucially important to desert people.

Most scholars think that the marks were part of the Nazca religion. They may have been footpaths followed during ritual processions. And although it's extremely unlikely that they were intended for extraterrestrials, many experts think it likely that the lines were oriented toward Nazca gods—perhaps a monkey god, a spider god, and so on, who could be imagined gazing down from the heavens upon likenesses of themselves.

Red Eyes Over Point Pleasant: The Mysterious Mothman

In 1942, the U.S. government took control of several thousand acres of land just north of Point Pleasant, West Virginia. The purpose was to build a secret facility capable of creating and storing TNT that could be used during World War II. For the next three years, the facility cranked out massive amounts of TNT, shipping it out or storing it in one of the numerous concrete "igloo" structures that dotted the area. In 1945, the facility was shut down and eventually abandoned, but it was here that an enigmatic flying creature with glowing red eyes made its home years later.

* * * *

"Red Eyes on the Right"

On the evening of November 15, 1966, Linda and Roger Scarberry were out driving with another couple, Mary and Steve Mallette. As they drove, they decided to take a detour that took them past the abandoned TNT factory.

As they neared the gate of the old factory, they noticed two red lights up ahead. When Roger stopped the car, the couples were horrified to find that the red lights appeared to be two glowing red eyes. What's more, those eyes belonged to a creature standing more than seven feet tall with giant wings folded behind it. That was all Roger needed to see before he hit the gas pedal and sped off. In response, the creature calmly unfolded its wings and flew toward the car. Incredibly, even though Roger raced along at speeds close to 100 miles per hour, the red-eyed creature was able to keep up with them without much effort.

Upon reaching Point Pleasant, the two couples ran from their car to the Mason County Courthouse and alerted Deputy Millard Halstead of their terrifying encounter. Halstead couldn't be sure exactly what the two couples had seen, but whatever it was, it had clearly frightened them. In an attempt to calm them down, Halstead agreed to accompany them to the TNT factory. As his patrol car neared the entrance, the police radio suddenly emitted a strange, whining noise. Other than that, despite a thorough search of the area, nothing out of the ordinary was found.

More Encounters

Needless to say, once word got around Point Pleasant that a giant winged creature with glowing red eyes was roaming around the area, everyone had to see it for themselves. The creature didn't disappoint. Dubbed Mothman by the local press, the creature was spotted flying overhead, hiding, and even lurking on front porches. In fact, in the last few weeks of November, dozens of witnesses encountered the winged beast. But Mothman wasn't the only game in town. It seems that around the same time that he showed up, local residents started noticing strange lights in the evening sky, some of which hovered silently over the abandoned TNT factory. Of course, this led some to believe that Mothman and the UFOs were somehow connected. One such person was Mary Hyre of *The Athens Messenger*, who had been reporting on the strange activities in Point Pleasant since they started. Perhaps that's why she became the first target.

Beware the Men in Black

One day, while Mary Hyre was at work, several strange men visited her office and began asking questions about the lights in the sky. Normally, she didn't mind talking to people about the UFO sightings and Mothman. But there was something peculiar about these guys. For instance, they all dressed exactly the same: black suits, black ties, black hats, and dark sunglasses. They also spoke in a strange monotone and seemed confused by ordinary objects such as ballpoint pens. As the men left, Hyre wondered whether they had been from another planet. Either way, she had an up-close-and-personal encounter with the legendary Men in Black.

Mary Hyre was not the only person to have a run-in with the Men in Black. As the summer of 1967 rolled around, dozens of people were interrogated by them. In most cases, the men showed up unannounced at the homes of people who had recently witnessed a Mothman or UFO sighting. For the most part, the men simply wanted to know what the witnesses had seen. But sometimes, the men went to great lengths to convince the witnesses that they were mistaken and had not seen anything out of the ordinary. Other times, the men threatened witnesses. Each time the Men in Black left a witness's house, they drove away in a black, unmarked sedan. Despite numerous attempts to determine who these men were and where they came from, their identity remained a secret. And all the while, the Mothman sightings continued throughout Point Pleasant and the surrounding area.

The Silver Bridge Tragedy

Erected in 1928, the Silver Bridge was a gorgeous chain suspension bridge that spanned the Ohio River, connecting Point Pleasant with Ohio. On December 15, 1967, the bridge was busy with holiday shoppers bustling back and forth between West Virginia and Ohio. As the day wore on, more and more cars started filling the bridge until shortly before 5:00 p.m., when traffic on the bridge came to a standstill. For several minutes, none of the cars budged. Suddenly, there was a loud popping noise and then the unthinkable happened: The Silver Bridge collapsed, sending dozens of cars and their passengers into the freezing water below.

Over the next few days, local authorities and residents searched the river hoping to find survivors, but in the end, 46 people lost their lives in the bridge collapse. A thorough investigation determined that a manufacturing flaw in one of the bridge's supporting bars caused the collapse. But there are others who claim that in the days and weeks leading up to the collapse, they saw Mothman and even the Men in Black around, on, and even under the bridge. Further witnesses state that while most of Point Pleasant was watching the Silver Bridge collapse, bright lights and strange objects were flying out of the area and disappearing into the winter sky. Perhaps that had nothing to do with the collapse of the Silver Bridge, but the Mothman has not been seen since . . . or has he?

Mothman Lives!

There are reports that the Mothman is still alive and well and has moved on to other areas of the United States. There are even those who claim that he was spotted flying near the Twin Towers on September 11, 2001, leading to speculation that Mothman is a portent of doom and only appears when disasters are imminent. Some believe Mothman was a visitor from another planet who returned home shortly after the Silver Bridge fell. Still others think the creature was the result of the toxic chemicals eventually discovered in the area near the TNT factory. And then there are skeptics who say that the initial sighting was nothing more than a giant sand crane and that mass hysteria took care of the rest.

Was It the Japanese or Extraterrestrials?

A little over a month after the attack on Pearl Harbor and America entering the war, fears were high across the cities of the nation. In Los Angeles, on February 24, 1942, searchlights scanned the sky for an object reported to have been floating over the city.

* * * *

ANTI-AIRCRAFT GUNS WERE aimed, air raid sirens wailed through the streets, and wardens put the city on blackout.

Everyone thought that a Japanese attack was in the works, but their fears were never fully confirmed. What was in the sky that night is disputed to this day. Some claim they saw one giant craft slowly lurching over the city, while others claim to have seen a fleet of smaller crafts flying over. Whatever it was, it was soon fired upon by anti-aircraft guns in a barrage of explosions that should have destroyed any modern craft of the day. The craft was shot at for nearly four hours as it moved across the city, causing several buildings to be destroyed and the death of six people in the defense effort.

News reports took off with the story the next morning, claiming it had been a Japanese attack that ended with the fighter planes being destroyed. No evidence was ever found to confirm the account and Japan continues to claim that there was no attack. The next day, the *Los Angeles Times* ran a picture of the object lit with searchlights, only causing more confusion. The U.S. government stated that the panic was caused by "war nerves." Many believe what they saw was extraterrestrial.

Sounds for Space

This gold record doesn't climb to the top of the charts—it sails through deep space.

* * * *

Voyagers Abroad

SOMEWHERE IN SPACE, two unmanned scientific probes—Voyagers 1 and 2—move through previously unexplored areas of our solar system. NASA launched the twin probes in 1977, hoping they would uncover valuable information about Jupiter, Saturn, and the outer planets. And indeed they have: The spacecraft found active volcanoes on Jupiter's moon and a surprising amount of structure in Saturn's rings. Voyager 2 launched first and is the only spacecraft to have visited Uranus and Neptune. But Voyager 1, moving at a brisk 38,000 miles per hour, has probed the farthest.

Hit Record

Aboard each Voyager is a 12-inch, gold-plated copper disc called The Golden Record. This disc is a time capsule of sorts, containing a hodgepodge of sounds and images intended to convey to extraterrestrials what life on our planet is like. That's right: It's an LP designed for aliens—on the infinitesimally small chance that they exist and that The Golden Record will reach them in the distant future. And that they can figure out how to play it.

Astronomer Carl Sagan curated the project along with his future wife, Ann Druyan, and a committee of scientists. The group selected 115 images to depict human life, as well as sounds from nature, including birdsongs, crickets, wild dogs, and wind. They also included music from various cultures, including Peruvian panpipes, an Indian raga, the song "Johnny B. Goode" by Chuck Berry, greetings in 55 languages, and printed messages from U.S. President Jimmy Carter and UN Secretary-General Kurt Waldheim.

As Sagan noted, the likelihood of an advanced civilization encountering the record is so rare as to be nearly impossible, yet "the launching of this 'bottle' into the cosmic 'ocean' says something very hopeful about life on this planet." In 2008, Voyager 1 slipped out and away from our solar system—who knows who (or what) will find its cargo?

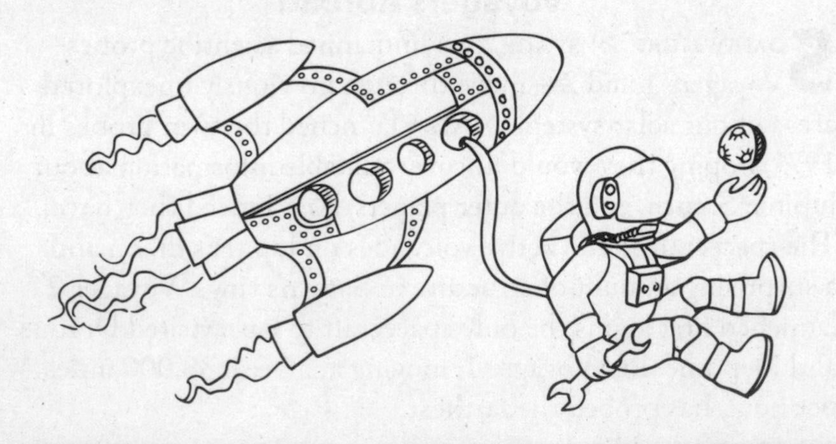

✳ Chapter 6

Game Changers: Consumer Inventions

From Pie Plate to Toy: The Frisbee

For something that now seems so familiar, the flying disc has an interesting story.

✳ ✳ ✳ ✳

William Russell Frisbie

IT ALL STARTED with W. R. Frisbie, a bakery manager. In 1871, he bought a Connecticut baking business, renamed the firm Frisbie Pie Co., and got to work baking pies. When he passed away in 1903, his son Joseph P. Frisbie donned the baker's hat. By the time of Joe's death in 1940, the Frisbie Pie Co. had become a great regional success. One example of Joe's savvy marketing: metal pie plates stamped with "Frisbie's Pies," so that anyone keeping the plate remembered the name.

Recreational Yalies

Yale University is just a quick flick of the disc away from the Frisbie Pie Co.'s old New Haven bakery. Like any self-respecting college students, Yalies both loved a good feed and liked to find creative ways to amuse themselves and annoy others. After porking out on Frisbie's pies, they flung the empty tins around for fun, quickly noting that they flew better if thrown with a quick flick of the wrist to impart spin. If you threw them with a spin, you could play catch with them. You could also accidentally bonk Professor Stuffshirt upside the head, so they learned to yell "Frisbie!"—as golfers holler "Fore!"—to warn of an incoming hazard.

Morrison and Franscioni

Nope, that isn't a personal-injury law firm from Jersey. Walter "Fred" Morrison was a World War II air combat veteran working for a bottled gas company in California in the late 1940s. He brought the idea for the frying plate to his employer, fellow vet Warren Franscioni. They started experimenting on the side.

The partners soon learned that a streamlined plastic disk was the ideal configuration. They began to make and market the Flyin' Saucer—attempting to capitalize on budding American interest in UFOs—but the product didn't take off. In the meantime, the gas company tanked. Franscioni rejoined the Air Force and was relocated to South Dakota.

Wham-O

Morrison renamed the Flyin' Saucer the Pluto Platter. In 1957, he was demonstrating the projectile and caught the attention of a small slingshot company named Wham-O. Impressed, the Wham-O people offered to market the Pluto Platter, and they knew what they were doing. The toy sold well. Not long thereafter, Wham-O cofounder Rich Knerr was giving out Pluto Platters at East Coast universities to build brand awareness and demand. At Yale, he saw students chucking metal pie tins around. When a tin was headed for a noggin, the Yalies yelled "Frisbie!," as tradition and safety demanded. (Irony: That same year, Frisbie's Pies shut down.)

Knerr soon renamed the Pluto Platter the Frisbee. Whether he changed the spelling for trademark reasons or thought that's how it was spelled, sales took off. Along with the Hula Hoop, the Frisbee became a Wham-O cash cow. Morrison got royalties; Franscioni did not. He died in 1974 while still considering legal action. Wham-O's official histories of the Frisbee do not mention him.

Video Games

It's hard to believe, but Pong*—the arcade game that featured a two-dimensional version of table tennis—is more than 45 years old. The game looks almost laughably simple in the 21st century, consisting of nothing more than a couple lines to represent "paddles," and a dot for the "ball." But when it was released,* Pong *was a big hit; and it helped to launch what is now a multi-billion-dollar industry.*

* * * *

ALTHOUGH PONG WAS the first video game to find true success with consumers, it was not the first video game to ever be created. In the 1950s, computers were just starting to make their way into mainstream life, having been used mostly for military applications after their initial creation. Companies like Remington Rand and IBM purchased the new technology, and universities, government entities, and businesses began using computers on a regular basis. As much for fun as for training and research purposes, computer scientists began designing games. These games were short-lived, and usually discarded once the research was over; but ambitious programmers were always trying to improve on previous designs.

In 1952, two British computer scientists, Christopher Strachey and Alexander Douglas, created a simple checkers game and tic-tac-toe game. These were the first games to incorporate a monitor screen. This was followed two years later by the first game to include graphics that updated in real time, a billiards game that was created by William Brown and Ted Lewis at the University of Michigan. And in 1958, a game dubbed *Tennis for Two* was designed by physicist William Higinbotham (who was also notable for being on the team that developed the first nuclear bomb).

But the earliest computer games were less concerned with war on Earth, instead focusing on war in space. In fact, in 1962, computer scientist Steve Russell designed *Spacewar*, the first game that could be played on multiple computer installations. The game featured two spaceships, controlled by two different players, battling it out in the gravity well of a star. Players' ships would be destroyed if they were hit by the other player's torpedoes, or if they accidentally ran into the star. *Spacewar* also made use of an early version of a gamepad, with each player using a control device with switches for maneuvering their ship, and a button to launch torpedoes.

Spacewar's popularity spread throughout the programming community, and it led to the creation of the first two commercial arcade video games, *Galaxy Game* and *Computer Space*, which debuted in 1971. Both games built off the initial premise of *Spacewar*, and they became the first coin-operated games available commercially. Although moderately successful, *Galaxy Game* and *Computer Space* never quite became runaway hits. However, they marked the beginnings of what would soon be a wildly popular industry.

Ready for the Masses

Meanwhile, an engineer named Ralph Baer had been hard at work on his own ideas for this new industry. In 1966, Baer designed a prototype of a multiplayer, multi-program video game system that could be plugged into a standard television set. He called his system "The Brown Box," and sold it to television company Magnavox, who renamed it the Odyssey. The Odyssey went on sale in September 1972 as the first home video game console, and included a dozen games, with ten more that could be purchased separately.

The Odyssey's programming was only able to render three spots and a line, greatly limiting the graphic abilities of the system. It was sold with plastic overlays that could be attached to the television set to make the games more visually interesting, as well as including accessories like cards and dice. Still, home gamers grew bored with the system quickly, and the Odyssey eventually fizzled out.

Enter *Pong*

Despite its failure, the Odyssey helped to prime the video game industry. In 1972 the two engineers who created *Computer Space*, Nolan Bushnell and Ted Dabney, formed their own video game development company. They named Atari after a Japanese word often used in the board game Go. Bushnell and Dabney hired computer scientist Allan Alcorn to create a new arcade game, and Alcorn, building off of a table tennis game that had been included with the Odyssey system, designed *Pong*. Atari released the game in 1972, expanding the release further in March 1973, when it became an arcade favorite.

In 1975, after advancements and improvements in technology, Magnavox released the Odyssey 100 and Odyssey 200, and Atari released a home version of *Pong*, which was a huge hit with consumers. But these were still "ball and paddle" type systems, with limited capabilities. Atari was determined to do better, so in 1977, it released the Atari 2600, which featured removable cartridges and multi-colored games. The immensely popular console sold more than 27 million units worldwide, ushering in a new era of video gaming.

The late 70s and early 80s introduced video game fans to some of the most popular games in history, including *Space Invaders*, *Asteroids*, *Pac Man*, *Donkey Kong*, and *Q*bert*. By 1982, coin-operated video games were hauling in $7.7 billion in the United States, making arcade video games the most popular form of entertainment in the country.

In 1985, the Nintendo Entertainment System arrived in the U.S. from Japan. It included better graphics, color, and sound than previous consoles. Nintendo introduced the world to games like *Super Mario Bros.* and *The Legend of Zelda*, and in 1989 released its popular Game Boy handheld video game device. The Game Boy often came with another now-classic game, *Tetris*. Companies like Coleco and Sega also jumped into the game console market, creating rivalries between gaming systems, with each trying to outdo the others with better games, faster speeds, and cheaper prices.

At the same time, more and more families began adding computers to their households, like the Commodore 64 and the IBM PC. So it may have been inevitable that video games made the leap from television consoles to computer monitors.

Online and Beyond

As early as the 1980s, online gaming started to become popular, through the use of dial-up bulletin board systems. PC games on floppy disc were next to follow, which soon led to the rise of 3-D graphics, sound cards, and eventually CD-ROMs. PC games also led to massively multiplayer online role-playing games, or MMORPGs, where players from anywhere around the world can participate in the same game together. The proliferation of computers radically changed the world of gaming, with computer programmer Sid Meier, who designed games like *Formula 1 Racing* and *Magic: The Gathering*, calling the IBM PC one of the most important innovations in the history of video games.

Games made the move to social media platforms and smartphones, with the popular mobile device game *Angry Birds* bringing in $200 million in 2012. But that's not to say that video game consoles have died out; quite the contrary. Although the Atari brand has faded into the background in recent years, the Microsoft Xbox, Sony Playstation, and Nintendo Switch continue to provide hours of fun to kids and adults everywhere.

Radio

It's hard to imagine living without radios. They're in our homes and cars, piped into stores, integrated with clocks, even streamed over the internet. But in the early 20th century, radios were a brand-new invention that fascinated the public and became the first electronic mass medium, decades before the television became popular.

❋ ❋ ❋ ❋

THE HISTORY OF the radio only goes back to the late 1800s, when German physicist Heinrich Rudolph Hertz began to experiment with the properties of electromagnetic waves. He discovered that certain frequencies, which are on what is now called the "radio spectrum," could be transmitted through the air. Thanks to his discovery, these radio waves were originally called "Hertzian waves," and we still refer to the oscillation frequency of sound waves, once called the cycle per second, as a Hertz. The radio spectrum includes frequencies between 3 Hz and 3000 GHz (gigahertz).

Hertz published his findings from his radio wave experiments between 1886 and 1888, catching the attention of many scientists, including Frederick Thomas Trouton, William Crookes, and Nikola Tesla, who conducted their own experiments with the newfound electromagnetic radiation. Some of these experiments were quite impressive, like the 1894 demonstration by Indian Bengali physicist Jagadish Chandra Bose in which he ignited gunpowder and rang a bell from a distance using radio waves. Bose was convinced that this "invisible light," as he called it, would one day be used to transmit messages without wires.

The Pragmatic Italian

The same year, a young Italian inventor named Guglielmo Marconi noticed that even though many scientists were toying with radio waves, none of them seemed to be searching for a practical use for them. Why not attempt to use them, as Bose suggested, for wireless communication? So Marconi gathered the research of previous inventors and used it to help him construct portable transmitters and receiver systems. Initially, Marconi was only able to transmit signals a half a mile, which other scientists had predicted would be the distance limit for radio waves. But the young inventor was undeterred, tweaking his system until it was able to transmit up to two miles, even over hills. In 1896, Marconi was granted a British patent for his radio system, and in 1901, he became the first person to transmit radio signals across the Atlantic Ocean.

For several decades, Marconi's successful radio was used mainly to transmit signals between ships at sea, usually through the use of Morse code, and his radio even played a crucial part in the rescue of the survivors of the *Titanic* disaster. But in 1900, a Brazilian priest and inventor by the name of Roberto Landell de Moura changed radio forever when he transmitted a human voice over a five-mile distance. Several years later, de Moura moved to the United States, where he was granted patents for a "Wave Transmitter," "Wireless Telephone," and "Wireless Telegraph," building a foundation for future radio advancements.

AM: Amplitude Modulation

On Christmas Eve 1906, the very first voice and music broadcast was made from Brant Rock, Massachusetts, just south of Boston, by inventor Reginald Fessenden. Fessenden had created a new type of radio transmitter that produced continuous wave signals of constant amplitude and frequency, improving on "damped wave" transmitters which produced a great amount of electromagnetic interference. This new transmitter was called amplitude modulation, now known as AM broadcasting.

Fessenden experimented with an AM radio broadcast, playing Christmas music and reading Bible passages for any radio amateurs or ships nearby that might be listening. The broadcast led to other experimental radio programs, including that of Californian Charles Herrold, who, in 1912, became the first radio operator to provide a regularly scheduled program for a small audience of local amateur radio enthusiasts.

Although radio was beginning to gain widespread notice, it was mostly a hobby taken up by a few curious listeners, who, before there were mass-produced radio receivers, had to make their own equipment at home. And the advent of World War I stalled advancements in the medium, as the government halted radio transmissions during the war, resulting in a decade of little progress.

So it wasn't until 1920, after the war ended, that radio began to move forward again. On August 31 of that year, the first radio news program was broadcast from the Detroit, Michigan, station 8MK. And in October, station 2ADD out of Schenectady, New York, began broadcasting regular Thursday-night concerts that could be heard within a 100-mile radius. Around the same time, sporting events, including college football, were beginning to be broadcast, as well. By the end of 1921, about 30 radio stations had popped up in the United States, most originating from amateur operations.

But radio was starting to interest more than just those who wished to build their own equipment; the general public was fascinated by the new medium. By the mid-1920s, the vacuum tube, which controlled electric current flow, was invented, greatly improving radio receivers and transmitters.

Commercial Radios

The Westinghouse Company, which had received a commercial radio license in 1920, began to advertise the sale of radios to the public. General Electric and the Radio Corporation of America (RCA), also began selling radios, and by 1922, a million American homes had radios and the number of radio stations in the U.S. had swelled to 550.

The 1930s and 40s are considered America's "Golden Age" of radio, and surprisingly, the Great Depression helped bolster its popularity. A family could make a one-time investment in a radio, and be promised years of entertainment to come. Even those who had to give up cars or expensive furniture held on to their radios, using them as a means of escape from the realities of a harsh life.

FM: Frequency Modulation

Despite the struggles within the country, another leap forward in radio technology was made in the 1930s in the form of frequency modulation, or FM, technology. Invented by American engineer Edwin Armstrong in 1933, FM broadcasting transmits sound through changes in frequency, which results in a clearer sound with less static.

But regardless of the better sound quality, Armstrong's invention got caught up in red tape and lawsuits with radio giants RCA and NBC, stifling the progression of the new technology. It would not be until the end of the 1970s that AM radio would cease its dominance of the airwaves. Finally embracing the clearer sound quality of Armstrong's FM radio, 70 percent of the global radio audience was tuning in to FM broadcasts by 1982.

The creation of FM radio may have been one of the most significant advancements of the medium in the 20th century, but other developments, like the pocket transistor radio, created in 1954, and satellite radio, launched in the early 2000s, continued to improve the quality and convenience of radio throughout the decades.

While some things have changed, the basic concept of radio—a way for people to stay connected to the world around them—has remained the same from the very first broadcasts.

Air Conditioning

If you've ever been to a state like Arizona, where temperatures in July in Phoenix average over 100 degrees, you've probably wondered how on earth humans could've lived in such a place in the days before air conditioning.

✻ ✻ ✻ ✻

WHILE MANY OFTEN joke that the soaring temperatures in states like Arizona and Nevada are a "dry heat," the same can't be said for other sweltering locations, like humid Houston, Texas, or Savannah, Georgia. And even northern locations like Boston, New York, and Chicago can reach uncomfortably warm temperatures in the middle of summer, hitting at least 90 degrees several times a year.

Before air conditioning, people dealt with the heat in many different ways. Homes were often built with air flow in mind, with high ceilings to allow the warmer air to rise, and windows placed on opposite sides of a room so opening them would create a cross-breeze. Homes in New Orleans were commonly built "shotgun" style: rooms were lined up in a single row, with doors at each end of the house to allow a breeze to flow through. Families would spend time on porches outdoors to escape the stifling interior of their house, sometimes snoozing on screened-in "sleeping porches."

In big cities like New York, residents could lounge, and even sleep, on fire escapes. And some people got even more imaginative, hanging wet laundry just outside doorways so the breeze would carry cooled air into the home.

While these methods may have provided some relief from relentless heat and humidity, one physician in 1800's Florida felt that they weren't enough. John Gorrie, who attended the College of Physicians and Surgeons in Fairfield, New York, moved to the warm Gulf Coast town of Apalachicola, Florida, in 1833. The 30-year-old doctor was studying tropical diseases, and embraced the popular theory of the time that warm, humid air contributed to the spread of illnesses like malaria. In fact, this theory was so commonplace that the name "malaria" was derived from the medieval Italian for "bad air."

Gorrie felt that cooling the air in hospital rooms would help prevent the spread of disease, so he came up with the idea of suspending buckets of ice from the ceiling. But in warm Florida, the only way to obtain ice was to have it shipped from northern states, an expensive and complicated endeavor. In an attempt to cut down on costs and time, Gorrie designed a machine to create ice, using a compressor that could be powered by water, steam, wind-driven sails, or even a horse.

Gorrie was granted a patent for his invention in 1851, but it never really caught on as a marketable device. Still, it was the first attempt anyone had ever made to find a mechanical way to artificially cool an indoor space. It would take another half century of hot, humid, sweaty summers before the idea of mechanical cooling would come to the forefront again.

In the summer of 1902, the Sackett-Wilhelms Lithographing and Publishing Company in Brooklyn, New York, had a problem. Magazine pages that were printed by the company kept wrinkling, due to the high humidity in the plant. They sought the help of the Buffalo Forge Company, an equipment manufacturing business. The task of solving this problem luckily fell to engineer Willis Carrier.

Coils That Cool

Carrier developed a mechanical system that used coils that could both cool and remove moisture from the air by cooling water, becoming essentially the world's first modern air conditioner. Carrier continued to refine and improve his invention, and his "Apparatus for Treating Air" was granted a U.S. patent in 1906. Carrier was the first to recognize the relationship between dew point—the temperature at which air must be cooled to create water vapor—and humidity, helping engineers design more efficient air conditioning systems.

But the first air conditioners were still much too big, bulky, and loud to be used in homes, so most people got their first taste of indoor cooling in public buildings. In 1904, an air conditioning system was used to cool the 1,000-seat auditorium in the Missouri State Building at the St. Louis World's Fair, marking the first time air conditioning was used for the general public, and providing hot attendees with a much-needed cooldown. By the 1930s, with the advent of Hollywood movies, Americans had a new refuge for escaping summer heat: movie theaters. Carrier, who founded the heating, ventilating, and air conditioning (HVAC) company Carrier Corporation in 1915, developed an air conditioning system for movie theaters that became popular with the general public, especially during the struggles of the Great Depression.

Greatest Contribution of the Century

In the late 1930s, air conditioning technology was improving, resulting in smaller, quieter systems, and the first window air conditioner made its debut. Still too expensive for anyone but the upper classes, window units were seen as a luxury, found mostly in fancy hotels. But over the next decade, improvements in manufacturing made window air conditioners more and more affordable, and by the late 1940s, many American homes were enjoying the comfort of indoor cooling. The technology was beginning to catch on in other countries, as well, with British politician Sir Sydney Frank Markham writing in 1947, "The greatest contribution to civilization in this century may well be air conditioning—and America leads the way."

This "great contribution" eventually became popular in the corporate world in the 1940s and 50s. Originally hesitant to spend money on air conditioning, employers soon changed their minds when studies began proving that employees worked more efficiently in comfortable temperatures. By the end of the

1950s, nearly 90 percent of employers believed that air conditioning created a positive effect on employee productivity, with one study showing that typists were 24 percent more efficient in a cooled space.

Nation of Addicts

By the 1960s and 70s, air conditioner use was soaring in the United States, with southern states like Florida and Arizona seeing huge population increases. But the road to cool air wasn't always smooth; surprisingly, the life-changing technology had critics as well as converts. In 1979, *TIME* magazine writer Frank Trippett lamented that Americans were "all but addicted" to the refreshing technology, and his point wasn't without merit: with only 5 percent of the world's population at the time, the United States was consuming more air conditioning than all other countries combined. When the 1979 energy crisis hit, lawmakers realized that American energy consumption needed to be dialed back. The Energy Department established the Appliance and Equipment Standards Program, which implemented minimum energy conservation standards for appliances such as refrigerators, washing machines, and of course, air conditioners.

Air conditioners were struck with another blow in the 1990s, when chlorofluorocarbon coolants, which had been used as air conditioner refrigerating fluid since the 1930s, were linked to the depletion of the ozone in the Earth's stratosphere. Although the fluids were safe and non-flammable, they were eventually phased out in favor of hydrofluorocarbons (HFCs), which do not deplete ozone; unfortunately, HFCs are a greenhouse gas that can affect climate change. But don't worry about ditching your central AC just yet: recent research is discovering new, non-toxic, energy efficient coolants, and these will one day replace HFCs and make air conditioning much more environmentally friendly.

Today, air conditioning is extremely common in the United States, with around 88 percent of all new homes constructed including central air conditioners. In the Southern states, the stats are even higher, with 99 percent of new homes including the much-appreciated feature. Even Canadians enjoy air conditioning in their cooler locale, with 55 percent of homes including one. While Frank Trippett may have been correct in his assumption of Americans being addicted to air conditioning, it's hard to imagine choosing to sleep on the fire escape.

Indoor Plumbing

There are a few basic essentials most of us look for when buying a house or renting an apartment: electricity; a way to heat in the winter and cool in the summer; and, of course, indoor plumbing.

* * * *

PLUMBING SEEMS DECEPTIVELY simple, with water appearing or disappearing at the turn of a knob or flush of a toilet. But when we consider the intricate labyrinth of pipes and sewers that make up our water and sanitation systems, the reality of their complexity becomes apparent. And even though we rarely think twice about our indoor plumbing, anyone who's dealt with a backed-up toilet or a clogged drain can immediately recognize that it's definitely a worthwhile convenience. So how did this modern marvel come about?

As water is necessary for life, our earliest ancestors obviously had to find ways to obtain it. And the more water available, the larger the civilization could be. Thousands of years ago, "plumbing" simply consisted of humans digging wells and carrying water to their dwellings, or, like ancient Mesopotamians, using slaves to carry water in large pots. Ancient Chinese civilizations were adept at digging very deep wells for drinking water, which they would top with a wooden frame and a pot with a simple pulley system.

Ancient Engineers

The basis for modern plumbing began sometime between 4000 and 3000 BC, when humans in the Indus Valley in South Asia created a surprisingly efficient water and sewer system. Archeologists have uncovered evidence in the Indus Valley of a sewer network that emptied into either nearby bodies of water or into cesspools, which were then drained and cleaned. Homes had private toilets and baths, and cities included public baths with underground drains. Cities channeled rainwater and saved it for later use, and wastewater was used for irrigation. There is even evidence of copper pipes in the location, making the Indus Valley truly ahead of its time.

By 3000 BC, the Egyptian king Menes began building irrigation ditches, canals, and dams, even diverting the course of the Nile River to create the city of Memphis. Egyptians became experts at diverting and draining water, and began fitting bathrooms with metal fixtures such as lead basin stoppers. And as in the Indus Valley, copper pipes have also been found in Egypt, in, of all places, the tombs in pyramids. Egyptians believed that those who died were simply moving from one life to the next and would therefore need the same things that the living needed. This included food, clothing, and bathrooms, which have been discovered in some Egyptian tombs.

In ancient Crete, Greeks created an underground water supply system out of clay pipes, and are credited with building the first flushing toilet. But it was the ancient Romans who were considered the masters of plumbing and sanitation, constructing aqueducts to supply the city with large amounts of water.

Construction on these aqueducts began around 270 BC, and the remnants of many can still be seen in the city of Rome today. Pipes, made of lead or terra cotta and sealed in concrete, brought water to homes and public baths. The Romans were also known for constructing the Cloaca Maxima (which literally means "greatest sewer"), considered a marvel of ancient engineering. But for all of the ingenuity displayed by ancient peoples, unsanitary conditions, horrible smells, and disease were still constant worries.

A Little Discretion Please

In Europe in the middle ages, natural waterways, like underground rivers or streams, were used to carry wastewater. Ignoring the sophistication of the Romans' designs, the water often ran near, or even in the middle of, busy streets, and was only occasionally covered. Open sewers were extremely unpleasant, so in 1370, the city of Paris constructed its first closed sewer; meanwhile, London made use of the subterranean River Fleet as a sewer, and consequently the area was lined with prisons and low-quality housing. As cities grew, residents made use of outhouses, or "pail closets," and waste would be collected by local authorities every day, who would either incinerate it or use it for composting. But there were also some who eschewed these practices altogether, using buckets or pails to collect human waste and then dumping it into the streets, where it could be washed away by rain showers. Streets were even constructed with raised stepping stones, so pedestrians wouldn't walk through the sewage.

In the 16th century, Sir John Harington, a writer and courtier under Queen Elizabeth I, invented England's first flushing toilet, which he dubbed "Ajax"—a reference to "jakes," which was a slang word for toilet. The toilet had a valve to let water out of the tank and a wash-down design to empty the bowl, and the waste was then released into a cesspool. Harington presented his invention to the queen, but reportedly she never used it, deeming it too loud. Since the queen set the example for the populace of the day, one has to wonder if toilets would've gained popularity more quickly had she had used it!

The Water Closet

By the 17th and 18th centuries, water systems expanded, with pumping stations often providing water throughout cities. In 1775, Scottish inventor Alexander Cummings created a prototype for the modern toilet, including an s-bend, an overhead cistern, and valves, and the invention was installed in "water closets." But only wealthy residents could afford such luxury, with most still making use of "pail closets" well into the 19th century. And with rapid population growth in many cities, the problem of sewage and polluted streets continued to cause outbreaks of diseases like cholera and typhoid. The need to provide residents with clean water and sanitary sewage systems was never more apparent.

As water systems in cities became more complex, cleaner, and more efficient, the rich were the first to reap their benefits. In 1829, the Tremont Hotel in Boston, which welcomed guests including Davy Crockett and Charles Dickens, became the first hotel to install indoor plumbing, running water, and toilets. Water was stored in a tank on the roof and fed by gravity down to the taps. Although guests had to leave their rooms to use the toilets and bathtubs, which were on the first floor and in the basement, the novelty of running water made the Tremont one of the most luxurious hotels in the country. The Tremont even beat the White House to indoor plumbing. President Andrew Jackson had running water installed in 1833.

Happy Are the Masses That Flush

By the mid-1800s, many large cities, including London, Paris, Hamburg, Chicago, and New York, had sewage treatment plants to improve the quality of their water. And by 1840, the water closet had become more widespread in Europe, with most middle-class families using one; the flush toilet then made its way to America, Australia, and other parts of the world, forever changing the way humans live.

The combination of better sewage treatment and indoor plumbing had a dramatic effect on public health. By the middle of the 20th century, incidents of waterborne diseases plummeted, and life expectancies in parts of the world with indoor plumbing greatly increased. Unfortunately, around two billion people in the world still live without adequate plumbing and sanitation, including more than a million in the United States.

Indoor plumbing is seen as such a vital advancement in human civilization that the Memorial Sloan-Kettering Cancer Center in New York City has stated that indoor plumbing does more to improve the health of residents in impoverished countries than even building new hospitals.

Clearly, indoor plumbing is a system that literally changed the world for the better. It might be hard to believe that there's still room for improvement, but experts continue to make advancements that have made our water use more efficient and cost-effective. Innovations like low-flow toilets and showerheads, hands-free faucets that can be activated by motion, and appliances that use less water to save money are just a few ways that indoor plumbing keeps getting better.

The Washing Machine

If you've ever lived in a home or apartment without a washing machine, you know how inconvenient laundry day can be. Piling dirty clothes into a basket or bag; dragging it all into a dark basement laundry room or driving across town to a laundromat; hoping no one steals any of your clothing when you're not looking—washing clothes is never much fun.

* * * *

BUT NEXT TIME you're schlepping that bag full of clothes around, consider what our ancestors had to do to clean their clothing. Before machines, ancient peoples would take clothing to streams or rivers, where they'd rub them with sand or rocks and rinse them in the water. Later, the Romans created a type of soap containing ash and animal fat that could be used to scrub clothes. It may sound less appealing than the fresh-smelling detergents we're used to today, but it was a huge improvement over the work of ancient Roman "fullers." Fullers collected laundry from the citizenry and cleaned it in a strange mixture of water and urine (used for its ability to bleach linens due to its high ammonia content). It was then "agitated" by stomping feet, before being rinsed and wrung out.

Whatever Works

In the Middle Ages, flowers and fragrant plants were added to wash water, to keep clothes smelling fresh for as long as possible. A good move, since clothes were washed very infrequently! By colonial times, people would wash clothes by boiling them in large pots. And in the 1800s, washerwomen would collect clothing and linens and wash them in streams, or take them to wash houses, which had running fountains of water and made the process a bit less laborious. But all of these washing methods took up precious time, and doing the laundry could be an all-day task.

Because of the time and labor involved, inventors were always coming up with new ways to wash clothes. But the question of who actually invented the washing machine isn't entirely clear. A patent for a "washing machine" was issued as far back as 1691 in England, and German professor, botanist, and inventor Jacob Christian Schäffer published a washing machine design in 1767. American Nathaniel Briggs was issued a "clothes washing" patent in 1797, but in 1836 the U.S. Patent Office burned down, destroying any record of his idea.

But early designs shared many of the same features: Enclosed basins with grooves or paddles would hold the laundry, and the person washing the clothes would use a stick to agitate the contents, helping to remove dirt. Later, it became common to use a wooden drum that could be cranked by hand, tumbling the clothes within. This design was improved with the introduction of metal drums, which allowed for a fire to burn beneath, keeping the wash water warm.

By the mid-1800s, cities in the United States were experiencing a population boom, and the Industrial Revolution was in full swing. Two more inventors, James King in 1851 and Hamilton Smith in 1859, were granted patents for washing machines. These designs would then be improved upon by the religious Shaker community. The Shakers built large wooden machines meant for commercial use, and their design was so popular that it won a gold medal at the Centennial International Exhibition in 1876.

We Love You, Alva

While commercial washing machines were often powered by steam and belts, home machines were hand powered. But the advent of electricity at the end of the 1800s began to change the way Americans lived, and the first electric washer, with the brand name Thor, was introduced for commercial use in 1908. Invented by Alva J. Fisher, the Thor washing machine was a drum with a galvanized tub. The brand is still available today.

The Maytag Corporation set their sights on home consumers, debuting an electric wooden-tub washing machine in 1907, and the Whirlpool Corporation was created in 1911 to manufacture electric wringer washers. By 1928, almost a million washing machines had been sold in the U.S., but when the Great Depression hit, families could no long afford the luxury of such appliances. So in the early 1930s, laundromats began popping up in many towns, the first one opening for business in Fort Worth, Texas, in 1934.

Meanwhile, washing machine design continued improving, and by the end of the Great Depression, 60 percent of homes with electricity also had a washing machine. While these machines were powered by electricity, their controls were still manual.

It wasn't until 1937 that the first automatic washing machine made its debut. During World War II, many manufacturers put washing machines on the back burner, focusing on the war effort; however, some appliance companies took this opportunity to research potential advancements to the machines, knowing that the end of the conflict would signal a new era in American industry.

Their work paid off: By the late 1940s and early 1950s, washing machine technology was vastly improved, with many machines including features that are still in use today. Even during the material shortages of the Korean War, appliance manufacturers kept up with high demand for their products, and by 1953, automatic machines outsold manual electric machines.

In the decades that have followed, washing machine advancements have made doing laundry easier, cheaper, and more efficient. Microcontrollers—tiny computers on integrated circuits—and other computer-controlled systems have made washing machines "smarter," and able to automatically adjust water level or agitation speed depending on load size. Even when we have to haul our clothes to a basement laundromat, laundry day has never been more convenient.

Microwave Ovens

It's happened to all of us at one time or another: We brew a nice, hot, cup of coffee or tea, settle into a desk chair to work or start watching some TV, and a half hour later realize we've failed to take a single sip. Only now the steaming hot drink is a lukewarm cup of disappointment. Fortunately, around 96 percent of American homes include a microwave; simply place the cup inside, hit a few buttons, and within seconds, that caffeinated beverage is ready to sip once more.

⁕ ⁕ ⁕ ⁕

MICROWAVE OVENS ARE hard to beat for their quick convenience, something that's appreciated in our hectic, modern times. But surprisingly, these handy appliances may have never come about had it not been for World War II and a peanut cluster bar.

At the end of the 1930s, just as the world was spiraling into war, a British physicist named John Randall was working on a way to improve the radar capabilities of the Allies. Along with a team of colleagues, Randall created a cavity magnetron, which produced small electromagnetic wave vibrations called microwaves. The cavity magnetron gave the Allies a radar advantage, allowing them to see smaller objects using a smaller radar set that could be installed in the nose of a plane.

But when the war ended, magnetrons were no longer in much demand. One company that had been producing them, Raytheon Manufacturing Company, was determined to find a more practical use for the technology. Engineer Percy Spencer, who had earned several patents while working with magnetrons, began testing the devices and trying to improve their power. One day in 1946, as he got ready to leave the lab for a lunch break, he stuck his hand in his pocket and found the peanut cluster bar he'd brought for a snack had melted.

Maybe We'll Stick to Grilling

It was no secret that food could be heated with radio waves; in fact, during the 1933 World's Fair in Chicago, Westinghouse had featured a shortwave radio transmitter that cooked steak and potatoes between two metal plates. It was a fun demonstration for the fair crowd, but no one thought much of it.

Spencer, however, began to wonder about the possible applications of this interesting phenomenon. Soon after the melted peanut cluster episode, Spencer placed an egg underneath the magnetron tube; the egg quickly exploded. Next, the engineer experimented with corn kernels—the resulting popcorn was shared with the entire office. Spencer verified his findings by aiming the microwaves from a magnetron into a metal box with food inside. The temperature within the food rose rapidly, and the microwave oven was born.

A Tough Sell

Within a year, the first commercial microwave, called the "Radarange," was available. Weighing almost 700 pounds and costing more than $2,000, it was marketed to restaurants and airlines and its magnetron tubes had to be continuously cooled with water. The massive machine did not sell well, and never made much of a name for itself. Still convinced that this invention was useful, in 1955 Raytheon began licensing its microwave technology. The Tappan appliance company sold the first microwave to home consumers. However, at $1,295 (around $11,000 in today's dollars) per appliance, microwaves were far too expensive for most home cooks. Again, the launch of the microwave flopped.

It would take another decade before the microwave began to make its way into more American kitchens. In 1965, Raytheon acquired Amana Refrigeration, which had been known for manufacturing freezers, refrigerators, and air conditioners. But now they added a heat source to their repertoire: a compact version of the Radarange. The smaller microwave could fit on a countertop and cost a more reasonable $495. Although this was still a major expense (coming in around $4,000 in today's dollars), consumers showed much more interest, and other companies began jumping on the microwave bandwagon. Sharp introduced an oven with a rotating turntable, providing more even heating; and Litton Industries improved on the original design of the oven, fueling more sales.

"Beware the Microwave Oven"

In 1970, about 40,000 microwaves were sold in the U.S. By 1975, one million ovens were sold over the course of the year. Clearly, the technology was a hit; but some consumers were still concerned about one thing: radiation. Microwaves work by passing microwave radiation through food, causing molecules to rotate or vibrate, releasing energy. This dispersed energy raises the temperature of the food. Understandably, using an appliance that emits radiation does sound like a disconcerting practice. In 1974, the *New York Times* even cautioned consumers to "beware the microwave oven," asserting that there was no way to know whether the technology was safe.

But the warnings didn't prevent microwaves from flying off the shelves. By 1986, a quarter of American households had a microwave, and by the late 90s, more than 90 percent of U.S. homes had one. Other parts of the world have embraced the convenience of quick cooking, as well: 95 percent of Australians and 88 percent of Canadians use them, and even the French, known for their rich cuisine, high culinary standards, and leisurely meals, find ways to use the machines, with 65 percent of homes including a microwave. Today, we know that the radiation emitted by microwaves is negligible, and that none of the radiation passing through our food can be transferred to our bodies. In fact, some experts assert that microwaves make food even safer to eat, since microwave rays can kill bacteria.

Microwaves are sò small, cheap, and commonplace now that it's hard to believe they were once huge behemoths that could cost as much as a car. But for those who are curious, an original Radarange is still found on the merchant ship NS *Savannah*, a nuclear-powered passenger/cargo ship that was deactivated in 1971 and is now moored in Baltimore, Maryland. Although the ship is not open to the public, tours can be requested—no word on whether microwave popcorn is provided!

A Toy So Fun It's Silly

Is Silly Putty a toy or an industrial compound? Both, actually.

* * * *

WHAT'S NOT TO like about Silly Putty? It bounces, it stretches, it breaks. It's good, clean fun. But surprisingly enough, Silly Putty didn't begin its career as a toy. It's actually a by-product of war research.

During World War II, the Japanese military began a series of conquests in southeast Asia with the intention of cutting off the Allies' supply of rubber. In response, the United States began to look for ways to develop their own synthetic rubber products. Because, after all, it's hard to fight an effective war without rubber.

Enter James Wright. While working at General Electric in 1943, Wright combined boric acid with silicone oil. The result was a polymerized substance that could bounce and stretch and had a high melting temperature. Unfortunately, it couldn't replace rubber and thus served no purpose for the military.

General Electric was determined to find a use for Wright's invention. The company bounced the putty across the country, hoping to drum up interest. In 1949, the putty landed in the hands of a New Haven, Connecticut, toy-store owner named Ruth Fallgatter. Fallgatter teamed up with marketing guru Peter Hodgson to sell the product through her toy catalog as a bouncing putty called Nutty Putty. It was an instant hit. In 1950, Hodgson christened the new toy "Silly Putty" and went on to sell the product in the now-famous egg-shaped containers. Originally marketed to adults, the product ultimately found success with kids ages 6 to 12—especially after 1957 when a commercial for Silly Putty debuted on the *Howdy Doody* show.

Silly Putty has done a brisk business over the years, with more than 300 million eggs sold since 1950. Silly Putty also has practical uses—it is useful in stress reduction, physical therapy, and in medical and scientific simulations. It was even used by the crew of *Apollo 8* to secure tools in zero gravity. Because there's nothing silly about being hit in the head with a wrench when you're in space.

Refrigerators

Refrigerators are one of those things we all take for granted in the 21st century. After all, most of us have probably never known life without one. The ability to stock a fridge with our choice of fresh food, which remains safe and useable for days or weeks, is such a simple concept; yet the advent of the refrigerator truly changed human lives.

* * * *

BEFORE THIS APPLIANCE made its debut in the world, humans relied on ice for their preservation needs. By 1000 BC, the Chinese had figured out how to cut and store ice from frozen rivers and lakes. Five hundred years later, Egyptians and Indians realized that water in pots left outside on cold nights would freeze, resulting in homemade ice. And Hebrews, Romans, and Greeks would collect fallen snow and save it for later use. All of this ice or snow would be packed together, insulated with straw or wood, and stored in an "ice house," where it could remain frozen for months. The ice was then used to preserve food or to cool drinks.

By the 17th century, Europeans had discovered that when saltpeter, or potassium nitrate, was dissolved in water, it had a cooling effect and could create ice. This method became very popular in France to make iced drinks, liqueurs, and frozen juices. Europeans also preserved ice by salting it, wrapping it in flannel, and storing it underground.

For centuries, humans relied on variations of these methods to store food. They would place food in cool cellars, outdoors, or in springhouses, where cold stream water would keep food cool. Or preserved ice and food was placed in "ice boxes," which were insulated containers made of wood or metal. But even with these various approaches, illness from spoiled food remained a huge problem, especially during warmer summer months.

In the early 1700s, the first concept of artificial refrigeration was born when a Scottish scientist by the name of William Cullen discovered that rapidly heating diethyl ether in a partial vacuum, causing it to evaporate, actually absorbed heat from the surrounding air and created a cooling effect. Cullen never found a practical application for his discovery; but years later, his idea would serve as the basis for modern refrigeration.

A Refrigerator in Name Only

But interestingly, the very first "refrigerator" would not use Cullen's ideas, but rather harken back to the days of the ice box. In 1802, cabinetmaker, engineer, and farmer, Thomas Moore, was looking for a way to transport butter to his farm. He designed a container consisting of a tin box inside a cedar tub, with a gap between the two for ice. He then insulated the whole thing with rabbit fur and woolen cloth, and called it a "refrigiratory," but in 1803 changed the name to "refrigerator" and had it patented. Although this particular refrigerator never gained popularity, it did have one notable fan: President Thomas Jefferson, who not only signed the patent, but purchased his own refrigerator from Moore and used it for several years.

Vapor Compression

But with more people moving to cities and away from sources of fresh foods, inventors were hard at work searching for a practical use for Cullen's decades-old discovery. In 1805, American inventor Oliver Evans designed a refrigeration machine that used vapor-compression, building on Cullen's original idea. Although Evans never moved past the design stage, his concept was sound; vapor-compression refrigeration is still widely used today in air conditioners and refrigerators.

Meanwhile, in England, famed British scientist Michael Faraday, known best for his work with electricity and magnetism, was running his own vapor-compression experiments by liquifying gasses using high pressure and low temperature. And in 1834, American Jacob Perkins, who was living in Great Britain at the time, built the first working refrigeration system, using the vapor-compression idea first developed by Evans. Patented in 1835, the machine never became a very big commercial success; but it did earn Perkins the title of "father of the refrigerator."

The path laid out by these early inventors and scientists cleared the way for many improvements and new discoveries around the world. Scottish-Australian James Harrison patented a refrigeration machine that used ether, ammonia, or alcohol in 1856, and within five years a dozen of his machines were found in breweries and meatpacking houses in the Australian state of Victoria. In 1859, French engineer Ferdinand Carre designed a refrigerator that used a mixture of ammonia and water, which was patented in 1860. And in 1876, Carl von Linde, a professor at the Technological University Munich in Germany, created an improved process for liquifying gasses, which led to the use of not only ammonia, but also sulfur dioxide and methyl chloride as refrigerants.

The Quest for Cold Beer

By the late 1800s, refrigeration was quickly gaining on the old ice box method. Not only was the new technology more reliable, but finding clean ice was becoming a problem. As a result of pollution and sewage dumping, using ice from frozen lakes and rivers could be a health hazard; and businesses such as meatpacking plants and dairies all looked to mechanical refrigeration for a more sanitary solution.

Breweries, especially, embraced refrigeration, looking to provide their customers with a quality product year-round. In fact, breweries were the first major industry to extensively use mechanical refrigeration, with the first fridge installed at S. Liebmann's Sons Brewing Company in Brooklyn, New York, in 1870. Within two decades, almost every brewery in the country had a refrigerator. The meatpacking industry followed suit, installing the first refrigerator in Chicago in 1900. And dozens of other businesses began to see the benefits of these machines, including sugar mills, florists, chocolate factories, bakeries, and tea companies. Even textile mills, oil refineries, and metalworkers found a use for refrigeration, and fur companies used refrigerated warehouses to store clothing to prevent moth damage.

With the success of mechanical refrigeration in industries, engineers turned their attention to the home market. Fred W. Wolf, an inventor from Fort Wayne, Indiana, created the first electric refrigerator for domestic use, which could be mounted on top of an ice box. Although his creation never sold well, Wolf is also credited with inventing the ubiquitous ice cube tray, which was immediately popular and is still included in many refrigerators today.

In 1916, Alfred Mellowes invented a self-contained refrigerator with a compressor in the bottom cabinet. The design was a hit, and two years later, Mellowes' successful company was bought out by William C. Durant, the founder of Frigidaire, who began mass-producing the machines. Meanwhile, the

Kelvinator company introduced the first fridge with automatic control, and by 1923 they held 80 percent of the American market for electric refrigerators. In Sweden, technology students Baltzar von Platen and Carl Munters invented an "absorption refrigerator," which used a heat source to drive the cooling process, and appliance manufacturer Electrolux began marketing it to worldwide success.

Too Big and Too Expensive

Although mechanical refrigeration was definitely catching on in the 1920s, there were still a few snags. First was the cost: a fridge could cost more than $700, which was more expensive than even the price of a Model-T Ford car! Another issue was safety, as the machines often used either sulfur dioxide or methyl formate as refrigerants, both of which are highly toxic. And lastly, all of these machines required a separate area for the mechanical parts to be installed, usually in a basement or adjacent room, making them bulky and not always practical.

Fortunately, improvements over the decade were swift, leading to safer, smaller, and more affordable machines. Freon, which is a stable and nonflammable type of gas or liquid, was introduced to the refrigeration market in the 1930s and replaced the more corrosive chemicals that had previously been used. And General Electric introduced the "Monitor-Top" model in 1927, which contained the compressor assembly in a ringed container above the refrigerator cabinet. The Monitor-Top became the first refrigerator to see truly widespread use, selling more than a million units. New refrigerators were becoming cheaper, as well, with prices dropping into the $100 and $200 range. By the mid-1930s, around 6 million refrigerators had been sold to consumers.

After World War II, separate freezer compartments, often called a "deep freeze," became popular as frozen foods were introduced to supermarkets. Over the next few decades, refrigerators saw more advancements, like automatic defrosting and ice making. Freon, which was found to be damaging to the ozone layer, was replaced with more environmentally friendly refrigerants, and efficiency was improved. And of course, decorative colors—like pastels in the 60s, avocado green in the 70s, gold in the 80s, and (thankfully) the more neutral white, black, and stainless steel that endure today—were introduced to match interior design schemes.

Refrigerators have come a long way since the old-fashioned ice box, today even incorporating internet, Wi-Fi, voice control, and smart apps that help you control your fridge from your phone. There are even refrigerators that can order groceries for you and have them delivered to your door. It seems the only thing they can't do is prepare your food for you—but perhaps that feature is in the works.

Television Changed Everything

Think of some of the biggest moments in history: the moon landing; the Challenger *disaster; the fall of the Berlin Wall; O.J. Simpson's race from the police and subsequent trial. If you witnessed any of these events, chances are good that it wasn't in person. Like millions of other Americans, you watched these events, and many more like them, on television.*

* * * *

A RUSSIAN SCIENTIST AND inventor, Constantin Perskyl, is said to have coined the word "television" in 1900, combining the Greek word *tele*, meaning "far," with the Latin *visio*, meaning "sight." But television's inception began even before the turn of the century, with inventors attempting to mechanically scan and transmit images to a screen as early as the mid-1800s. One of these inventors was a German university student named Paul Julius Gottlieb Nipkow. He created a rotating disk with a pattern of holes in a spiral pattern. Images would be transmitted to the disk, which divided a picture into a linear sequence of points.

This "Nipkow disk" became the inspiration for other inventors, including Scottish engineer John Logie Baird, American inventor Charles Francis Jenkins, and Japanese engineer Kenjiro Takayanagi. Baird created a system using a Nipkow disk that could televise silhouette images, which he demonstrated at Selfridge's Department Store in London in 1925. And on January 26, 1926, in what has often been considered the first official demonstration of television, Baird transmitted the face of his business partner, Oliver Hutchinson, using a system consisting of two Nipkow disks. A year later, Baird was able to transmit a signal between London and Glasgow, a distance of more than 438 miles. The inventor then established the Baird Television Development Company Ltd., which made the first transatlantic television transmission in 1928.

Meanwhile, Jenkins was also using the Nipkow disk to transmit images, which he demonstrated for the public in 1925. Jenkins was granted a U.S. patent for his device, which transmitted the image of a toy windmill over a five-mile distance from a radio station in Maryland to his lab in Washington, D.C. These early mechanical televisions created by Baird and Jenkins impressed the public, but were primitive and rudimentary compared to what was soon to come.

In 1926, Takayanagi demonstrated his television system, consisting of a Nipkow disk scanner and a cathode ray tube display, at Hamamatsu Industrial High School in Japan. The addition of the cathode ray tube made Takayanagi's invention the first "all-electronic" television. And whereas Baird and Jenkins had been unable to achieve more than 48 lines of resolution with their systems, Takayanagi created a system that improved the resolution to 100 lines. For his contributions to the early advancements in the medium, the inventor earned the nickname, "the father of Japanese television."

Show Me the Money

Back in America, inventor Philo Taylor Farnsworth gave the first demonstration of an electronic television to the press in 1928. The 22-year-old had needed the help of financial backers to get his invention off the ground, who demanded to know when they'd start seeing some dollars returned from their investment. So, fittingly, Farnsworth's first transmission on his electronic television was the image of a dollar sign. Farnsworth went on to improve his system, later transmitting live images of people.

Television technology was immediately embraced by inventors and the public alike. The first television stations launched in 1928, including W2XB in Schenectady, New York, and Jenkins' own station, W3XK, in Washington, D.C. These stations broadcast "radio movies," which featured silhouetted actors performing the scenes in radio shows. W2XB, now called WRGB, is still in operation today, giving it the distinction of being the oldest continually operating television station in the world.

By 1934, most mechanical televisions were phased out in favor of electronic televisions, although a few mechanical stragglers were found in universities until 1939. Farnsworth made history again when he demonstrated an all-electronic television using a live camera on August 25, 1934. Throughout the rest of the decade, cameras, as well as televisions, began to improve, with a team from British conglomerate EMI creating and patenting more sensitive camera tubes. These cameras were able to broadcast the first live street scene, a feat that Farnsworth would later duplicate for the 1939 New York World's Fair.

The World's Fair also marked the first time televisions became commercially available to the public. The large boxes with 12-inch screens cost around $450 dollars—about a third of a family's yearly income at the time. Needless to say, although the public viewed the technology with awe and admiration, few could afford an early television set. But those who could were treated to such sights as boxing matches, news reports, and the famous images of Adolph Hitler announcing the opening ceremonies of the 1936 Berlin Olympics.

Color Television

Although early televisions most commonly broadcast in black and white, inventors had been experimenting with color televisions since the late 1800s. Baird demonstrated the world's first color transmission in 1928, and made the first color broadcast ten years later from his television studio to a screen at London's Dominion Theater. Other early innovators of color television include Mexican inventor Guillermo Gonzalez Camerena, who patented his "trichromatic field sequential system" color television in 1940, and Hungarian engineer Peter Carl Goldmark who the same year demonstrated a color television system to the Federal Communications Commission.

Broadcasters began testing out color transmissions in the 1940s, but the advent of World War II slowed down the television market's evolution. But the strong post-war economy created a television boom, with sets selling for a more reasonable $100. By the 1950s, average Americans could finally afford television sets, and shows like *The Ed Sullivan Show*, *Howdy Doody*, and *I Love Lucy* ushered in television's first golden age. The RCA Corporation developed the first all-electronic color television system, which was introduced in 1953, although most consumers would continue to have black and white sets until the early 1970s.

Digital Television

Televisions themselves would change little over the next decade, although accessories like remote controls, game consoles, and VCRs made TVs more versatile. But by the 1990s, as high-performance computers became more prevalent, electronics companies began to develop digital television. Up until this time, televisions used analog signals to broadcast content, which could create electronic noise and static, sometimes resulting in a "snowy" picture. But digital technology eliminated this issue, and by 1994, DirecTV debuted a digital satellite platform, with cable providers like Tele-Communications, Inc. and Time Warner launching digital cable broadcasts.

Digital television marked the first major breakthrough in the medium since the color TVs of the 50s. It was far superior to analog systems, and many countries began transitioning their broadcasting from analog to digital by the year 2000. Berlin became the first major city in the world to completely shut off analog signals in 2003. Today, Australia, India, Mexico, Japan, and most of Europe have terminated all of their analog signals, whereas the United States and Canada still have some lingering analog signals.

It's doubtful that the early television pioneers—Baird, Jenkins, Farnsworth, and others—could have envisioned what television has become in the 21st century. "Smart TVs" merge televisions and internet to create a more interactive and controlled experience; streaming content means that we no longer need to wait to watch our favorite shows; and those original 12-inch screens have morphed into 40-, 50-, and 60-inch screens (and even larger, if you have the room!). Few inventions have had as much of an impact on the world as television; it's safe to say that our lives would be quite different without it.

One-Channel Wondering

How come there wasn't a channel one on your television?

⁕ ⁕ ⁕ ⁕

HAVE YOU EVER wondered which programs Bruce Springsteen saw in 1992 as he surfed the channels available on his newly installed cable and began muttering a new song under his breath? Was it the smarmy cast of *L.A. Law* that so repulsed him? Or was it the predictable mystery of *Matlock* that pushed him over the edge? Or perhaps it was the vampiric visage of Ron Popeil? This is a question for philosophers to debate; we'll probably never know the answer. But there's one thing that we can say for sure: In all of his channel flipping, the Boss never took a look at channel one—he couldn't have. Channels start at two and go up from there.

It wasn't always this way. In April 1941, New York's two stations began broadcasting on channels one and two, respectively. And within a year, channel one was a going concern over more regions of the country. But World War II had a chilling effect on the fledgling medium. For the next several years, the country devoted its resources to more pressing needs. By the time commercial broadcasting was ready to resume in 1946, new technological developments had changed both radio and television. Competition for the airwaves was fierce—stations could broadcast farther, faster, and on higher frequencies than ever before.

Everyone wanted a piece of the big pie in the sky. A series of congressional hearings were held to apportion the broadcast spectrum, and by 1947, the Federal Communications Commission (FCC) had awarded a total of thirteen channels to the television networks. Channel one was designated a community channel for stations with limited broadcasting range because it had the lowest frequency.

But there was trouble in this television paradise. As the number of broadcasters increased, the airwaves began to get crowded, especially in larger metropolitan areas. Frequencies started to overlap, causing chaos and complaints when viewers found their quiz shows scrambled with the nightly news, or vice versa.

The FCC took steps to reduce this interference. In 1948, the organization decided to free up space by disallowing broadcasts on the lowest frequency—channel one. That bandwidth would instead be devoted to mobile land services—operations like two-way radio communication in taxicabs. Commercial television retained channels two through twelve. As the FCC's plan went into effect, television manufacturers simply stopped putting a one on the tuning dial; the millions of people who bought their first television sets in the 1950s barely even noticed its absence.

Since then, our options for televised entertainment have multiplied at a staggering rate. Bruce Springsteen's fifty-seven-channel cable package sounds quaint to contemporary subscribers who have hundreds of stations to choose from. But even with this nearly unlimited number of channels available in our living rooms, we'll never again have a channel one. As to whether there's anything worth watching? We'll leave that up to you and Bruce.

Spectacles and Contact Lenses

If you're like the majority of Americans—three fourths, to be exact—who need some sort of vision correction, it's probably hard to imagine life without glasses or contacts. We use our eyesight to process information, navigate the world, marvel at nature's beauty, and assess situations; just try to walk through an unfamiliar room in the dark and the value of good eyesight will be immediately obvious.

* * * *

IN ORDER TO see an image clearly, light rays must focus on the retina of the eye, while the pupil and cornea shrink, focus, and curve the image so it can be processed by the brain. If any parts of the eye have irregularities, the resulting image will be blurry. For nearsighted individuals, the shape of the eyeball causes light to focus in front of the retina; for those who are farsighted, the light focuses behind the retina. In order to correct this, some type of lens is needed to refocus the light. That's where spectacles and contact lenses come into play.

Devices to assist vision have been around for a surprisingly long time. As long ago as 700 BC, the Assyrians figured out that polished crystal could magnify objects. Ancient Greeks and Romans filled transparent spheres with water to create a magnifying effect. And legend has it that the Roman Emperor Nero used a polished emerald (although it may have been a less impressive, more common green mineral) to watch gladiator fights in arenas at a distance, to help with his nearsightedness.

Sometime between the years 1000 and 1250, humans began to use convex-shaped glass to magnify words when reading. These "reading stones" were often used by monks as they pored over documents. Philosophers and scholars, such as English theologian Robert Grosseteste and Franciscan Friar Roger Bacon, wrote about such stones in the mid-13th century, just before the first spectacles were invented.

No one is quite sure who should be credited for the invention of the first eyeglasses, but we do know that they first appeared near Pisa, Italy, around 1290. Consisting of two glass or crystal stones surrounded by a frame, these spectacles were held to the eyes with the aid of a handle. By 1300, the guild of crystal workers in Venice had created regulations for crafting "discs for the eyes," and soon this new invention was making its way throughout Europe.

As eyeglasses gained popularity, factories in other parts of Italy began to create them too. Manufacturers in Florence soon realized that not everyone needed or wanted the exact same type of magnification in their spectacles, so they came up with a grading system to create lenses of different strengths. The eyeglasses they created for nearsightedness, farsightedness, and for reading were in high demand; by the 15th century, spectacles were sold throughout Europe and had made their way to China, India, and the Middle East. The demand for eyeglasses increased even more as newspapers began publication, such as the *Oxford Gazette* in 1665. Owning a pair of spectacles was seen as a symbol of intelligence, wealth, and status, and these "status symbols" soon crossed the pond and debuted in America.

Nose Pinchers

Up until the early 1700s, spectacles were designed to either be held by the hand or to perch on the bridge of the nose, a style known as "pince-nez." But these both had drawbacks: what if the wearer needed both hands free? What if the "pince-nez" spectacles didn't quite sit right on the nose, and kept falling off? Eyeglass wearers came up with all manner of contraptions to keep their eyeglasses steady, including securing them around the head with ribbons or attaching long handles to make them easier to maneuver. But around 1730, a London optician named Edward Scarlett came up with the idea of adding rigid sidepieces to eyeglasses. This allowed them to rest securely atop the ears. Surprisingly, this innovation was not immediately popular, and many people continued to use other styles that were considered more "fashionable." Fortunately, Scarlett's idea eventually caught on, and today we can wear our eyeglasses without worrying that they'll topple off our nose.

Of Course Leonardo Already Thought of That

But even without the threat of falling spectacles, some people would prefer to forgo devices on the face altogether. Contact lenses sit directly on the cornea, and not only correct vision, but provide clearer peripheral vision and eliminate the problem of rain or fog on glass lenses. The idea of lenses that sit directly on the eyes goes all the way back to Leonardo da Vinci, who noticed the way that water could help to focus images when looking through it. He envisioned some sort of contraption that could hold a water-filled vessel over the eyes; the idea was later echoed by French philosopher Rene Descartes, who suggested water-filled glass tubes that could be held in contact with the eyes.

Obviously neither of these ideas was very practical, but they did illustrate an understanding of how light appears different when viewed though different mediums. It wasn't until 1888 that contact lenses became a reality. Two inventors, ophthalmologist Adolf Gaston Eugen Fick and physiologist August

Müller, each created lenses made of blown glass. Although they worked fairly well to correct vision, they also cut off the oxygen supply to the cornea and could only be worn for short periods of time.

Blown-glass contact lenses were the only option until the 1930s, when acrylic was developed. American optometrist William Feinbloom created the first lenses made of a combination of glass and plastic in 1936, and in 1939, the first all-plastic lenses were introduced by Hungarian optometrist Istvan Gyorffy. These first contact lenses were large "scleral" lenses, which sat on the sclera, or the white, of the eye and created a space between the lens and the cornea, which was filled with a dextrose solution. Even the plastic scleral lenses were a bit unwieldy, and not always very comfortable. But shortly after World War II, "corneal" lenses were developed, which sat only on the cornea and not across the entire expanse of the eye. These greatly improved lenses could be worn for up to 16 hours a day, and grew in popularity throughout the 50s and 60s. President Lyndon Johnson even wore contacts on television in 1964, becoming the first president to do so.

But these acrylic contact lenses still came with drawbacks. The worst disadvantage was that they choked off the oxygen to the cornea, causing infections, swelling, and sometimes nerve damage. So in the 1970s, chemist Norman Gaylord developed a material called siloxane-methacrylate that allowed oxygen to pass through, which he marketed as "Polycon." Lenses made of Polycon, which were called rigid gas permeable lenses, could be safely worn all day while allowing oxygen to reach the surface of the eye.

Again, though, there was a drawback: rigid lenses, while gas permeable, were uncomfortable at first, and required an adjustment period. But in the late-1950s, two Czech chemists, Otto Wichterle and Drahoslav Lim, had published an article entitled, "Hydrophilic gels for biological use" in the journal *Nature*.

They had developed a prototype of a contact lens made of a material that formed a hydrogel in water. Six years later, the rights to their idea was licensed to eye health company Bausch & Lomb, who began to manufacture hydrogel lenses in the United States. These lenses soon outsold rigid gas permeable lenses, thanks to their convenience and comfort.

Over the next few decades, soft lenses improved even more, with the introduction of silicone hydrogel lenses, which boast even better oxygen permeability. New polymers and materials have improved lenses to the point that some can be worn for extended periods of time, sometimes even up to 30 days.

Eyeglasses and contact lenses have come such a long way since their inception that some choose to wear them simply as fashion statements. But for those who need glasses or contacts to correct poor vision, these inventions are no longer merely status symbols; they are nothing short of remarkable.

A Clean Slate: The Cutting Edge in Classroom Technology

The introduction of the blackboard revolutionized teaching practices. Chalk it up to a Scottish teacher's desire to get the word out.

* * * *

FEW THINGS SEEM as mundane or uninspiring as a classroom blackboard: a slate surface covered in dusty, white text detailing tonight's reading assignment or the date of next week's test. What could be less revolutionary? But when James Pillans, headmaster of the Old High School of Edinburgh, Scotland, began using one in his classroom in 1801, he sent ripples of excitement throughout the educational world.

Before that time, teachers had no way of displaying written material to an entire class; they also had no way of mechanically reproducing copies of material written on paper. Consequently, teachers spent a considerable amount of time writing out assignments, dictating them to the class, or painstakingly copying them down on each student's small slate.

The Writing Is on the Wall

Pillans's innovation was considered a revolutionary piece of technology that would—and did—transform the educational experience. In fact, in 1841, Josiah F. Bumstead, a writer and educator, said that the inventor of the blackboard "deserves to be ranked among the best contributors to learning and science, if not among the greatest benefactors of mankind." Teachers could now work out equations, diagram sentences, and write out the week's new spelling words for the entire class at one time. And students could now experience the dread of going up to the board to work out those impossible math problems while the whole class watched!

The first recorded use of the blackboard in the United States was likely by George Baron at the United States Military Academy at West Point in 1801. By the mid-1800s, a blackboard could be found in virtually every schoolroom in America—even the isolated one-room schoolhouses that dotted the Western frontier.

New Changes Afoot

Of course, many of today's classrooms have since replaced their chalkboards with whiteboards. Teachers are also increasingly relying on digital technologies that enable them to pass out and collect assignments electronically. Pillans's innovation may be on its way out, but there is no denying that it served teachers and students well for more than 200 years. Its impact won't be erased anytime soon.

Cell Phones

There's a scene in the 1987 movie Wall Street *in which Michael Douglas, in his Oscar-winning turn as wealthy corporate raider Gordon Gekko, strolls down a beach while talking on his cell phone. But this was not one of the sleek phones we're used to today: this phone was a Motorola DynaTAC 8000X, a 13-inch-long, two-pound monstrosity also known as "The Brick." Today, watching Gordon Gekko heft around his giant cell phone is almost laughable; but at the time, The Brick was seen as a status symbol, used only by those who could afford its $3,995 price tag. Since then, cell phones have not only undergone a makeover, but they have drastically changed the way we communicate and live.*

* * * *

BEFORE THE CELL phones we know today, portable, two-way radio transceivers, or walkie-talkies, were created for use during World War II. Both Canadian inventor Donald Hings and Canadian-American engineer Alfred J. Gross played a part in the development of the devices, with teams at telecommunications company Motorola manufacturing the first handheld transceivers for the U.S. military. But these sorts of devices were not limited to military use; two-way radios also found a place in police cars and taxis, so officers and drivers could communicate with each other or with a central base location.

Just after the war, in 1946, Bell Laboratories developed a mobile telephone system in St. Louis, Missouri, that allowed users to place and receive calls from automobiles. Within two years, AT&T had introduced "Mobile Telephone Service" to 100 different towns and highway corridors.

While it was a novel idea, the equipment used for such calls weighed 80 pounds, and only three radio channels were available, limiting calls to only three users per city. Only about 5,000 customers signed up for the service, which cost $15 per month (more than $150 in today's dollars). Needless to say, the system never quite caught on with the general public. But AT&T didn't give up: in 1965 they launched their "Improved Mobile Telephone Service," with more radio channels and smaller user equipment, hoping the obvious name might draw in customers. Unfortunately, state regulatory agencies limited the number of customers to 40,000, and even with more radio channels, some callers had to wait more than 30 minutes before they could place a call.

A Dream That Wouldn't Die

Even with these early fails, the idea of making phone calls from cars, trains, or any area without pay phones was a dream that wouldn't die. And surprisingly, the basis for our modern cell phones began long before the iPhone was even a glimmer in Steve Jobs' eye. In fact, as far back as 1947, Bell Labs engineers Douglas H. Ring and W. Rae Young had an idea for using hexagonal cells to provide service for mobile phones in cars. But at the time, the technology to create such a system didn't exist. Two decades later, engineers Richard H. Frenkiel, Joel S. Engel, and Philip T. Porter, also with Bell Labs, expanded on Ring's and Young's ideas, suggesting antennas that transmitted and received in three different directions to reduce interference.

The Call Handoff

By 1967, cell phone technology was definitely taking shape, but there was still a problem: cell phone users had to remain in a single coverage area. If the cell phone left the area serviced by the base station, the user could no longer continue the call. Enter another Bell Labs engineer, Amos Edward Joel, who, in 1970, developed a "call handoff" system so a call could be continued from one area to another. Joel's idea, along with those of the other Bell Labs engineers, completed the blueprint for the cellular system we know and love today. On April 3, 1973, Motorola engineer Martin Cooper placed the first cell phone call to Engel at Bell Labs, using a prototype of the DynaTAC, proving that the dream of cell service could be a reality.

But even though the technology had been developed, it would take more than 10 years before the Federal Communications Commission would approve a request from AT&T for cellular service. Finally, in 1983, the Advanced Mobile Phone System (AMPS) was launched in the United States. This "1G"—or first generation—system was an analog system, which translated audio into electronic pulses. And for those who were anxious to make use of the new system, the Motorola DynaTAC 8000X was introduced on March 6, 1983. "The Brick" debuted to surprisingly strong demand, despite the steep price tag, 10-hour charge time, and measly 35-minute talk time.

Throughout the 1980s, cell technology advanced and mobile phones got (thankfully) smaller, although they would still be considered bulky and cumbersome by today's standards. In the 1990s, with the emergence of 2G phones, digital cellular was introduced. While analog phones had been increasingly popular, they were also unencrypted, which meant that conversations were susceptible to eavesdropping and could even be accidentally picked up by scanners or baby monitors. But digital technology, which breaks audio down into binary code and then reassembles it, is easier to encrypt and therefore much less vulnerable to curious ears.

The Smartphone

In 1993, what could be considered the world's first "smartphone," the IBM Simon, was introduced. The phone had a calendar, address book, clock, notepad, and email service, along with being a phone, fax machine, pager, and personal digital assistant (PDA). It even included a keyboard and a touchscreen. But the battery of the one-pound Simon only lasted an hour, and with newer, lighter, and smaller phones introduced almost at the same time, this original smartphone met its demise within six months.

Second generation phones also introduced a new form of communication that would soon become more commonplace than phone calls themselves: SMS (Short Message Service), or text messages. The first person-to-person SMS message was sent in Finland in 1993, and by the late 1990s, text messages became a preferred form of communication, especially among the younger generations. Finland also introduced the world to the first paid and downloadable media content for cell phones, a ringtone service called Harmonium.

Considering that the original realization of cellular phones took almost 40 years to implement, the last two decades have seen amazingly fast advances in the technology. Third generation technology, which was first rolled out in Tokyo in 2001, began using packet switching as opposed to circuit switching for data transmission. Packet switching allows data to be broken up and sent through whichever channels are optimum, before being regrouped at the destination, which allows for faster and more efficient data transfer. This became crucial for 3G phones, which allowed for media streaming like online videos and music, and access to the internet.

Today, we've reached the fourth generation of cellular networks (with the promise of 5G on the horizon), which offer speeds up to 10 times faster than previous 3G phones. This means better video quality, the ability to stream television and movies, and an all-IP (Internet Protocol) network. This all-IP network sends all data over the public internet instead of a telephone network. It's hard to believe that cell phones have morphed from Gordon Gekko's heavy, expensive brick into the pocket-sized minicomputers that can do everything from play videos to log onto social media to send emails. Oh, and they can still make phone calls, too.

Teflon

What would the world be without polytetrafluoroethylene?

* * * *

ANYONE WHO LOVES to cook knows the frustration of cleaning a pan covered with stuck-on food. Eggs, cheese, and meat are often some of the worst culprits, resulting in pans that require a soak in the sink and an extra dose of elbow grease to clean. The proteins in these foods form molecular bonds with the metal in hot pans, causing extra stickiness. But that's not the only reason bits of food are left behind: No matter how smooth and shiny metal pans may look, they are actually covered with microscopic ridges, dips, and holes. Like a rock climber who uses small indentations in the rock for traction, food holds on to these tiny undulations for dear life, making cleanup extra difficult.

Of course, there are ways to prevent your food from sticking to metal pans. One way is to simply keep the food constantly moving in the pan, so the proteins are never in contact with the metal long enough to form bonds. Another is to use hot oil in the pan, which fills in all those crevices and holes so food is less likely to stick to them. But even these methods don't always work perfectly, so for those of us who are not professional chefs, there's another option: polytetrafluoroethylene.

Or, in plain English, Teflon.

Slippery Serendipity

Like many wonders of science, the story of Teflon begins with an accidental discovery. In 1938, chemist Roy Plunkett was working for DuPont, the company that also developed materials like nylon, Lycra, and neoprene. Plunkett had been assigned the task of researching new and safer refrigerants to replace the dangerous ammonia and sulfur dioxide that was in use at the time. One day, he was experimenting with a substance called tetrafluorethylene, or TFE, which he stored in small cylinders that were then frozen and compressed.

On April 6, 1938, Plunkett took one of his cylinders and opened the valve to release the TFE, but he was surprised when nothing came out. The cylinder still weighed the same as when it was first filled, so he knew nothing had leaked out. Curious, he decided to saw the container in half and see what was inside. To his surprise, the cylinder was coated with a waxy, white, slippery substance. Tests showed that this was a polymerized perfluoroethylene, also known as polytetrafluorethylene (PTFE), created when the iron from the storage cylinders reacted with the TFE at high pressure.

Plunkett was fascinated by this unintentional creation, and he decided to do more research on the unusual substance. He discovered that it was non-corrosive, chemically stable, and had an extremely high melting point. But perhaps the most interesting thing about PTFE was its slipperiness. In fact, it was found to be one of the slipperiest substances known to man! DuPont was so intrigued by PTFE that they transferred the study of it to their central research department, where scientists with experience in polymer research could learn more about it. The substance was patented in 1941, and in 1945, DuPont copyrighted the name Teflon.

The slippery Teflon wasn't immediately considered useful for the kitchen; in fact, at first, it was mostly used for industrial and military applications. One of its first uses was to coat valves and seals in pipes containing uranium during the Manhattan Project. Then, in 1954, the wife of a French engineer named Marc Gregoire suggested that he use the slippery substance he'd been using on his fishing tackle on her cooking pans. Gregoire wisely followed his wife's idea, creating the first PTFE-coated cooking pans. He sold them under the brand name "Tefal," combining the words "Teflon" and "aluminum," and opened a factory in the Paris suburbs. The facility manufactured 100 frying pans a day in its first year of operation. Today, in the United States, Gregoire's brand is marketed under the name "T-Fal."

Oh Happy Pan Day

Back in the United States, inventor Marion A. Trozzolo founded Laboratory Plasticware Fabricators in 1957, a company that manufactured utensils including a Teflon-coated stirring rod. He started experimenting with coating frying pans with the slippery substance, and in 1961 he debuted the first PTFE-coated pan in America, the Happy Pan. Trozzolo's pan enjoyed short-lived success, as other manufacturers quickly caught on to the popularity of Teflon-coated cookware and introduced better offerings. But one of the first Happy Pans still holds a place of honor in an exhibit at the Smithsonian Institution, and Trozzolo was nominated for a spot in the Plastics Hall of Fame.

Today, many other cookware manufacturers, including Calphalon, All-Clad, and Swiss Diamond International, sell Teflon-coated pans. But the substance is used for much more than just cookware: Teflon is still used as an industrial coating for pipes, hoses, and machine parts, ensuring that things run smoothly. It is used to coat clothes irons, hair styling tools, and windshield wipers. It makes a great stain repellant, so it is often used to keep carpets and furniture looking brand new. And it is used in the automotive, semiconductor, and space industries.

But there's still one question that many cooks ask: How does Teflon, one of the slipperiest substances on earth, stick to a pan? The answer goes back to all those cracks and crevices in metal pans. The same ridges and holes that cause food to stick act like an adhesive layer for the Teflon. Before a pan is coated with the substance, it is first blasted with sharp iron grit to make its surface even more rough and uneven. The Teflon is then sprayed or brushed on, and baked in a 700-degree oven to lock it in place. The Teflon binds into place with the metal, and the pan is ready to cook up a fluffy (and nicely non-sticky) cheese omelet.

It may have been an accidental invention, but there's no doubt that life would be different without Teflon. As for Plunkett, the man who stumbled onto this slippery discovery, he received an immediate promotion from DuPont and went on to direct their production of Freon before his retirement in 1975. He was inducted into the Plastics Hall of Fame in 1973, and the National Inventors Hall of Fame in 1985.

An Indispensable Machine

Vending machines are a part of modern living, but they've been around much longer than you think.

⁕ ⁕ ⁕ ⁕

VENDING MACHINES SEEM to be distinctly modern contraptions—steel automatons with complex inner workings that give up brightly packaged goods to anyone with a few coins to spare. The first modern versions were used in London in the 1880s to dispense postcards and books. A few years later, they were adopted in America by the Thomas Adams Gum Company for dispensing Tutti-Frutti-flavored gum on subway platforms in New York City. The idea of an automated sales force caught on quickly, and vending machines were soon found almost everywhere. The idea perhaps reached its peak in Philadelphia with the Automat, which opened in 1902. These "waiterless" restaurants allowed patrons to buy a wide variety of foods by plunking a few coins into a box.

Today, we think of vending machines as an everyday part of our lives. Americans drop more than $30 billion a year into them, and Japan has one vending machine for every 23 of its citizens. All kinds of products—from skin care items, pajamas, and umbrellas to DVDs, iPods, ring tones, and digital cameras—can be bought without ever interacting with a salesperson.

As high-tech as all that may be, the most remarkable thing about vending machines lies not in the modern era but in the distant past. A Greek mathematician and engineer named Hero of Alexander built the very first vending machine in 215 BC! Patrons at a temple in Egypt would drop a coin into his device. Landing on one end of a lever, the heavy coin would tilt the lever upward and open a stopper that released a set quantity of holy water. When the coin slid off, the lever would return to its original position, shutting off the flow of water.

Velcro

One day in 1941, a Swiss engineer named George de Mestral went for a walk with his dog in the Jura Mountains, a sub-alpine mountain range north of the Western Alps. In the midst of what was no doubt breathtaking scenery, de Mestral was bothered by one issue: walking through the woods always resulted in pant legs (and a dog) full of burrs. Originating from a plant commonly known as burdock, these prickly seeds clung tenaciously to fabric and fur, requiring painstaking removal after he returned home. But de Mestral was curious: what was it about these burrs that made them so sticky and gave them the ability to cling to different materials?

* * * *

USING A MICROSCOPE, he examined one of the annoying little offenders and discovered it was covered in tiny hooks. The hooks were what allowed the burrs to attach themselves to passing fabric or animal fur and the reason they were so hard to simply brush off.

De Mestral then wondered if this unique characteristic of the burdock seed could somehow be replicated for a more useful purpose. If he could somehow create a synthetic form of this burr, he would have a whole new type of fastener, something not quite a button and not quite a zipper. So he set off to find a fabric manufacturer that was willing to help him make his vision a reality.

His quest took far longer than he anticipated, because many of the fabric manufacturing companies he visited were quite skeptical of his idea. It took him eight years, but he finally found a manufacturer in Lyon, France, that was intrigued by what he was trying to accomplish. With the help of this fabric manufacturer, De Mestral used two strips of fabric made of cotton and nylon, one covered in thousands of tiny hooks and one covered in thousands of tiny loops, and created a fastener that easily stuck together, and wouldn't unstick until it was ripped apart. It was exactly like the sticky, tenacious burrs!

De Mestral patented his invention in 1955, and then created a company, called Velcro, to manufacture his new product. The name was taken from the French words *velour*, or "velvet," and *crochet*, or "hook," in a descriptive portmanteau that summed up the invention. However, de Mestral's innovative product met with some difficulty right off the bat, as the tiny hooks in the material had to be created by hand, limiting the amount that could be made. Wanting to find a way to mass-produce the fasteners, de Mestral devised a modified version of barber's clippers that could precisely cut the hooks to the proper specifications and made it possible to manufacture the fasteners much more quickly.

Now possessing the invention and means to mass-produce it, all de Mestral needed was a customer base. Unfortunately, this new hook-and-loop fastener didn't take off as quickly as the inventor had hoped. People who were used to their buttons and zippers didn't really see a need for the sticky stuff, and most clothing manufacturers weren't interested in taking a chance on it. But de Mestral was about to get a lucky break, thanks to the American space program.

Velcro Blasts Off

In the early 1960s, NASA astronauts were just beginning to head into space, and they realized they had a small, albeit frustrating, problem. Floating in orbit, without the aid of gravity, the astronaut's pens, food packets, clipboards, and other equipment simply floated around their spacecraft. Although they attempted to secure the objects, nothing seemed to work well. Nothing, that is, until they heard about de Mestral's new fastener. The simple hook-and-loop fabric was easy to incorporate into spacecraft, and it kept everything secure and easy to find. Suddenly, de Mestral's invention was a space age fabric-of-the-future, and the public took notice.

Soon, the hook-and-loop fastener was showing up in all kinds of different places. Hospitals used it for blood pressure cuffs and on patient gowns; car manufacturers added it to floor mats to keep them in place; couch slipcovers and drapes featured the fasteners for easy attachment and removal; and surprisingly, the fabric was even showing up in high fashion, with designer Pierre Cardin particularly smitten with the product.

By the late 60s, Velcro fasteners had become especially popular in the shoe industry, where the fastener caught on like wildfire. And by the 1980s, sneakers with Velcro were everywhere. Anyone who grew up during the decade boasted at least a pair or two. But by this time, de Mestral's patent had expired, leading to cheaper, lower quality versions of his hook-and-loop fastener. While we tend to use the name "Velcro" for any hook-and-loop fastener, the word is actually a brand name; and Velcro began to work hard to prove that their version was the best, coming up with new ways to market their product to appeal to consumers.

Today, de Mestral's innovative product is used for so much more than sneakers. It's used for wall hangers, car organizers, bag closures, and even toys; it's also found in industrial markets, like construction and transportation, and is commonly used in the medical industry. It's amazing to think that something as small as a burr on a pant leg would one day inspire a product that is now used around the world.

Temper Foam

Memory foam mattresses: Some people love them, some people hate them; but they certainly have a long list of "pros" in their favor. These mattresses never need to be flipped, they hold their shape well, and they don't sag. They also conform to your body shape, which can help align your spine and prevent stiff, achy mornings. Plus, they are less likely to hold on to dust mites, so sleeping on one instead of a traditional mattress can help relieve allergies. Oh, and there's one other "pro" in memory foam's favor: this interesting material was invented by NASA.

✻ ✻ ✻ ✻

THINKING ABOUT THE National Aeronautics and Space Agency conjures up images of space shuttles and rockets and stars; space travel is probably the last thing on our minds when shopping for a new mattress. But when temper foam was first developed by aeronautical engineer Charles Yost in 1966, space travel most definitely was on his mind. Yost had already proven his value to NASA four years earlier, when he had helped to design a recovery system for the Apollo command module. This time, NASA asked Yost to develop a way to keep astronauts cushioned against the stressful forces of blastoffs.

Yost created a new material that he called "slow spring-back foam," because the material molded to fit whatever pressed against it, and then slowly regained its shape once the pressure was removed. It was also heat sensitive, so the foam would soften when it was in contact with heat from a body, while cooler areas remained firmer. NASA installed the material in space shuttles to give astronauts a more comfortable ride into space and to protect them against the jarring forces encountered on blastoff.

Sitting Down on the Idea

But that wasn't the last of it: NASA's Ames Research Center in Moffett Field, California, thought this "slow spring-back foam" would work well in airplane seats, as well, especially since its ability to evenly distribute body weight would make long flights more comfortable for passengers. The foam was considered safer than other materials used for seating as its high energy absorption meant that it provided better impact protection in the event of an accident.

NASA was so impressed with the new material that Yost immediately realized he had something bigger than just a cushion for space shuttles and airplane seats. So in 1969, he founded his own company, Dynamic Systems, Inc., to market what was now called "temper foam." Since then, the material has been used in a myriad of different ways. But what makes this squishy foam so special?

Temper foam has a unique property called "viscoelasticity," meaning it exhibits both viscous and elastic characteristics. A viscous material moves depending on how high or low its viscosity is. A liquid like water has a very low viscosity, so it is easily moved; honey, on the other hand, has a much higher viscosity so it moves more slowly. Elastic simply means that a material can be stretched but still return to its original shape. So temper foam's viscosity allows it to move when a force is applied, and its elasticity helps it to return to its normal form.

But this interesting material has even more unique features. Viscoelastic materials respond differently to gradual pressure versus sudden pressure, being much more reluctant to change shape with sudden pressure. For instance, if you punch a memory foam mattress, the material will feel stiffer than if you lie down on it and allow the foam to slowly change shape around you. This property makes it useful both for cushioning and for absorbing impact. Temper foam also exhibits a phenomenon known as "hysteresis," which is when the value of a physical property lags behind changes in the effect causing it. In temper foam's case, this is displayed by the long amount of time it takes to return back to its original shape when pressure is removed, and it helps to dissipate and absorb energy.

The Memory Foam Mattress

One of the most common uses for temper foam is in memory foam mattresses, which have been equally revered and reviled since their introduction in 1991. One of the biggest complaints from consumers when these mattresses were introduced was their tendency to feel much too hot. And interestingly, this, too, is a side effect of another unique characteristic of temper foam which is related to its viscosity.

When you heat up a viscous liquid, like honey, it becomes much less viscous; the same is true for a memory foam mattress. Warm body temperature causes the viscous material to become much more flexible, which results in a squishy mattress that cradles a sleeper. But at the same time, the foam isn't ventilated in any way, so all that trapped body heat and cushy cradling begins to feel warm. Fortunately, this problem has been addressed with "reticulated foam," which has an open-celled structure that allows for ventilation, or by adding beads of gel to the temper foam, which have a cooling effect.

An Extra Dose of Comfort

Since we spend about a third of our lives in our beds, most of us prefer a comfortable sleeping space, so these improvements have made memory foam mattresses extremely popular. But mattresses are only one of many uses for temper foam. It's widely used in the medical field to make orthopedic seating pads, wheelchair inserts, and mattress pads that prevent bed sores. It's also used in prosthetic limbs to provide cushion between the skin and prosthesis. Temper foam has found a place in the sports world, as well, being used to cushion the helmets of football players and race car drivers. And the military uses the material as a shock absorber for vehicles and aircraft ejection seats. There's no doubt about it: this unique material has made the world a more comfortable place for all of us.

Chemical Batteries

Watches, flashlights, laptops, smoke detectors, hearing aids—what do they all have in common? They all use batteries. We use billions of these chemical power packs every year, which can be as small as a pin or as large as a bookcase, and without them, life would be vastly different.

* * * *

OF COURSE, THE idea of portable energy is nothing new. After all, even prehistoric humans would create torches to burn and carry in the dark, which we could consider a very primitive version of the modern flashlight. By the industrial revolution, humans had figured out steam power, which created its own portability by powering locomotives (which, in turn, carried people from place to place—portable power which made people portable!). But burning fuel, whether it be wood or coal, could be dirty, dangerous, and time-consuming. So, as curious minds are wont to do, some scientists began researching a better way to make energy portable.

From Volta to Volts

One of these scientists was Alessandro Volta, an Italian physicist and chemist who was fascinated with electricity. One day, Volta's friend, fellow scientist Luigi Galvani, was dissecting a frog that was attached to a brass hook. When he touched the frog's leg with an iron scalpel, the leg twitched, making Galvani believe that the frog's leg was driving the reaction with "animal electricity." But when Volta heard about the incident, he disagreed, thinking the reaction was caused by the two different metals in contact with the moisture contained in the frog's leg.

To test his theory, Volta created a stack of alternating zinc and silver discs, layered with brine-soaked paper in lieu of frog's legs. When he connected a wire to each end of the stack, a steady current of electricity flowed. What's more, Volta discovered that he could alter the amount of current produced by using different types of metals, like copper instead of silver, and by making the stack larger or smaller. After unveiling this electric stack to the Royal Society of London in 1800, it became known as a "voltaic pile," the first true battery ever invented.

The key to Volta's battery was the use of two different metals. One of these metals acted as the negative anode, which released electrons, while the other metal acted as the positive cathode, which accepted electrons. Volta's brine-soaked paper was what is now known as an electrolyte, which is a chemical medium that allows electricity to flow between the anode and cathode. While the voltaic pile may have been a bit primitive and cumbersome, the same basic principles are used for batteries today.

Volta's battery was immediately seen as a useful device by the scientific community, although it did have some flaws. One was the tendency for the electrolyte brine to leak out and cause short circuits, due to the weight of each disc squeezing out the liquid. Scottish chemist William Cruickshank realized this could be fixed by simply laying the pile sideways in a box instead of stacking all the components on top of each other.

Another issue was the short battery life of the copper and zinc voltaic pile, which was about an hour. There were two reasons for this: First, the current running through the electrolyte solution caused hydrogen bubbles to form on the copper, and second, impurities in the zinc caused many short circuits to occur.

By the mid-1830s, scientists were coming up with new ways to construct batteries that eliminated these problems. One of these batteries was the Daniell cell, created in 1836 by English chemist John Frederic Daniell. His design consisted of a copper pot filled with copper sulfate, surrounding a porous earthenware container filled with sulfuric acid and a zinc electrode. Because the Daniell cell used two different electrolyte solutions, the hydrogen produced by the first was consumed by the second, giving the hydrogen no chance to build up on the copper. Around the same time, inventor William Sturgeon discovered that amalgamated zinc, which contains mercury, didn't have the same short circuit problems caused by pure zinc. With these changes, batteries became practical devices, used not only for experimentation, but also to power telegraphs and doorbells.

The Battery Is Dead Again?

While many other inventors, including John Dancer, Johann Christian Poggendorff, and William Robert Grove, created their own versions of batteries, they all had one drawback: once the chemical reactions powering them were spent, the battery was permanently drained. So in 1859, French physicist Gaston Plante invented the first rechargeable battery, the lead-acid battery. This battery consisted of a lead anode and a lead dioxide cathode immersed in sulfuric acid, which could be recharged by passing a reverse current through the battery. Amazingly, more than a century later, these types of batteries are still in use, mostly to power automobiles.

Two more advancements, the dry cell battery and the alkaline battery, were invented by the end of the 19th century. In 1886, German inventor Carl Gassner created a battery with an ammonium chloride electrolyte mixed with plaster of Paris to hold it immobile. This meant that the battery could be used in any orientation without spilling or leaking, making it a truly portable energy source. These dry cell batteries, which could literally be held sideways or upside down and continue to function, eventually led to the invention of our modern flashlights. And by the end of the century, Swedish engineer Waldemar Jungner had invented a battery with nickel and cadmium electrodes in an alkaline solution of potassium hydroxide. Alkaline batteries had a longer life than their acid counterparts, but they were also much more expensive, keeping them out of reach of most general consumers.

Portable Power for the Masses

But in the mid-20th century, Canadian chemical engineer Lewis Urry was determined to make alkaline batteries affordable for the public. His battery, which was made up of a manganese dioxide cathode, a powdered zinc anode, and an alkaline potassium hydroxide electrolyte, went on the market in 1959, and was the forerunner to the popular Duracell and Energizer batteries we use today.

Over the last few decades, advancements in battery technology have made them more environmentally friendly and longer lasting. Nickle-metal hydride batteries were introduced in the late 1980s as a safer replacement for nickel cadmium batteries, since cadmium can release toxic fumes. And in 1991, Sony began marketing lithium-ion batteries, which have a very high energy density and can be repeatedly recharged. These batteries have become staples in laptops, phones, and other gadgets, and can also be used to power electric cars.

Batteries have become a vital part of our daily lives, and chances are, we all use something powered by a battery at least a few times a day. And to think, if it hadn't been for a twitching frog leg, we might be much farther behind in our quest for efficient portable energy than we are now!

Social Media

When we think of social media, we automatically think of websites like Facebook, Twitter, Instagram, and other computer-mediated technologies. But the origins of social media go back much further than even the first computer. In fact, humans have been finding ways to connect for centuries.

⁕ ⁕ ⁕ ⁕

IF WE THINK of social media as a way to keep in touch with others, we could say that the very first "social media" was simply the postal service, whose earliest incarnation was in Persia around 550 BC. This idea would, of course, eventually spread throughout the world. It may be hard for younger generations to comprehend, but for centuries, people could only share their favorite recipes or vacation stories through the mail!

The optical telegraph, or semaphore telegraph, was invented in 1792, which sent visual messages from line-of-sight towers that were placed between five and 20 miles apart. Obviously, the messages relayed had to be short and to the point; but they were still conveyed more quickly than if they were sent by horse and rider. When electric telegraphs came along in the 1830s, they were even more efficient, although still limited in scope. But in a way, these early forms of communication were primitive forerunners to today's text messages.

With the advent of telephones and radios coming to the masses beginning in the late 1800s, communication between distances became a simpler process. Being able to pick up a phone and instantaneously speak to someone in another location was an amazing advancement for humans. Now those recipes and vacation stories could be shared over the phone; and radio allowed people to stay up-to-date on the news or listen in on a sporting event. Humans were connected in ways that had never been possible before.

Computers Up the Game

But even telegraphs, telephones, and radios would soon seem outdated as computers made their debut and technology began taking huge leaps forward. By 1969, ARPANET was established. This precursor to today's internet was used to connect dozens of universities in the United States. Although first envisioned during the Cold War as a way to maintain communications in the event of a Soviet attack, it soon became a place where users on opposite sides of the country could exchange information and ideas. This was one of the first true versions of social media.

In 1973, the first computerized bulletin board system, named Community Memory, debuted in Berkeley, California. The system was originally located on a teleprinter at a record store, where people could post messages and also look through messages to find specific information. Other terminals were added throughout the city, and it became a way for people to ask for restaurant recommendations, organize study groups or carpools, or share poetry. In fact, Community Memory was home to the very first "internet personality," a poet who went by the name of Benway. The system was shut down in 1975, but it was just the foreshadowing of what could be accomplished with social media.

In 1979, computer scientists Tom Truscott and Jim Ellis created Usenet, which allowed anonymous users to read and post messages. Usenet helped to popularize many of the social media terms we know today: for instance, messages were called "posts," and users could read messages in their "news feed." This was also the birth of terms like "FAQ" and "spam," which we still commonly use today.

While Usenet is still active, the anonymity of the service is less appealing to those who wish to use social media to keep in touch with friends and family.

The Social Circle

The desire to connect to others soon led to the creation of networking websites. "Classmates" debuted in 1995, as a site where users could search for people they knew from elementary school, high school, or college. This was followed by "Six Degrees" in 1997, a site that employed the idea of "six degrees of separation" by allowing users to invite family, friends, and friends of friends to join the site. This idea of "social circles," where a user can not only be connected to a friend or family member in their direct social circle but also to someone in their friends' or family members' social circles, became the basis for more successful social media sites down the road.

One of these was LinkedIn, which launched in 2003 and has become the most popular social media site for business professionals. Job seekers can post resumes, and employers can post job availability, giving both groups a head's up as to what's going on in the job market. A far cry from old-school job searches that required circling ads in the classifieds! Today, LinkedIn features about 590 million registered members in 200 countries; so even if your job search is international, social media can play a part in helping you get hired.

And Then It Really Took Off

Another site that debuted in 2003 was MySpace, which became the largest social media site in world from 2005 to 2009. But in 2004, Harvard College student Mark Zuckerberg launched "TheFacebook," originally conceived as a student directory with photos and information about each user. But the site was such a success at Harvard that it soon spread to other universities in the United States and Canada. In 2005, Zuckerberg bought the domain name "facebook.com" and dropped the "the" from the site's name. Facebook membership was expanded to include employees of Apple and Microsoft, and Facebook's operations were moved to Palo Alto, California.

Finally, in September 2006, Facebook was officially made available to anyone over the age of 13 with a valid email address. By 2008, Facebook surpassed MySpace in worldwide visitors, and by January 2009, Facebook was ranked the most used social networking service in the world. Today, more than 2.2 billion monthly active users access the site, making it the most popular social media site in the world.

Twitter, Instagram, YouTube, Pinterest, and Reddit are just a few of the many other social media sites that make it easier to share pictures, videos, stories, and opinions (so many opinions!) with people around the world. With the click of a few buttons, hundreds of people can see us hula dancing in Hawaii or skiing in the Alps. And to think, just a few decades ago, we would've had to mail those vacation pictures to friends and family, and then waited days or weeks for them to tell us they "liked" them.

The Long Extension Cord of History

Hybrid and electric cars are so twenty-first century . . . If they hadn't been invented in the 1800s.

* * * *

EVERYONE UNDERSTANDS THAT VHS tapes surpassed Betamax, but similar older rivalries are a bit more obscure. Automotive history buffs know that early cars were electric, and they were beaten out by the internal combustion engine because of logistical issues that still exist today.

A Current Affair

Electric cars are the environmentally friendly alternative to fossil-fuel vehicles, at least by most measures. For consumers with an international frame of mind, electrics also represent a way to work toward energy independence from the chaotic Middle East. They're quiet—electric luxury cars sometimes have engine sounds piped back in—and often they don't need traditional transmissions or many of the other trappings of fossil-fuel vehicles. But these pros and cons are cutting edge compared to the factors affecting consumers in the late 1800s.

Plugged In to Safety

Electric cars—still often called "horseless carriages" at the time—outsold fossil-fuel cars by a factor of 10 to 1 in 1890. Internal combustion engines were still frightening to many people, so choosing electric was a matter of safety and comfort. Today, refueling at a gas station is very structured and sterile, but carrying gasoline around in the 1800s probably didn't give anyone confidence. At the time, new car garages were built far from their corresponding houses in case of explosions or fires.

And gas cars needed to be hand-cranked to start their engines, which required physical strength and then catlike reflexes to get out of the way of the spinning crank starter. In 2007, *Top Gear* hosts James May and Jeremy Clarkson visited England's National Car Museum and drove what many consider to be the first modern-layout car, the 1916 Cadillac Type 53. But in the same visit, they drove an antique car with a crank starter, which Britons call the starter handle. May joked, with sincerity, that the crank starter nearly broke his arm.

The Feminine Mystique?

So there were great reasons to choose electric in the early automobile age, despite the fact that electrics cost more at the time and were less practical—does that sound familiar? The crank starter issue alone meant that many who drove electric cars were women, and the cars developed a reputation as something of a "woman's car." Automotive journalist Gary Witzenburg wrote in 2013 about the way some manufacturers "bolted dummy radiator grilles on the fronts" to help stanch this stigma. Presumably, like the brick facade on a McMansion, this didn't fool anyone.

Ford Every Revenue Stream

Henry Ford worked closely with Thomas Edison to introduce an electric car for consumers. He felt strongly about electric motors and believed this to be the way forward for city travel and trucking. But their deal fell through in a similarly modern way, as the media were given different excuses and postponed release dates with little concrete information. Henry Ford told the *New York Times* in 1914, "The problem so far has been to build a storage battery of light weight which would operate for long distances without recharging."

Edison was no stranger to generating, so to speak, his own buzz. He crusaded against alternating current technology, insisting that his own direct current was safer and better. In the course of this public relations war, Edison electrocuted more than a dozen animals in public demonstrations. His intentions were good, but this barbaric campaign just didn't work. Alternating current (AC) worked at higher voltages than direct current, but the relative danger of AC was mitigated after electrical engineers in Hungary invented the electrical transformer. Edison is largely looked back on as a sore loser as well as an inventive genius.

So electric-car skeptics doubted that Edison could help to make the affordable electric car he was shilling—he and Ford claimed that multiple prototypes existed, but just a couple were ever seen. The project dragged on for years and overstayed the public's attention span. Eventually Ford's electric car dreams went up in environmentally friendly smoke: Edison couldn't provide a battery suitable for the car, and he covertly swapped a much heavier battery in its place. The bromance ended on this sour note.

The Decline of Production Electrics

Costly consumer electric cars were made for decades beginning in the 1880s, and the only revolutionary thing about Edison and Ford's plans, decades later, was that their electric car would hypothetically be more affordable. In the meantime, high-end customers continued to buy and drive electric cars in small but steady numbers. What led to the near disappearance of electric car technology was partly Ford's own doing: His assembly-line Model T brought the price of a gasoline car down to a level a more average American could afford. A few years after the introduction of the Model T, prolific inventor Charles Kettering developed the electric starter motor. Production cars no longer needed to be cranked by hand, opening the market to anyone previously discouraged by the hand crank.

Progress Marches On

But the electric car wasn't simply bested by the gas car. As technology continued to improve, the electric car stood still while progress marched on. Even today, affordable, durable, practical batteries are an obstacle to many electric-car development programs—imagine the issues faced 100 years ago before many Americans even had electricity in their homes. In 1925, half of American homes had electricity, which probably trickled out the way internet service did in the 1990s and 2000s: cities first, everyone else much later. And these folks lived further apart than their city counterparts, meaning they needed vehicles that could carry them more miles than an electric car could manage. Filling up a gas tank takes a matter of minutes; charging an electric car took hours, even back then. Electric cars were slower, too.

In the Interim

Electric concept cars were made on and off throughout the decades between the decline of early electrics and the revival with cars like the Toyota Prius in the 1990s. A modified Chevrolet Corvair known as the Electrovair was shown in 1964, with the Electrovair II in 1966. The Electrovair II was packed with over $150,000 worth of batteries that still amounted to a range of 80 miles at most—you'd be better off wearing the Heart of the Ocean while you tooled around on an electric scooter.

It was inventor Charles Kettering who inadvertently led Americans back to environmentally friendly electric vehicles. Along with the electric starter motor, he invented both leaded gasoline and freon coolant. These two inventions meant that Kettering almost singlehandedly punched a hole in the Earth's ozone layer, and it was the discovery of this hole by scientists that really galvanized modern environmentalism.

The Rest Is Modern History

Toyota introduced the Prius in Japan in 1997 and began introducing it to the rest of the world in 2000. The idea of a hybrid vehicle combines the safe modern use of internal combustion with the environmental savings of a rechargeable battery, and the women driving electric cars in the 1890s would probably love to know the part they played in this very long story. One notable early adopter was fictional Rory Gilmore (from the television series *Gilmore Girls*), whose very ugly early Prius was a graduation gift from her grandparents. You can still see these first-generation Priuses—rarely—on highways today. And they are where America's love affair with alternative fuel sources began all over again in the 21st century.

SMS (Texting)

On December 3, 1992, Richard Jarvis, an employee of British telecommunications company Vodafone, was attending a holiday party when he received a message from his friend Neil Papworth. The message simply said, "Merry Christmas," and under most circumstances, it would've barely garnered a glance. But this wasn't just any message: it was the very first text message ever sent.

* * * *

PAPWORTH, A COMMUNICATIONS engineer with IT services company Sema Group, had to type the message on his personal computer, since cell phones didn't have keyboards yet; and Jarvis's brick-sized Orbital 901 mobile phone couldn't send a reply. But this humble message was the advent of a form of communication that has now become a part of our daily lives.

Long before cell phones became a reality, electric telegraphs used signals to send text messages. By the late 19th century, radio waves made wireless telegraphy possible. By 1933, the German postal service had introduced the "telex" network, a network of mechanical typewriters that could be used to send messages. While these may not seem anything close to what we consider a "text message" today, the idea of sending quick, short messages that bypass the necessity of a post office is not new.

The idea for a short message service (SMS) that could one day be used for mobile phones began to take shape in 1984. At that time, German engineer Friedhelm Hillebrand was working for Deutsche Telekom, and French telecommunications engineer Bernard Ghillebaert was employed by France Télécom. Together, Hillebrand and Ghillebaert developed a proposal for the Global System for Mobile communications (GSM) standard. But it would be six more years before the world's first GSM call was made on July 1, 1991, in Finland.

In the meantime, Hillebrand worked on setting up a format for SMS messages. He sat down at his home typewriter and began typing out random sentences, then counting up every letter, punctuation mark, and space. He discovered that almost always, 160 characters was all that was needed to convey a message, and proposed that this should be the character limit for SMS messages.

A year and a half after the first GSM call in Finland, Papworth sent his "Merry Christmas" message, probably not realizing what an impact this technology would soon have. The Finnish Radiolinja GSM operator was the first network to launch person-to-person SMS service in 1994, with their competitor, Telecom Finland, launching the service in 1995. Strange to say, this made Finland not only the first country in the world to offer SMS text messaging, but also the first to offer it from competing services.

Texting Arrives in America

The first text messaging service in the United States was provided by American Personal Communications (APC), the first GSM service in America. Forty-nine percent of APC was owned by Sprint Telecommunications Venture, and on November 15, 1995, operating under the name Sprint Spectrum, they launched their service in Washington, D.C. and Baltimore, Maryland. Vice President Al Gore made the first phone call to inaugurate the service, dialing up Baltimore mayor Kurt Schmoke.

When it first began, text messaging was not the popular communication choice it is today. Slow to take off, GSM customers sent an average of only 0.4 messages a month. Perhaps the method of inputting text, which was called "multi-tap," was a hindrance. With multi-tap, each number on a phone is connected to three or four letters, and a user must tap each number until the letter that is needed appears. While simple enough to understand, this method of inputting letters was time consuming and not very efficient. It could be frustrating to scroll through letter after letter just to type a single word, so it is understandable that SMS text messaging wasn't an immediate hit.

But not long after SMS debuted in the U.S., inventor Cliff Kushler created "T9," which was short for "text on 9 keys." With T9, predictive text technology suggests common words after a single key press. And with more use, T9 becomes familiar with the words and phrases most often used by the texter, offering better suggestions with each text. The technology made text messaging just a little bit simpler and less frustrating, and is now licensed on hundreds of millions of cell phones every year.

By 1997, the Nokia 9000i Communicator was introduced, which had a full keyboard for typing SMS text messages, and keyboards became popular throughout the late 90s and early 2000s. Up until the end the century, SMS text messages could only be sent between texters who used the same network, such as AT&T or Verizon, but in 1999, messages could at last be sent between different operators. All of these changes helped to make texting more popular, and by 2000, Americans sent an average of 35 text messages per month.

The use of text messages exploded in the early 2000s, with networks and cell phones themselves getting more efficient and faster all the time. By 2007, when Apple introduced the iPhone, texting surpassed phone calls as Americans' preferred form of communication: an average of 218 text messages per month versus 213 phone calls. Text messaging was used to raise funds for United Way, to vote for singers on American Idol, and by presidential candidate Barack Obama, all ushering in a new era of connectivity.

Today, more than seven trillion texts are sent worldwide every year, although internet-based mobile messaging, such as Facebook Messenger and WhatsApp, are leading to a decline in SMS in some parts of the world. But that very first "Merry Christmas" proved to be more world-changing than anyone could've ever thought possible.

Music Synthesizers

If you've ever listened to music from the 1980s, you've no doubt noticed the obvious inclusion of electronic music. Synthesizers, which are electronic musical instruments that convert signals to sounds, became cheaper and more widely available, and thus more popular during the decade. But these instruments have been used for far more than just 80s dance music.

* * * *

ALTHOUGH UNDOUBTEDLY THE most popular, the synthesizer wasn't the first electronic musical instrument in the world. In fact, as soon as humans began to understand electricity, they started searching for ways to use it in musical applications, often in a novel fashion.

In the mid-1700s, a Czech theologian, natural scientist, and music lover by the name of Prokop Divi took great interest in the phenomenon of electricity. He created a musical instrument he named the "Denis d'or," an elaborate keyboard that was said to be able to mimic the sounds of harps, lutes, and wind instruments. Very little is known about the Denis d'or, which some consider the first electrical musical instrument, because after Divi died in 1765, the instrument was sold, taken to Vienna, and was never heard of again.

Sparks Fly

A much better known early electronic musical instrument was the "clavecin électrique," created in 1761 by the French Jesuit priest and scientist Jean-Baptiste Thillais Delaborde. This instrument, which can still be found in the national library of France, was an electronic carillon played with a special keyboard. It also had the added effect of producing sparks during live performances, clearly demonstrating its electronic nature; but it's probably safe to say that not everyone wants to play an instrument that doubles as a fire hazard.

The late 1800s and early 1900s saw an increase in the invention of electric instruments, such as the "musical telegraph," created by electrical engineer Elisha Gray in 1876. Gray, who was also one of the first to create a telephone prototype, constructed an instrument powered by electromagnets with sound that was transmitted over a telephone line. Other creations included the Telharmonium organ, patented in 1897; the Theremin, invented in 1920; the Trautonium, created in 1928; and the 1929 Hammond organ, which used mechanical and electrical parts and became popular for blues and jazz music.

Forklift Not Included

The Hammond Organ Company also debuted an electronic keyboard in 1939, the Novachord, that was commercially available for three years until sales were stopped due to World War II. But the keyboard was significant because of its use of an envelope controller, a circuit (or software) that tells the keyboard what sort of sound to produce depending on which instrument it is mimicking. Often considered the world's first synthesizer, the Novachord contained 163 vacuum tubes and weighed 500 pounds.

In 1955, RCA acoustical engineers Harry Olson and Herbert Belar created the Mark II Music Synthesizer, which was the first programmable music synthesizer. It was operated by feeding the synthesizer a punched paper tape, similar to a player piano. But the Mark II, which is housed at the Columbia-Princeton Electronic Music Center in New York City, was created more for research than performance. Its array of components and vacuum tubes took up an entire room, and it could only be played by programming music into it, so spontaneous jam sessions were out.

By the 1960s, synthesizers were becoming more compact, but were still mostly confined to studios due to their lack of portability. But in 1963, engineer Paul Ketoff created the Syn-Ket (for Synthesizer-Ketoff), a transistor-based instrument that was small enough to sit on a table. The Syn-Ket had three, two-octave keyboards and three sound modules that could be controlled individually and combined into a single output. Composer John Eaton was impressed with the instrument, and used it for more than a thousand performances between 1966 and 1974. But the Syn-Ket was never marketed commercially, so it never caught on in mainstream music.

Enter the Moog

In the mid-1960s, one of the engineers who had worked on the behemoth Mark II, Robert Moog, designed a synthesizer that was exhibited at the Audio Engineering Society convention in 1964. Much more musical instrument and much less machine than the Mark II, Moog's synthesizer could be intuitively used by musicians after only a bit of trial and error, and by the later part of the decade, hundreds of popular recordings had been made using his synthesizers. However, these instruments were still modular, meaning they were composed of several different parts to control different functions. But, to the appreciation of traveling bands everywhere, even more mobile, far less cumbersome instruments were on the horizon.

In 1970, Moog designed the Minimoog, a non-modular synthesizer with integrated components. Though this integration made the instrument less flexible than its larger counterparts, its portability was a hit with musicians, and 12,000 units were sold. But the Minimoog had one big drawback: it was monophonic, so it was only able to produce one tone at a time. Several synthesizers were created that were able to play two tones simultaneously, but by 1976, polyphonic synthesizers, which could play several tones at the same time, were being designed, and in 1977, the first polyphonic synthesizer with a microprocessor controller debuted.

Digital Synthesizers

By the 80s, sound engineers were experimenting with digital sound synthesis, leading to the first stand-alone digital synthesizer, the Yamaha DX-7, introduced in 1983. The instrument used frequency modulation synthesis, developed by composer John Chowning in the late 60s, which could alter the sounds made by the synthesizer. The patent was licensed exclusively to Yamaha, leading to a spate of Yamaha synthesizers throughout the 1980s. But manufacturers like Roland and Korg developed their own synthesizers to great success, as well.

With so many new electronic instruments flooding the music market, musicians decided they needed a standardized interface with which all of the instruments could communicate. They called it Musical Instrument Digital Interface, or MIDI. A MIDI link can control instruments, computers, and other audio equipment, so a single performer can play multiple instruments with the stroke of a few keys.

By the end of the 20th century, electronic dance music was all the rage, and software synthesizers, which use computer programs to create music, became more and more popular. Throughout the years, some musicians have worried that the synthesizer, with its ability to create sounds of many instruments, may put traditional musicians out of work. There was even an effort to ban the instrument in Great Britain in 1982, but obviously it was unsuccessful.

Although the concern is not without merit—for instance, some theater productions in New York and London have reduced their number of musicians by half, opting for synthesized music to fill in the gaps—synthesizers have also introduced new and innovative ways to make music, perhaps providing some musicians who otherwise would've stayed in the wings with the opportunity to showcase their talents.

Who Invented the Computer Mouse?

Douglas Engelbart. And here's an extra-credit question: When did he do it? Is your guess the 1990s or the 1980s? If so, you're wrong. The correct answer is the 1960s.

❋ ❋ ❋ ❋

ENGELBART GREW UP on a farm, served in the Navy during World War II, then obtained a Ph.D. He wound up working at the Stanford Research Institute, where he pursued his dream of finding new ways to use computers. Back in the 1950s, computers were room-filling behemoths that fed on punch cards.

Engelbart believed that computers could potentially interact with people and enhance their skills and knowledge; he imagined computer users darting around an ethereal space that was filled with information. Most folks couldn't envision what he had in mind—no one thought of a computer as a personal machine, partially because the models of the day didn't even have keyboards or monitors.

Engelbart set up the Augmentation Research Center lab in 1963 and developed something he called the oNLine System (NLS). Today, we would recognize the NLS as a series of word-processing documents with hypertext links, accessed via a graphical user interface and a mouse.

On December 9, 1968, after years of tinkering, Engelbart presented his new technology at the Fall Joint Computer Conference in San Francisco. Engelbart's mouse—with a ball, rollers, and three buttons at the top—was only slightly larger than today's models. It was called a mouse because of its tail (the cord that connected it to the computer), though no one remembers who gave it its name. We do know, however, that engineer Bill English built the first mouse for Engelbart.

To Engelbart's disappointment, his new gadgets—including the mouse—didn't immediately catch on. Some curmudgeons in the audience thought that the ninety-minute demo was a hoax, though it received a standing ovation from most of the computer professionals in attendance. Eventually, Engelbart got the last laugh: His lab hooked up with one at UCLA to launch the ARPANET in October 1969. The ARPANET, as any computer geek knows, was a precursor to the Internet.

Engelbart's mouse patent expired in 1987, around the time the device was becoming a standard feature on personal computers. Consequently, he has never received a dime in royalties. But he was never in it for money—Engelbart's motivation was to raise humanity's "Collective IQ," our shared intelligence.

In November 2000, President Clinton presented Engelbart with the National Medal of Technology, the highest honor the nation can award a citizen for technological achievement. Engelbart may not have gotten rich from his computer mouse, but at least he gained a measure of lasting fame.

The Segway

In 2001, mastermind Dean Kamen, a self-taught physicist and established inventor, developed the Segway, claiming he could make walking "obsolete." Kamen had previously developed a phonebook-size portable dialysis machine, a nonpolluting engine, and more than 150 other patented contraptions.

* * * *

- Kamen's Segway is officially referred to as a "human transport device."
- Segway PTs ("personal transporters") have electric motors that drive the apparatus at speeds up to 12.5 miles per hour.
- The Segway is designed with "redundant technology." This means the device features duplicates of its important pieces

of hardware. If one function fails, an internal computer uses the duplicate function to keep the machine stable long enough for the rider to hop off safely.

* Segways respond as if they're controlled by the rider's thoughts alone. The secret on newer Segways is in the control shaft, which sways in sync with the rider if he or she wants to turn.
* You operate a Segway like this: Turn it on, step onto the two-wheeled platform, grip the waist-high handle, lean forward, and off you go.
* It is illegal to use a Segway on streets, roads, or highways. They are allowed on sidewalks and bike lanes.
* The use of Segways is not permitted in public areas in the United Kingdom, but they are legal in most places in the United States. They are also legal in Austria, the Czech Republic, France, Greece, Hungary, Italy, and Portugal.
* Most Segways can travel about 12 miles before they need to be recharged.
* Despite a lot of media attention, Segways haven't sold especially well. In 2006, *Time* magazine reported that the company had sold fewer than 25,000 units since the device was unveiled.

He Made His Mark

Indoor cats and their owners should give thanks to Ed Lowe, the inventor of Kitty Litter.

* * * *

Stumbling on Paydirt

BORN IN MINNESOTA in 1920, Ed Lowe grew up in Cassopolis, Michigan. After a stint in the U.S. Navy, he returned to Cassopolis to work in his family's business selling industrial-strength absorbent materials, including sawdust, sand, and a powdered clay called fuller's earth. Due to its high concentration of magnesium oxide, fuller's earth has an extraordinary ability to rapidly and completely absorb any liquid.

Back in those days, domestic kitties did their kitty business in litter boxes filled with sand, wood shavings, or ashes. One fateful morning in 1947, a neighbor of Lowe's, Kaye Draper, complained to him about her cat tracking ashes all over the house. She asked him if she could have a bag of sand from his company's warehouse.

Instead, Lowe gave her a sack of fuller's earth. Draper was so pleased with the results that she asked for more. After a while, her cat used only fuller's earth—it was the first Kitty Litter-using critter in the world.

Sensing that he was on to a good thing, Lowe filled ten brown bags with five pounds of fuller's earth each and wrote "Kitty Litter" on them. He never explained exactly how he came up with the name, but it was certainly an inspired choice.

The Idea Catches On in Catdom

Initially, convincing pet shop owners to carry Kitty Litter proved to be a challenge. Lowe's suggested price of 65 cents per bag was a lot of money at that time—the equivalent of about $5 today. Why would people pay so much for cat litter, the shop owners asked, when they could get sand for a few pennies? Lowe was so sure Kitty Litter would be a success that he told the merchants they could give it away for free until they built up a demand. Soon, satisfied customers insisted on nothing but Kitty Litter for their feline friends, and they were willing to pay for it.

Lowe piled bags of Kitty Litter into the back of his 1943 Chevy and spent the next few years traveling the country, visiting pet shops and peddling his product at cat shows. "Kitty Litter" became a byword among fastidious cat owners. The *Oxford*

English Dictionary cites this advertisement from the February 9, 1949, issue of the Mansfield, Ohio, *Journal News* as the phrase's first appearance in print: "Kitty Litter 10 lbs $1.50. Your kitty will like it. Takes the place of sand or sawdust."

The Kitty Litter Kingdom

By 1990, Lowe's company was raking in almost $200 million annually from the sale of Kitty Litter and related products. He owned more than 20 homes, a stable of racehorses, a yacht, and a private railroad. He even bought up 2,500 acres of land outside of Cassopolis, where he established the Edward Lowe Foundation— a think-tank dedicated to assisting small businesses and entrepreneurs. Lowe sold his business in 1990 and died in 1995. As far as anyone knows, he never owned any kitties himself.

Slippery Slopes

From our nation's first oil well came slime with healing properties.

* * * *

SOME OF THE best inventions are the ones that come about accidentally. In 1859, while visiting the boom town of Titusville, Pennsylvania, chemist Robert Chesebrough stumbled upon petroleum jelly. Chesebrough was in the region, home of the famous Drake Well, to strike crude oil deals with oil-drilling bigwigs. As the 22-year-old owner of an "illuminating oil" business, Chesebrough was one "slick" customer with his eyes set firmly on future prizes. He wasn't aware of it yet, but he was about to realize success beyond his wildest dreams.

While touring Titusville's oil fields, Chesebrough noticed a worker scraping waxlike goo from a pump rod. Curious, he asked the man what he was doing. "Scraping off rod wax," came the reply. It turns out the substance was a by-product of the drilling process that would cause the pumps to foul if allowed to accumulate.

Then the man told Chesebrough something that piqued his interest. When a worker would burn or cut himself, a dollop of this stuff applied to the wound would "fix it right up." Chesebrough was intrigued. *What magical healing ingredients might be in this oil?* he wondered. A new business venture was about to emerge.

Over the next decade, Chesebrough modified naturally occurring rod wax and used himself as a guinea pig to test its effectiveness. He became convinced of the product's curative properties and brought it to market.

By 1875, the invention was selling at the rate of one jar per minute. By the 1880s, the substance was a staple in American homes. These days, people refer to it not as rod wax but as Vaseline. Chesebrough had found his success in the slimiest of places. We all share in his rewards.

The Little Disc That Could

Digital downloads may have hurt the traditional record store, but the CD remains the world's most popular medium for audio recordings. Not only that, but the technology also changed the way digital data is stored.

* * * *

COMPACT DISCS FIRST hit the American market in 1982. After little more than a decade, the format dominated the marketplace, quickly replacing the vinyl records and cassette tapes that had been the mainstay for decades. Although music purists and staunch audiophiles maintain that vinyl has a richer or "warmer" sound, the compact disc has superior advantages—notably that repeated playback causes virtually no wear to the surface of the disc; the recordings sound the same after 10 plays or 10,000 plays.

The compact disc was actually invented by James T. Russell in the late 1960s. Russell patented the technology in 1970.

The first CD player, Sony's CDP-101, was released in Japan in October 1982 and arrived in the United States the following spring. The compact disc quickly became a breakthrough format for most popular music. The 1985 album *Brothers in Arms* by Dire Straits was the first compact disc to sell over one million copies.

The CD ultimately lent itself to more than just music. The same digitally encoded data that made music storage possible also worked as a storage device for computers. In 1990, the CD-ROM (read-only memory) and recordable CDs (for music and data) were introduced. And while the MP3 may have stolen the spotlight from the music CD, the discs will live on through volume, if nothing else; to date, more than 200 billion discs have been sold worldwide.

✳ Chapter 7

Humans: Warts and All

Warts

Does tape clear up a plantar wart? What about vitamin C?

✳ ✳ ✳ ✳

MOST PEOPLE HAVE one of these ugly bumps at one point or another. Warts are caused by the human papilloma virus (HPV) and are contagious. That's why an initial wart can create a host of others. Common warts are the rough-looking lesions most often found on the hands and fingers. The much smaller, smoother flat warts can also be found on the hands but might show up on the face, too. Warts that occur on the soles of the feet are called plantar warts and can sometimes be as large as a quarter. Genital warts, which have become a more common problem, develop in the genital and anal areas. If you suspect that you have a genital wart, see your doctor; do not try the remedies suggested here.

No one knows why warts occur and disappear and later recur in what appears to be a spontaneous fashion. For example, some women say they develop warts when they become pregnant, but the gnarly lumps disappear soon after they have their babies. A medical mystery also surrounds the fact that researchers have yet to find a way to get rid of warts for good. The solution may lie in developing a wart vaccine, but an approved, safe vaccine has yet to be created. That leaves the wart sufferer with two options: Having a dermatologist treat the warts or trying a few methods on their own. As for home remedies, some people swear by certain tactics, while others will never have any success with them. And it seems that in some cases, prevention may be the best medicine. Here are some tips to help you be wart-free.

Be sure it's a wart. First and foremost, before you try any type of treatment, know whether your skin eruption is a wart or another condition. Warts (except the small, smooth flat wart) commonly have a broken surface filled with tiny red dots. Moles, on the other hand, are usually smooth, regularly shaped bumps that are not flesh colored (as flat warts can be). A rough and tough patch that has the lines of the skin running through it may be a corn or a callus. There is also a chance that the lesion is skin cancer. You may be able to recognize skin cancer by its irregular borders and colors. But if you have any doubt, ask your doctor. In addition, if you have diabetes, circulation problems, or impaired immunity, do not try any home therapy for wart removal; see your doctor.

Don't touch. The wart virus can spread from you to others, and you can also keep reinfecting yourself. The virus develops into a wart by first finding its way into a scratch in the skin's surface—a cut or a hangnail, for instance. Even shaving can spread the flat warts on the face. Inadvertently cutting a wart as you trim your cuticles can cause an infection. So keep viral travels to a minimum by not touching your warts at all, if possible. If you do come in contact with the lesions, thoroughly wash your hands with soap and hot water.

Stick to it. An effective treatment for warts that's cheap and doesn't leave scars is adhesive tape. In fact, a 2002 study found that tape therapy eliminated warts about 85 percent of the time, compared to a standard medical treatment using liquid nitrogen, which was only successful on 60 percent of warts. Wrap the wart with four layers of tape. Be sure the wrap is snug but not too tight. Leave the tape on for six-and-a-half days, and then remove the tape for half a day. You may need to repeat the procedure for about three to four weeks before the wart disappears. You can try this on a plantar wart, but be sure to use strips of tape that are long enough to be properly secured.

Try castor oil. The acid in castor oil probably does the trick by irritating the wart. It works best on small, flat warts on the face and on the back of the hands. Apply castor oil to the wart with a cotton swab twice a day.

"C" what you can do. Vitamin C is mildly acidic, so it may irritate the wart enough to make it go away. Apply a paste made of crushed vitamin C tablets and water. Apply the paste only to the wart, not to the surrounding skin. Then cover the area with gauze and tape.

Heat it up. One study found that having patients soak their plantar warts in very hot water was helpful because it softens the wart and may kill the virus. Be sure the water is not hot enough to cause burns, however.

Don't go barefoot. Warts shed viral particles by the millions, so going shoeless puts you at risk for acquiring a plantar wart. The best protection is footwear. Locker rooms, pools, public or shared showers, even the carpets in hotel rooms harbor a host of germs—not just wart viruses. You can catch any of a number of infections, from scabies to herpes simplex. Never go barefoot; at the very least, wear a pair of flip-flops.

Keep dry. Warts tend to flourish in an environment that's damp, especially in the case of plantar warts. That's why people who walk or exercise extensively may be more prone to foot warts, says the American Academy of Dermatology. So change your socks any time your feet get sweaty, and use a medicated foot powder to help keep them dry.

Cover your cuts and scrapes. The wart virus loves finding a good scratch so it can make its way under your skin. By keeping your cuts and scrapes covered, you'll help keep the wart virus out.

Take precautions with over-the-counter preparations. The Food and Drug Administration (FDA) has approved wart-removal medications made with 60 percent salicylic acid, but most common over-the-counter remedies contain 17 percent. While the stronger formulas may work well for adults (except for those who have sensitive skin), they are not recommended for children. Salicylic acid works because it is an irritant, so no matter which strength of solution you use, try to keep it from irritating the surrounding skin. If you are using a liquid medication, do this by smearing a ring of petroleum jelly around the wart before using the medication. If you're applying a medicated wart pad or patch, cut it to the exact size and shape of the wart. Apply over-the-counter liquid medications before bed and leave the area uncovered.

The Human Lint Trap

It's called the belly button, and its primary duty seems to consist of gathering copious amounts of fuzz.

* * * *

The Experts Tackle BBL

IT'S AN AFFLICTION that embarrasses most people. Some call it "dirty" and "gross"; others simply find it mysterious. When it is discussed, it's usually late at night, behind closed doors. Yes, we're referring to belly button lint. But if there's one thing we've learned in our weekly belly button lint support group, it's that this accumulation of fuzz is natural. Still, each evening as we shamefully dislodge another tuft of blue-gray lint, we wonder just where it comes from.

Fortunately for humanity, not one but two scientists have taken on the Herculean task of identifying the source and nature of belly button lint. In 2001, Australian researcher Dr. Karl Kruszelnicki embarked upon a massive survey of nearly 5,000 people in order to identify the risk factors for belly button lint (BBL). What did he learn? The typical BBL sufferer is male, middle-aged, slightly paunchy, and has a hairy st and an "innie" navel. Kruszelnicki suggested that BBL is minute fibers that are shed by the clothes we wear every These fibers are channeled by abdominal hair into the bel button, where they collect until they are extracted. Dr. K o that the reason most BBL is a blue-gray color is that blue je rub the most against the body.

The Steinhauser Study

Dr. Kruszelnicki's research was a landmark study in BBL, but it wasn't quite detailed enough for some people. Enter Austrian chemist Dr. Georg Steinhauser, who decided that it was necessary to spend three years of his life chemically analyzing more than five hundred samples of BBL, mostly of his own making. Along the way, Steinhauser discovered that BBL isn't merely fibers from clothes, as Kruszelnicki had believed, but also includes bits of dead skin and fat.

Steinhauser went even further, establishing a list of practices to discourage the development of BBL. Shaving the abdomen seems to be the most foolproof method, though this strategy is, of course, temporary. Wearing older clothes may also help, because they have fewer loose fibers than new duds. Additionally, a belly button ring appears to have some effect in preventing BBL.

But it's another Australian man (what is it with Aussies and belly button lint?) who has taken BBL research to a whole new level. Graham Barker has been collecting his own BBL—which he calls "navel fluff"—in jars since 1984, earning himself a spot in Guinness World Records. Thanks to Barker's courage, it is now safe for those afflicted with BBL to come out of the closet and show their lint-filled bellies to the world.

Cholesterol Confusion!

You know the old saying, "You can't judge a book by its cover"? Well, you can't judge your cholesterol level by your cover, either.

⁂ ⁂ ⁂ ⁂

YOU'VE PROBABLY HEARD about someone who runs daily and eats a lean, healthful diet but is suddenly struck by a heart attack. The culprit: undiagnosed clogged arteries because of high blood cholesterol. But it's a common misconception that high cholesterol is always the result of poor diet. Although diet can play a significant role in boosting (and lowering) cholesterol levels, other factors may also be involved.

Our genes partly determine how much cholesterol our bodies make, so high blood cholesterol often runs in families. Dietary cholesterol found in eggs and other foods seems to make very little difference, despite the infamous "Is Your Breakfast Killing You" rash of headlines in the mid 20th century. Other potential causes include:

- Excess weight. Being overweight boosts cholesterol levels; losing weight can bring down bad LDL cholesterol and increase good HDL cholesterol.
- A sedentary lifestyle. Being inactive is a major risk factor for high cholesterol. Increasing physical activity will help lower LDL cholesterol and raise HDL levels.
- Age and gender. Cholesterol levels tend to rise naturally as we get older. Before menopause, most women have lower total cholesterol levels than men of the same age. But after menopause, women's LDL levels frequently rise.

Facts on Left-Handedness

If you're of the sinister persuasion, read on. Being a southpaw has its good and bad sides.

⁕ ⁕ ⁕ ⁕

- There is no standard for what constitutes left-handedness, making research into handedness difficult.
- Left-handed adults find many workplaces inefficient or dangerous because they're designed for right-handed people.
- Around 10 percent of the population is left-handed.
- Famous lefties include Mark Twain, Whoopi Goldberg, Ronald Reagan, George H. W. Bush, Bill Clinton, Jay Leno, Julia Roberts, Oprah Winfrey, John McCain, and Barack Obama.
- More men than women are left-handed.
- Left-handers are more likely to stutter, have dyslexia, and suffer from allergies.
- International Left-Handers Day is August 13, and was first celebrated in 1976.
- Homosexuals are 39 percent more likely to be left-handed or ambidextrous.
- Lefties are three times more likely to become addicted to alcohol or other substances.
- The term *southpaw* was first used to refer to left-handers in the 1890s.
- At one time, teachers would force students to write with their right hands, even when left-handed. Luckily, that bias has (mostly) gone away.
- Lefties might be better in hand-to-hand combat.

* Researchers at Oxford University have found a gene for left-handedness.
* Scientists have found that handedness develops *in utero.*
* The hair of right-handed people swirls clockwise on the top of their head, but the hair of left-handed people can swirl in any direction.
* Lefties tend to perform better on IQ tests.

The Shocking Truth About Bad Hair Days

Researchers have actually studied the effects that ill-shaped locks can have on a person, and the results aren't pretty.

* * * *

A Horrific Way to Start the Day

SKY-HIGH FRIZZ, LITTLE sprigs of cowlick, the combover that won't comb over—no magic comb, curling iron, or straightening serum can fix this tress mess. It's only 8:00 a.m., but when your coif doesn't cooperate, a promising new day seems doomed. Oh, look: The cat just peed on your briefcase. What else can go wrong?

A whole lot, according to a Yale University "bad hair day" study. It seems that the effects of an unmanageable mane extend beyond what's in the mirror. The Yale research, headed by Dr. Marianne LaFrance in 2000, found a direct relationship between a bad hair day and psychological well-being.

"Interestingly, both women and men are negatively affected by the phenomenon of bad hair days," reported LaFrance. "Even more fascinating is our finding that individuals perceive their capabilities to be significantly lower than others when experiencing bad hair."

That's right—the study, commissioned by Procter & Gamble's Physique hair care line, found that bad hair lowers performance self-esteem, increases social insecurity, and intensifies self- criticism. It turns out that a bad hair day can spiral into a self-loathing, self-destructive, mangy mess of a pity party. No wonder you missed the train, spilled coffee on your boss, and dropped your keys through a drainage grate.

What to Do When Good Hair Goes Bad

The positive news is, there's more than one way to lock down wayward locks. For starters, get the very best haircut you can afford. "It's the cut that determines how easy your hair will be to style," counsels Beverly Hills hairdresser Nick Chavez. "A good one can go a long way in helping you avoid a bad hair day."

Next, use a shampoo and a conditioner that are designed to deal with your hair type. Got haystack hair? Go with a moisturizing formula. Your scalp is an oil slick? Get rid of the grease with an oil-controlling concoction. And there's a simple fix for staticky, flyaway, just-been-electrocuted hair: Rub it down with a dryer sheet. Bounce, Downy, Snuggle—basically, just grab whatever's in the laundry room. It'll keep hair from sticking together and make styling a lot easier.

But do you know what's even simpler? A fashionable hat.

Who Needs a Radio When You've Got Dental Fillings?

It's not beyond the realm of possibility that you can pick up radio stations via your dental work.

* * * *

Lucy's Strange Story

IF YOU'VE WATCHED enough *Gilligan's Island*, you know that it can happen on TV. And maybe you've heard that it happened in real life to Lucille Ball of *I Love Lucy* fame. According to Jim Brochu, author of *Lucy in the Afternoon*, the actress claimed that her dental fillings picked up radio signals as she drove to her home outside Los Angeles in 1942. She also claimed that the signals were later traced to a Japanese spy who was eventually taken into custody by law enforcement authorities, perhaps the FBI.

Who knows if Lucy's tale is true? Nobody's found documented evidence of a Japanese spy nest infiltrating California in 1942, and Lucy's FBI file contains no mention of such an event. (Yes, Lucy had an FBI file. At the urging of her grandfather, she had registered to vote as a Communist in the 1936 elections, so she had some 'splaining to do when she was investigated by the House Select Committee on Un-American Activities in 1953.) At one point, the Discovery Channel television show *MythBusters* devoted a segment to debunking Lucy's claims.

The Voices in His Head Were Real

Other people have claimed that they picked up radio signals via the metal in their heads, whether it was dental work or something else. The anecdotal accounts are easy to find but hard to verify. In 1981, however, a doctor in Miami wrote to *The American Journal of Psychiatry* to report that he had treated a patient who suffered from headaches and depression and complained of hearing music and voices. The patient was a veteran who had been wounded by shrapnel to the head during combat 12 years earlier. But after receiving successful treatment for the headaches and depression, the patient claimed that he still heard the mental music.

This led the doctor to sit down with the patient and a radio; the two of them listened to various stations, trying to find one that matched what the patient heard in his head. When they found what seemed to be the offending frequency, the doctor listened to the radio with an earphone while the patient described what he was hearing in his head. Although the patient couldn't hear the voices clearly, he passed the test convincingly; he was even able to tap out the rhythm and hum along with the songs that played. The doctor concluded that the shrapnel in this man's head was receiving radio signals and conducting the sound through bone to his ear.

So apparently, it is possible for the metal in your head to receive radio transmissions—but don't look for the American Dental Association to begin marketing the iTooth portable music player anytime soon.

Can the Cold Give You a Cold?

No, you won't catch a cold by running around in the frigid air while wearing only underwear or traipsing through town with wet hair. Sorry, Mom—an old wives.

* * * *

WHILE IT'S TRUE that colds are more prevalent during the nippy months from September to April, the cold temperatures are probably not to blame. These just happen to be the months when viruses are typically spread.

One study did conclude that cold temperatures might indeed give you a cold. Researchers at Cardiff Common Cold Centre in Wales asked 180 volunteers to sit with their feet in bowls of ice-cold water for twenty minutes. Over the next five days,

29 percent of the cold-feet volunteers caught a cold, compared to 9 percent of an empty-bowl control group. It's thought that cold temperatures can constrict the blood vessels of the nose, turning off the warm blood supply to white blood cells (the ones that fight infections).

However, most research continues to show that being physically chilled or wet really has nothing to do with catching a cold. We spend more time indoors during the winter, oftentimes exposed to sniffling coworkers who refuse to take sick days. (Are they still hoping to win a gold star for perfect attendance?) Before you know it, January rolls around and you're drowning in a mound of soft ply tissues and begging someone—anyone—to make you a batch of chicken noodle soup.

Who hasn't been there? That's why it's the "common cold." More than two hundred types of viruses can cause it. There are the rhinovirus (the leading cause of the common cold, made famous in Lysol disinfectant TV commercials), the respiratory syncytial virus, and lots, lots more. These nasty bugs lurk on telephones, cutting boards, computer mice, doorknobs, hand towels, and pretty much everywhere hands are meant to go.

So wipe those areas down with disinfecting sprays or wipes and be vigilant about cleansing your hands with antibacterial soap and water or hand sanitizer gel. Dr. Neil Schachter, author of *The Good Doctor's Guide to Colds and Flu*, says that people who wash their hands seven times a day get 40 percent fewer colds than the average person.

You know what else? It really can't hurt to throw on your ski mask and thermal snowsuit when the temperature dips below freezing. At least it will make your mom feel better.

A Bigger Brain Doesn't Translate to a Smarter Person

If you're someone who has an oversized noggin—and displays it like a trophy—we really hate to rain on your parade: You are not smarter than the rest of us. Scientific studies continue to show that size isn't everything where the human brain is concerned.

* * * *

History Lessons

SURE, IT MIGHT be easy to assume that a colossal cranium is capable of holding more intelligence—just by sheer mass. History suggests otherwise. William H. Calvin, a theoretical neurophysiologist and affiliate professor emeritus at the University of Washington School of Medicine, has pointed to notable periods in the historical timeline when the brain mass of ancient humans greatly increased, but toolmaking smarts did not seem to increase.

Although *Homo sapiens* in Africa 200,000 years ago had developed a brain size comparable to that of contemporary people, they continued to use the same crude, round-edged rocks for some 150,000 years before graduating to points, needles, harpoons, and hooks. You can't exactly say those bigger-brained primates were the sharpest tools in the shed.

Modern Science Weighs In

As for modern people, advancements in magnetic resonance imaging (MRI)-based scans are yielding pertinent data about the relationship between brain size and intelligence. (Before MRI, researchers had to measure the outside of a person's head to estimate brain size, or wait until that person died to get a measurement.) A 2004 study conducted by researchers at the University of California-Irvine and the University of New Mexico was one of the first to use MRI technology to show it's not overall brain size that counts, but brain organization.

How so? The researchers used MRI to get structural scans of the study participants' brains, and then compared those scans to respective scores on standard IQ tests. What they discovered was that human intelligence is less about total girth and more about the volume and specific location of gray-matter tissue across the brain. It appears there are several "smart" areas of the brain related to IQ, and having more gray matter in those locations is one of the things that makes us, well, smarter.

Undoubtedly, the relationship between brain size and intelligence will continue to be studied and debated, but some in the medical field now believe that brain size is purely a function of genetics and doesn't result in a greater intellect. Researchers at Harvard Medical School have even been able to identify two of the genes (beta-catenin and ASPM) that regulate brain size.

So if you've got a big head, don't be so quick to get a big head. It turns out that Albert Einstein's brain weighed only 2.7 pounds. That's 10 percent smaller than average.

Beefed Up

You're probably familiar with the terms "juiced," "roid-raged," "hyped," and "pumped"—all used to describe the effects of anabolic steroids. For better or for worse, steroids have invaded the worlds of professional and amateur sports, and even show business and beyond.

* * * *

Better Living Through Chemistry

ANABOLIC STEROIDS (ALSO called anabolic-androgenic steroids or AAS) are a specific class of hormones that are related to the male hormone testosterone. Steroids have been used for thousands of years in traditional medicine to promote healing in diseases such as cancer and AIDS. French neurologist Charles-Édouard Brown-Séquard was one of the first physicians to report its healing properties after injecting himself with an extract of guinea pig testicles in 1889. In 1935, two German scientists applied for the first steroid-use patent and were offered the 1939 Nobel Prize for Chemistry, but they were forced to decline the honor by the Nazi government.

Interest in steroids continued during World War II. Third Reich scientists experimented on concentration camp inmates to treat symptoms of chronic wasting as well as to test its effects on heightened aggression in German soldiers. Even Adolf Hitler was injected with steroids to treat his endless list of maladies.

Giving Athletes a Helping Hand

The first reference to steroid use for performance enhancement in sports dates back to a 1938 *Strength and Health* magazine letter to the editor, inquiring how steroids could improve performance in weightlifting and bodybuilding. During the 1940s, the Soviet Union and a number of Eastern Bloc countries built aggressive steroid programs designed to improve the performance of Olympic and amateur weight lifters. The program was so successful that U.S. Olympic team physicians worked with American chemists to design Dianabol, which they administered to U.S. athletes.

Since their early development, steroids have gradually crept into the world of professional and amateur sports. The use of steroids have become commonplace in baseball, football, cycling, track—even golf and cricket. In the 2006 Monitor the Future survey, steroid use was measured in eighth-, tenth-, and twelfth-grade students; a little more than 2 percent of male high school seniors admitted to using steroids during the past year, largely because of their steroid-using role models in professional sports.

Bigger, Faster, Stronger—Kinda

Steroids have a number of performance enhancement perks for athletes such as promoting cell growth, protein synthesis from amino acids, increasing appetite, bone strengthening, and the stimulation of bone marrow and production of red blood cells. Of course, there are a few "minor" side effects to contend with as well: shrinking testicles, reduced sperm count, infertility, acne, high blood pressure, blood clotting, liver damage, headaches, aching joints, nausea, vomiting, diarrhea, loss of sleep, severe mood swings, paranoia, panic attacks, depression, male pattern baldness, the cessation of menstruation in women, and an increased risk of prostate cancer—small compromises in the name of athletic achievement, right?

While many countries have banned the sale of anabolic steroids for non-medical applications, they are still legal in Mexico and Thailand. In the United States, steroids are classified as a Schedule III controlled substance, which makes their possession a federal crime, punishable by prison time. But that hasn't deterred athletes from looking for that extra edge. And there are thousands of black-market vendors willing to sell more than 50 different varieties of steroids. Largely produced in countries where they are legal, steroids are smuggled across international borders. Their existence has spawned a new industry for creating counterfeit drugs that are often diluted, cut with fillers, or made from vegetable oil or toxic substances. They are sold through the mail, the Internet, in gyms, and at competitions. Many of these drugs are sub-medical or veterinary grade steroids.

Impact on Sports and Entertainment

Since invading the world of amateur and professional sports, steroid use has become a point of contention, gathering supporters both for and against their use. Arnold Schwarzenegger, the famous bodybuilder, actor, and politician, freely admits to using anabolic steroids while they were still legal. "Steroids were helpful to me in maintaining muscle size while on a strict diet in preparation for a contest," says Schwarzenegger, who held the Mr. Olympia bodybuilding title for seven years. "I did not use them for muscle growth, but rather for muscle maintenance when cutting up."

Lyle Alzado, the colorful, record-setting defensive tackle for the Los Angeles Raiders, Cleveland Browns, and Denver Broncos admitted to taking steroids to stay competitive but acknowledged their risks. "Ninety percent of the athletes I know are on the stuff. We're not born to be 300 lbs. or jump 30 ft. But all the time I was taking steroids, I knew they were making me play better," he said. "I became very violent on the field and off it. I did things only crazy people do. Now look at me. My hair's gone, I wobble when I walk and have to hold on to someone for support and I have trouble remembering things. My last wish? That no one else ever dies this way."

Recently, a few show business celebrities have come under scrutiny for their involvement with steroids and other banned substances. In 2008, 61-year-old *Rambo* star Sylvester Stallone paid $10,600 to settle a criminal drug possession charge for smuggling 48 vials of Human Growth Hormone (HGH) into the country. HGH is popularly used for its anti-aging benefits. "Everyone over 40 years old would be wise to investigate it (HGH and testosterone use) because it increases the quality of your life," says Stallone.

"If you're an actor in Hollywood and you're over 40, you are doing HGH. Period," said one Hollywood cosmetic surgeon. "Why wouldn't you? It makes your skin look better, your hair, your fingernails. Everything."

Calling Hannibal Lecter

As the villainous cannibal from The Silence of the Lambs *would confirm, there are some choice cuts of meat on the human body. What exactly are the tastiest parts? Read on to find out.*

* * * *

A Crash Course on Cannibalism

THIS TOPIC COMES up all the time. After all, what if the plane carrying your national rugby team crashes in the mountains, as happened to the Uruguayan team in 1972? Or your wagon train becomes trapped for the winter within a desolate mountain pass, like the infamous Donner Party? You may need to know just which cuts of human flesh are the tastiest in such situations.

Okay, we're kidding. This topic almost never comes up. And since *Bon Appétit* has yet to publish its "Cannibal" issue—the headline would be easy to write: EAT ME—we're forced to go to primary sources to determine which parts of the human body are the most succulent. Fortunately, there aren't many of them, as it turns out.

According to archaeological evidence, human cannibalism has a long history that dates back to the Neanderthals. Despite the stereotype that cannibals only live on remote islands, or in the deepest jungles, evidence of cannibalism has been found in cultures on nearly every continent, including Europe and North America. However, most cannibalistic practices throughout history were of a ritual nature, and there were few food critics writing up snappy reviews of their human feasts. For that, we have to consult those individuals throughout history who dined on other humans for pleasure.

The current living expert on cannibalism—Armin Meiwes, the German cannibal who is serving a life sentence for devouring a willing victim—likened the taste of his "cannibalee" to pork. Meiwes prepared his meal in a green pepper sauce, with a side of croquettes and Brussels sprouts. Science seems to agree with Meiwes. Some Japanese researchers manufactured "an electromechanical sommelier," a kind of gastronomist robot capable of sampling wines, cheeses, meats, and hors d'oeuvres, and identifying what it has been fed. When one reporter stuck his hand in the robot's maw, the two-foot robot immediately identified it as prosciutto. When the accompanying cameraman offered his hand, the mechanical gourmand declared, "Bacon."

And the Tastiest Part Is ?

So, what is the tastiest part of the human body? That seems to be a matter of debate. Early 20th-century murderer and cannibal Albert Fish declared that the buttocks were the choicest cut, but latter-day cannibal Sagawa Issei disagrees, claiming that the thighs get that honor. In Fiji, where cannibalism was practiced until the late 1860s, men (women apparently were forbidden from partaking in this tasty treat) also favored the thighs (they also preferred the flesh of nonwhite women).

So if you really must know, there you have it: The thighs and buttocks are the prime cuts of a human. Just don't invite us to dinner.

Here a Yawn, There a Yawn, Everywhere a Yawn

"Contagious yawning," as it is commonly known, is one of the strangest quirks of the human body. Stranger still are some of the theories about why a simple yawn can spread from person to person like wildfire.

* * * *

Why Do People Yawn, Anyway?

YOU MAY THINK we yawn because we're tired or bored, or because oxygen levels in our lungs are low (that's the traditional medical explanation, after all). But did you know that babies yawn in utero? (They pick up the habit as early as eleven weeks after conception.) Fetuses don't take in oxygen through their lungs, and there's no way they are tired or bored—they sleep all day, and they certainly haven't viewed enough television yet to have problems with attention span. Olympic athletes have been known to yawn right before competing in events. Yawning also has been connected to certain conditions, including multiple sclerosis and penile erection. It's all pretty weird.

And Why Do You Yawn When You See Me Yawn?

Scientists don't fully understand why we yawn. Does involuntarily opening one's mouth wide serve any useful or healthful purpose? It's something of a mystery. We do know, however, that 55 percent of people will yawn within five minutes of seeing someone else do it. It's a phenomenon called "contagious yawning." Sometimes just hearing, thinking, or reading about a yawn is enough to make you unconsciously follow suit. (Did it work?) Again, scientists don't know exactly why, though they have paid it enough mind to conjure a few theories.

Some researchers hypothesize that contagious yawning is more common among the empathetic crowd. In other words, those of us who demonstrate a greater ability to understand and share other people's feelings are more likely to emulate their yawns. Taking that theory one step further, Dr. Gordon Gallup and researchers at the University of Albany say that empathetic or yawning evolved as a way to "maintain group vigilance." Gallup thinks yawning keeps our brains working at cool and alert levels. So in the days of early man, contagious yawning helped raise attentiveness and danger-detecting abilities of the group.

Even today, members of paratrooper regiments and airborne units report yawning together right before a jump. Could contagious yawning really be leftover hardwiring from the days of yore? Quite possibly. Other theories contend that contagious yawning may have been a more explicit form of early communication. The "herding theory" suggests humans might have used contagious yawning to coordinate their behavior. One member of the group would yawn to signal an event, as if to say, "Hey, let's to go hunt for a sabertooth tiger." And the other members in the group would yawn back to reply, "Yeah, let's go."

Humans aren't the only creatures that yawn. Foxes, sea lions, hippos, dogs, and cats are among the animals that do it. Recent studies have even demonstrated that some animals, like dogs and chimpanzees, may suffer from contagious yawning.

Amongst the Very Old at Heart

Was the average life expectancy in ancient times short, and basically everyone died by 30? Think again.

* * * *

IN THE 21ST century, there's no health headline scarier than declining life expectancy. What could more pointedly indict our culture as backsliding toward primitivity? After all, medical science has brought the average life expectancy in the U.S. from under 40 in 1850 to the mid 70s in 2000. Well, yes and no—it depends on how you measure.

Infant and Child Survival Rates

The greatest factor in overall life expectancy has always been infant and child mortality, and these numbers are powerfully affected by dozens of different factors in everyday life, not to mention the specific practices around childbirth and childrearing. In fact, this one statistic is so far-reaching that it can pretty accurately predict whether a national government will fail. The U.N. reports that world infant mortality, defined as a baby who dies during or after birth, dropped from nearly 16% in 1950 to about 7% in 1980 and about 5% by 2000.

Adult Life Expectancy

There have always been adults who lived a relatively long time by modern standards. Historians divide life expectancy for prehistoric people, ancient Egypt or Rome, and so forth into categories that try to account for child mortality. As a template, if an ancient child lived to age 10, 15, or adulthood, she could expect to live an additional 20, 30, 40 years, depending on the documents historians are able to parse. Different ancient groups had varying life expectancies that map to differences in lifestyle and location. The same is true today, when pockets of people who share an unusual longevity factor—cleaner air, lower stress, healthier staple foods—are studied as anomalies.

It's fun to consider the globetrotting explorers of the 1500s embarking on *Gulliver's Travels*-like adventures among groups and cultures with noteworthy numbers of healthy older people. More importantly, these anecdotes illustrate how important environmental and lifestyle factors could be even centuries before modern medical science emerged.

Old News

Today, reports of declining life expectancy cite the specific health dangers that plague our era, like coronary artery disease, HIV and AIDS in the developing world, and opioid overdose in the United States. Life expectancy has fallen sharply in specific nations like North Korea, where the stifling autocracy has created food shortages and medical crises along with the general risk factors like low general education or access to information. But any declines in the 21st-century developed world are small, especially compared to the almost history-ending severity of the waves of Black Plague that ravaged Europe in the late middle ages. As the U.S. government takes action to curb opioid addiction, for example, the lives saved will help to buoy the life expectancy overall. But public health researchers are rightfully concerned that the same socioeconomic factors that can lead to opioid addiction may also lead to a rise in infant mortality, and the U.S. already has one of the highest rates of infant mortality in the developed world—even as its single richest nation. This is a much more worrisome statistic in terms of life expectancy in the broad historical sense.

Does Listening to Mozart Make a Baby Smarter?

That depends on your definition of smarter.

⁕ ⁕ ⁕ ⁕

When researchers in a 1993 study had participants listen to a Mozart sonata, they found that those people scored slightly higher on spatial-reasoning tests for about ten or fifteen minutes. That's what the researchers tested, that's all they claimed, and their methods seemed sound.

But within a year, the *New York Times* wrote an article in which it summarized, breezily, that "listening to Mozart actually makes you smarter," and we were off to the races. In 1997, Don Campbell published a book called *The Mozart Effect: Tapping the Power of Music to Heal the Body, Strengthen the Mind, and Unlock the Creative Spirit.* Then came Campbell's *The Mozart Effect for Children.* Brand extensions of this powerful franchise are available to this day.

If the *New York Times* piece showed the suggestive power of the media, Campbell's books demonstrated the power of parental love. Here was a chance for a new generation of über-parents to achieve several desirable things simultaneously: They could help their kids become smarter, give them culture, and assuage their own residual guilt for having listened to Iron Maiden when they were young and impressionable.

Few separate studies have corroborated the limited findings of the original research. Other researchers have argued that Mozart doesn't make listeners smarter—it simply puts them in a better mood, which can translate to temporarily better scores on certain kinds of tests. Mozart's music also has been shown to cause significant—though again temporary—decreases in brain activity that leads to epileptic seizures.

In other words, there is little doubt that Mozart's music—considered to be both abstractly complex and aurally ingratiating—has a fleeting positive effect on people. But does this mean you are smarter for listening? And is it just Mozart? A composition by the Greek composer Yanni—whose cheesy fare you might know from infomercials—has been shown to have similar effects.

You'll have trouble finding many scientists who say Mozart makes anyone smarter. "Enjoyment arousal" is what one scientist calls it. That's certainly a good thing, but it's not enough to guarantee your children will go to Harvard—or will even prefer Mozart to their generation's version of Iron Maiden.

What Attracts Us to Some People but Not Others?

You've heard all the theories: Opposites attract. Opposites repel. It's all about the pheromones. Women like guys with fat pocketbooks. Guys like pinup models.

* * * *

WAIT A MINUTE—NOW we're actually on to something.

According Dr. Beverly Palmer—a professor emeritus of psychology at California State University, Dominguez Hills, and an expert in the science behind attraction, love, sexuality, and flirting—attraction is typically sparked by the sense of sight. And guess what? Stereotypical standards of beauty influence our perceptions of what we see.

This isn't to say that a great sense of humor isn't an appealing quality in the long run—especially for a guy with a "radio face"—but that initial fire isn't likely to be lit by a knock-knock joke; it'll be ignited by his broad shoulders or her long legs.

Or maybe it'll be set alight by a perfect waist-to-hip ratio (WHR) of 0.7. You see, some scientists think that they've boiled the laws of sexual attraction right down to an actual—science. For example, through studies of people's WHRs, psychologist Devendra Singh of the University of Texas at Austin has concluded that men are most enthralled by women with WHRs of about 0.7.

No need to do the math—these are women with waists that are significantly narrower than their hips. In other words, they have classic hourglass figures. And this brings us to *Playboy* models—an analysis of them (and Miss America contestants) shows that most of these stereotypically attractive women have WHRs of approximately 0.7.

So when it comes to initial attraction, it seems that what we're really looking for is good overall symmetry. And according to studies by New Mexico State University psychologist and researcher Victor Johnston, one of the best indicators of attractiveness is not only symmetry of the body, but also symmetry of the face. (You know who has an unusually perfectly symmetrical face? Denzel Washington. Enough said?)

It's true: Years of scientific research have revealed that universally, we seem to be bent on meeting people with symmetrical proportions. We're drawn to men with prominent foreheads and equally strong, square chins. We're captivated by women with large, bright, well-spaced eyes that are balanced by full, luscious lips.

But why? What makes us more attracted to the guy with the V-shaped torso and less attracted to the gal whose eyes are too close together? Evolution, baby.

Although human attraction is complex and not completely understood, many psychologists and biologists theorize that who—and what—we find attractive was hardwired into our brains by evolutionary needs. So why are we drawn to the strong jaw, glowing complexion, or gleaming white smile of the stranger across the bar? According to evolutionary biologists such as Randy Thornhill of the University of New Mexico, we're innately "programmed" to find those traits attractive because they advertise health, strength, fertility, dominance, and the ability to care for offspring.

So Denzel—any chance you're free for dinner?

Are Women More Emotional Than Men?

Get out your handkerchiefs, ladies. It's time for a chick flick, and we all know how emotional women get, don't we? Or do we? Are men really from Mars and women from Venus?

❋ ❋ ❋ ❋

According to psychologist Ann Kring of Vanderbilt University, men react just as strongly to a sentimental flick as women do. In 1998, she asked male and female volunteers to watch the same movie with electrodes attached to their non-dominant palms in order to measure sweating and an increase in body temperature. The men got damp palms, just like the women. What they didn't get, however, were damp eyes. Women, Kring noted, were more likely to show their emotions through facial expressions and tears.

This finding meshes with the observations of Dr. William Frey II, author of *Crying: The Mystery of Tears*. After age eighteen, Frey contends, women cry four times as often as men. This may be due to the surge of prolactin—a hormone that triggers lactation after childbirth—in women's bodies as they reach adulthood. Prolactin is found throughout the female body—in blood, sweat, and, yes, tears.

How can men and women feel the same but behave so differently? University of California–Irvine researchers Larry Cahill and Lisa Kilpatrick made some startling discoveries in 2006 when they studied the amygdalas of thirty-six male and thirty-six female volunteers via brain scans. The amygdala is a small cluster of neurons in the brain that processes strong emotions, such as fear and aggression.

In men, the right side of the amygdala is most active; the neurons there connect with areas of the brain that govern physical movement, vision, and other outward-oriented functions.

Women, in contrast, use the left side of the amygdala, which is connected to areas of the brain that monitor heart rate, blood pressure, and hormone levels.

How did these differences evolve? Cahill and Kilpatrick speculate that because women become pregnant, it might be more important for them to control what goes on inside their bodies rather than outside.

Just how significant are these differences in the context of everyday life? Linguist Deborah Tannen has spent nearly three decades analyzing how men and women speak. Armed with a tape recorder, she's eavesdropped on hundreds of conversations. In her bestselling book *You Just Don't Understand*, Tannen concludes that women use language to convey feelings while men use it to exchange information. These are broad generalizations, but her advice is valid: The keys to overcoming differences, she says, are patience, persistence, and compassion.

So walk a few blocks in your partner's shoes, be they stilettos or size-sixteen desert boots. And when you get to the movies, enjoy that sci-fi feature, the one about the Martian boy who meets the Venusian girl, and, well, you know the rest.

Are Old Men Grumpier Than Most People?

From the beginning of time—or roughly sixty years after the beginning of time—they've warmed our hearts with such expressions as, "Get off my lawn!" and "No! No! You're doing it all wrong! Get out of the way, I'll do it myself."

* * * *

THEY'RE GRUMPY OLD men, and they don't like you or anyone else, really. And they like it that way. And whether you like it or not, researchers confirm that grumpy old men are no joke; they're very real victims of their own physiology.

The English Longitudinal Study of Ageing surveyed nearly ten thousand people ages fifty-five and older and asked them to rate themselves in areas such as health, finances, and overall happiness. Across the board, men reported lower levels in quality of life than did women. Why? Are old men uncomfortable in those pants? Does the hair growing from their ears and nostrils irritate them?

The answer may be male menopause. Also known by its clinical name, andropause (or the snicker-inducing hypogonadism), male menopause is caused by testosterone deficiency. Some men experienced diminished testosterone production in their forties; full decline occurs in the early fifties.

The symptoms include flaccidity in a certain body part and an unwillingness to get it on. (See some connection here?) Irritability, depression, and loss of sleep are other signs. Treatments range from simple life changes—such as increased exercise, improved diet, and lowering stress—to chemical enhancements that include elevating testosterone levels through injections or supplements.

Identified by medical researchers as early as the 1940s, male menopause can be difficult for doctors to diagnose. It develops slowly over a long period, so it's tricky to pinpoint. And well, you try to get a grumpy old man who's sitting on a cold examination table wearing nothing but a paper gown to talk about his feelings.

But male menopause may yet find its way onto center stage. Baby Boomers, accustomed to whining until they get what they want, are now the generation complaining about growing old. In the process, the medical community is learning more about male menopause and how to identify and treat it effectively.

Yes, doctors are actually looking forward to the day when seventy-five-year-old men share graphic—and gruesome—details of their sex lives.

Life Before Alarm Clocks

Imagine how difficult it was to wake up before good old clock radios and other loud-sounding devices existed. Fortunately, there were some workarounds.

* * * *

The Knocker-Up

EVERYONE HAS A trick for waking up on time. Some people put the alarm clock across the room so that they have to get out of bed to turn it off; some set the clock ahead by ten or fifteen minutes to try to fool themselves into thinking that it's later than it is; some set multiple alarms; and some—those boring Goody Two-Shoes types—simply go to bed at a reasonable hour and get enough sleep.

We don't necessarily rely on it every day, and some of us definitely don't obey it very often, but just about everybody has an alarm clock. How did people ever wake up before these modern marvels existed?

Many of the tough problems in life have a common solution: hire someone else to do it. Long ago in England, you could hire a guy to come by each morning and, using a long pole, knock on your bedroom window to wake you up so that you would get to work on time. This practice began during the Industrial Revolution of the late eighteenth century, when getting to work on time was a new and innovative idea. (In the grand tradition of British terminology that makes Americans snicker, the pole operator was known as a "knocker-up.") There's no word on how said pole operator managed to get himself up on time, but we can guess.

The Alarm Clock Inside You

The truth is, you don't need any type of alarm, and you never did. Or so science tells us. Your body's circadian rhythms give you a sort of natural wake-up call via your body temperature's daily fluctuation. It rises every morning regardless of when you went to bed. Studies conducted at Harvard University seem to indicate that this rising temperature wakes us up (if the alarm hasn't already gone off).

Another study, conducted at the University of Lubeck in Germany, found that people have an innate ability to wake themselves up very early if they anticipate it beforehand. One night, the researchers told fifteen subjects that they would be awakened at 6:00 a.m. Around 4:30 a.m., the researchers noticed that the subjects began to experience a rise in the stress hormone adrenocorticotropin. On the other two nights, the subjects were told that they would get a 9:00 a.m. wake-up call—but those diligent scientists shook them out of bed three hours early, at 6:00 a.m. And this time, the adrenocorticotropin levels of the subjects held steady in the early morning hours.

It seems, then, that humans relied on their bodies to rouse them from the dream world long before a knocker-up or an alarm clock ever existed.

How Do I Become a Human Guinea Pig?

Ever want to do something for the good of all humankind? How about jogging on a treadmill with an oxygen mask strapped to your face? (Whew!) Or sleeping with a dozen electrodes attached to your body and a video camera watching? (Yikes!) Or eating a diet of specially prepared vegetarian food for three months? (Yum!) Or sniffing cotton balls saturated with the sweat of strangers of the opposite sex? (Phew!)

* * * *

ALL OF THESE are ways in which people have participated in clinical trials. Clinical trials give doctors insight into how we breathe, dream, absorb nutrients, and even sniff out our soul mates. Thousands of these trials take place every year. Want to be a volunteer? The government maintains a comprehensive database of these trials at ClinicalTrials.gov, which will point you in the right direction. If you live near a teaching hospital, you can call and ask if it has a department of clinical research.

Before you hand yourself over to eager medical students, however, you should know that in order to qualify as a genuine clinical trial, the experiment must have a set of rules, or protocols. Protocols define every step of a trial, including how volunteers are chosen, what kinds of treatment they receive, and how the results are measured.

If you qualify, you will be asked to sign a document testifying that you have given your informed consent. Don't take this lightly: You may very well be going where no man or woman has gone before. Though the vast majority of trials are safe, the effects of various medical treatments can't be guaranteed, even by the most careful researchers.

At the beginning of a study, volunteers are divided into several groups. One group, the control group, receives no treatment whatsoever. If the study involves medication, the control group will be given a placebo, which is usually a fake pill made of sugar and cellulose.

You've heard of the placebo effect? That's when the pseudo-pill-popping volunteers respond as if they are taking the real thing. It might be embarrassing to find out afterward that your amazing improvement was all in your mind, but the placebo effect is so important that a whole branch of clinical trials has evolved just to study it. Sort of like trials to test trials.

Will you be paid? Maybe. Payment, if offered, depends upon the length and complexity of the trial. It may be as little as ten dollars for a test that only takes an hour, but you could make several hundred dollars for trials that require hospital stays.

For most people, volunteering for a clinical trial is a one-shot deal. However, a small number of intrepid individuals actually turn it into a full-time job. A career as a professional human guinea pig may not be everyone's cup of antioxidant green tea, but it sure can provide some interesting conversation at Thanksgiving when Aunt Edna leans over and says, "So tell me, dear, what do you do for a living?"

Why Are There More Women Than Men?

Men are often referred to as the stronger sex, but women definitely outperform them when it comes to longevity.

⁂ ⁂ ⁂ ⁂

ACCORDING TO TOTAL population estimates in the CIA's *The World Factbook*, women outnumber men in most countries. Gender ratio estimates for 2018 say that there are 0.97 males for every female in the United States, 0.94 males for every female in Monaco, and 0.86 males for every female in the Ukraine.

Yet in newborn populations, men have a slight edge. There are about 107 boys born for every one hundred girls worldwide. If you're wondering where the boys are, they're not all having sex-change operations. It turns out that throughout different stages of life, men face higher mortality rates than women.

For example, between the ages of fifteen and twenty-four, men are four to five times more likely to die than women. Marianne Legato, a specialist in gender-specific medicine at Columbia University, says this is because men—particularly adolescents—are biologically more inclined toward risky behavior.

Researchers often refer to this increase in reckless and violent behavior as a full-on "testosterone storm." It gets guys in this pubescent age group into a whole lot of trouble—mainly death by car accident, homicide, suicide, or drowning.

After the age of twenty-four, the mortality rate for men and women tends to even out. But beyond the age of fifty, it's all downhill again for the misters. Research conducted at Harvard University Medical School shows that from the ages of fifty-five to sixty-four, men are once again more likely to die than women. This time, the deaths are often from tobacco and alcohol use and heart disease, in addition to car accidents and suicide.

In her book, *Why Men Die First,* Legato says that across national and cultural boundaries, men die an average of seven years earlier than women. Over a period of time, that slowly alters overall gender ratios.

In the United States, there are 1.05 males for every female in the under-fifteen age category, but only 0.73 males for every female in the sixty-five and older category. However, the gender gap is widest among those who live one hundred years or longer. Worldwide, female centenarians outnumber their male counterparts by a ratio of nine to one.

Ladies looking for love after hitting that century mark might consider a move to Qatar in the Middle East. In the sixty-five and older group, there are 2.92 males for every female. Who knows? You might nab yourself a nice spring chicken.

Would Men Still Live in Caves If It Weren't for Women?

Are women the reason behind human evolution?

* * * *

IT'S TRUE: WE'VE got women to thank for many of the greatest innovations of the modern age. We're talking about things like dishwashers (Josephine Cochran), windshield wipers (Mary Anderson), disposable diapers (Marion Donovan), liquid paper (Bette Nesmith Graham), the first computer language (Grace Murray Hopper), and Kevlar (Stephanie Kwolek), a synthetic material that's five times as strong as steel.

Researchers are now saying that women have always been major players in society. In their book *The Invisible Sex: Uncovering the True Roles of Women in Prehistory,* archaeologist J. M. Adovasio, anthropologist Olga Soffer, and writer Jake Page assert that early women drove the invention of language, agriculture, and the most useful tools, particularly string. They call it the String Revolution, and it opened the door to weaving the first clothes, baskets, food containers, and slings, as well as nets to catch game and fish for food. Talk about advancing humankind! If women were the primary weavers and fabric experts of prehistory, it's no wonder that they're still so good at decorating—and redecorating—with draperies, bed linens, and shabby-chic pillowcases. And it's probably no coincidence that the inventor of the Scotchgard line of stain repellents and cleaners was a woman (Patsy Sherman)—the ladies do like to keep their woven upholstery and carpets clean.

Yes, of the two sexes, it's the women who really know how to turn houses into comfy, cozy homes, complete with complementary color schemes and furnishings. Left to their own devices, most men would be content living in a tent or a cave—so long as those domiciles are equipped with a beat-up Barcalounger, mini-fridge, a few chip clips, and cable television with a remote control.

Maybe that's why some men still do live in caves. That's right—in places like northern China, southern Spain, and Tunisia, people continue to occupy natural mountain grottos and stone caves. Some of these earth dwellings have been outfitted with modern conveniences like electricity, running water, broadband connections, and Jacuzzis. Well okay, some women live there, too. But thanks to them, the guys have finally advanced beyond communicating with grunts and groans.

It's Okay to Awaken a Sleepwalker—Really

This is yet another old wives' tale that has been shredded by science. In fact, it's more dangerous not to awaken a sleepwalker.

* * * *

An Unsolved Puzzle

WE'VE ALL HAD the experience of waking up in the middle of the night to find ourselves drinking a giant Slurpee and singing Barry Manilow's "Copacabana" on the back porch of the neighbor's house. Wait, everybody has, right? Er, we meant that metaphorically.

Sleepwalking, or somnambulism, is one of the great medical mysteries. Anyone who has encountered a sleepwalker wandering around the house—or singing naked on the back porch—can attest that it is an eerie experience. Sleepwalking is listed in a group of sleep disorders known as parasomnia, and researchers aren't sure what causes it. They know that stress and irregular sleep patterns may contribute to episodes, and that children are far more likely to suffer from the condition than adults. They also know that the old wives' tale warning that a person awakened from a somnambulist daze may die is just that—an old wives' tale.

Some experts trace this myth back to the beliefs of various indigenous cultures that thought when a person slept, his or her soul left the body, and that if you woke up a sleepwalker, the soul would be lost forever. Others argue that the myth arose simply due to the distress and shock sleepwalkers sometimes experience when woken up.

Just Don't Use Cold Water

Though it is true that a sleepwalker may be distressed and disoriented upon being roused from a midnight stroll, there are no documented cases of sleepwalkers expiring from it. Indeed, sleep experts argue that not waking a sleepwalker can lead to more harm than waking one, especially if he or she is engaged in certain activities at the time (climbing, jumping, handling a knife, running the American government, etc.). In most cases, specialists suggest that it is best to gently guide the somnambulist back to bed.

There might be another reason to wake a sleepwalker. In 1982 an Arizona man named Steven Steinberg went on trial for killing his wife, who was stabbed twenty-six times with a kitchen knife. Despite overwhelming evidence and Steinberg's own admission that he had committed the crime, the defendant was unable to answer the simplest questions about the circumstances of his wife's death. Why? The man had killed his wife while sleepwalking. Steinberg was acquitted of the charges.

No Laughing Matter: Coulrophobia

It's a malady in which the sufferer has an abnormal or exaggerated fear of clowns, and experts estimate that it afflicts as many as one in seven people.

* * * *

Beware of the Man with the Red Nose

THE SYMPTOMS OF this strangely common affliction range from nausea and sweating to irregular heartbeat, shortness of breath, and an overall feeling of impending doom. Is the sight of Ronald McDonald more chilling than your Chocolate Triple Thick Shake? There could be a few reasons why.

The most common explanation for coulrophobia is that the sufferer had a bad experience with a clown at a young and impressionable age. Maybe the clown at Billy Schuster's fifth birthday party shot you in the eye with a squirting flower, doused your head with confetti, or accidentally popped the balloon animal he was making for you. Some of the most silly or mundane things can be petrifying when you are young. And though the incident may be long forgotten, a bright orange wig or bulbous red nose might be enough to throw you back into the irrational fears that plagued your younger days.

When Good Clowns Turn Bad

Who could blame you? If television and movies have taught us anything, it's that clowns often are creatures of pure evil. There's the Joker, Batman's murderously insane archenemy; the shape-shifting Pennywise from Stephen King's *It*; the human-eating alien clowns in *Killer Klowns from Outer Space*; and a possessed toy clown that comes to life and beats the bejesus out of a young Robbie Freeling in Steven Spielberg's *Poltergeist*.

Real-life serial killer John Wayne Gacy didn't do much to help the clown cause, either. Before authorities discovered the bodies of 27 boys and young men in his basement crawl space, Gacy was known as a charming, sociable guy who enjoyed performing at children's parties dressed up as Pogo the Clown or Patches the Clown. That ended when his crimes were discovered, but even on death row, he still had an unwholesome interest in clowning—he took up oil painting, and clowns were his favorite subjects.

Be Afraid—Be Very Afraid

It's enough to give anyone the heebie-jeebies. But some experts say there's more to coulrophobia than traumatic childhood events or pop-culture portrayals. Scholar Joseph Durwin points out that since ancient times, clowns, fools, and jesters have been given permission to mock, criticize, or act deviantly and unexpectedly. This freedom to behave outside of normal social boundaries is exactly what makes clowns so threatening.

A *Nursing Standard* magazine interview of 250 people ages four to sixteen revealed that clowns are indeed "universally scary." Researcher Penny Curtis reported some kids found clowns to be "quite frightening and unknowable." Seems it has a lot to do with that permanent grease-painted grin. Because the face of a clown never changes, you don't know if he's relentlessly gleeful or about to bite your face off. In the words of Bart Simpson: "Can't sleep; clown will eat me."

When Cold Burns

Frostbitten skin may feel like it's burning, but that doesn't mean you should try to cool it down.

* * * *

IF YOUR MOTHER told you to rub snow on frostnipped digits or appendages, chances are she was born before the 1950s—or else she was getting her treatment advice from a very old first-aid manual.

Rubbing snow on frostbitten skin had been standard treatment since Napoleon's army surgeon, Baron Larrey, proposed it in the frigid winter of 1812–13, during the retreat from Moscow. Larrey saw the disastrous results when soldiers defrosted their hands and feet by holding them over fires, and he decided that using snow was a better way to minimize infection.

The good doctor had it wrong, however. Physicians' discoveries during another military endeavor, the Korean War, followed by a landmark study a few years later, debunked the popular snow-on-frostbite treatment theory.

Frostbitten skin should not be thawed unless it won't be refrozen. Refreezing warmed frostbitten tissue is more harmful than leaving it alone. Of course, the soldiers did not have an option. They were stuck outdoors on the battlefield, and a great number of them ended up amputees.

Your Pet Goldfish Just Might Save Your Life

Research shows that gazing into fish-filled aquariums can help to reduce a person's stress, at least temporarily. That's why there are fish tanks in the waiting rooms of many doctors' and dentists' offices.

⁕ ⁕ ⁕ ⁕

IN THE 1980s, researchers at the University of Pennsylvania found that watching fish in an aquarium is far more effective at reducing stress than watching an aquarium without fish. In 1999, Nancy Edwards, a professor of nursing at Purdue University, discovered that Alzheimer's patients who were exposed to fish-filled aquariums were more relaxed and alert, and that they even began to eat a healthier diet. Another study showed that exposure to aquariums can contribute to decreased stress and hyperactivity among patients with Attention Deficit Hyperactivity Disorder.

Nobody is sure exactly why fish are so calming. The Purdue University study theorized that the combination of movement, color, and sound in an aquarium has a relaxing effect.

But it's not just fish that help reduce stress. Having a pet of any kind leads to better health. A 2007 study at Queen's University in Belfast, Northern Ireland, showed that dog owners tended to have lower blood pressure and cholesterol than non-dog owners. The study proposed that pet owners in general are healthier than the population on average and also suggested that having a dog is better for you than having a cat. However, the study didn't weigh in on how dogs and cats compare to fish.

And if you don't fancy keeping an aquarium, you can always just eat fish. Most fish contain a bunch of omega 3 fatty acids that are good for your heart and blood pressure.

Life on the Body Farm

When Mary Scarborough wrote the lyrics to "Old MacDonald Had a Farm," she probably didn't have a research facility in mind. In fact, one won't find cows, chickens, or pigs that go "oink" at the Body Farm—just scores of rotting human bodies.

* * * *

E-I-E-I-Oh, Gross

THE BODY FARM (officially known as the University of Tennessee Forensic Anthropology Facility) was the brainchild of Dr. William Bass, a forensic anthropologist from Kansas. Its purpose, however nauseating, is to help law enforcement agencies learn to estimate how long a person has been dead. After all, determining the time of death is crucial in confirming alibis and establishing timelines for violent crimes.

After 11 years of watching and learning about human decomposition, Bass realized how little was actually known about what happens to the human body after death. With this in mind, he approached the University of Tennessee Medical Center and asked for a plot of land where he could control what happens to a post-mortem body and study the results.

Bass's Body Farm drew the attention of readers when popular crime novelist Patricia Cornwell featured it in her 1994 book, *The Body Farm*. In it, Cornwell describes a research facility that stages human corpses in various states of decay and in a variety of locations—wooded areas, the trunk of a car, underwater, or beneath a pile of leaves—all to determine how bodies decay under different circumstances.

Reading the Body

According to Bass, two things occur when a person dies. At the time of death, digestive enzymes begin to feed on the body, "liquefying" the tissues. If flies have access to it, they lay eggs in the body. Eventually, the eggs hatch into larvae that feast on the remaining tissues. By monitoring and noting how much time it takes for maggots to consume the tissues, authorities can estimate how long a person has been dead. Scientists can also compare the types of flies that are indigenous to the area with the types that have invaded the body to determine whether the body has been moved. "People will have alibis for certain periods," says Bass. "If you can determine that the death happened at another time or location, it makes a big difference in the outcome of the court case."

But the farm isn't all tissue decomposition—scientists also learn about the normal wear-and-tear that a human body goes through. For instance, anthropologists look at the teeth of the victim to try to determine their age at the time of death. The skull and pelvic girdles are helpful in determining a person's sex, and scientists can also estimate how tall the person was by measuring the long bones of the legs or even a single finger. Other researchers watch what happens to the five types of fatty acids leaking from the body into the ground. By analyzing the profiles of the acids, scientists can determine the time of death and how long it has been at its current location.

Unfortunately, the perps are catching on. Some criminals try to confuse investigators by tampering with the bodies and burial sites, spraying the victim with insecticides that prevent insects (such as maggots) from doing their job.

Further Afield

At another facility at the University of New Mexico, scientists have collected over 500 human skeletons and store them as "skeletal archives" to create biological profiles based on what happens to bones over time. And in Germany, the Max Planck Institute for Computer Science has been working on a 3-D graphics program based on forensic data to produce more accurate likenesses of the victims.

Although other proposed farms never got off the ground due to community protest, since the inception of Bass's original Body

Farm, another farm has been established at Western Carolina University. Ideally, Bass would like to see body farms all over the nation. Since decaying bodies react differently depending on their climate and surroundings, says Bass, "It's important to gather information from other research facilities across the United States."

Ready, Set, Cycle!

From the sandbox to the sorority house, gal groups are powerful. Guys might even say magical. But can women exert a kind of chemical alchemy over one another?

⁕ ⁕ ⁕ ⁕

- A landmark 1971 study of 135 members of an all-female college dormitory showed that women who lived together tended to have menstrual periods within days of one another. In the study, roommates whose periods averaged more than six days apart in October were less than five days apart six months later.
- The theory of menstrual synchrony—called the "McClintock Effect" after the author of the study—was long accepted as established fact, especially among women. Martha McClintock, a graduate student who then became a University of Chicago psychologist, speculated that pheromones (chemical messengers received through the sense of smell) were responsible for the phenomenon.
- Recent studies have cast doubt on the theory of synchrony. In 1992, H. Clyde Wilson published a report accusing a number of studies—McClintock's included—of faulty research and shoddy methodology. And researchers have pointed out that synchrony is impossible when women have cycles of different lengths.
- In 2007, psychologist Jeffery Schank published a study involving 186 Chinese women who lived together in a college

dorm. Though he uncovered some interesting menstruation patterns among the women, he found no evidence that their cycles were in sync. The researchers reviewed McClintock's original study and went so far as to say that the results in 1971 could be chalked up to chance.

* The strongest evidence against menstrual synchrony is that, in all studies, women's cycle lengths continue to vary radically, even if their start dates get closer over a designated period of time.

Leave No Trace?

Movie bad guys are always trying to sand or burn off their fingerprints, but can it really be done?

* * * *

An Ancient Art

THE USE OF fingerprints as identifiers or signatures dates back thousands of years, when different ancient civilizations used fingerprints on official documents. By the 1300s, someone had already noted, in surviving documents, that all the fingerprints he had seen in his life were unique. Scientists now know that fingerprints emerge during the first trimester of human embryo development and last in the same form for the entire human lifetime. Older people can naturally have very shallow fingerprints that are hard to read, but the shapes of these fingerprints are still the same.

People who work in certain fields are more likely to sand their fingerprints off during their regular daily routines. Construction workers who build with brick, stone, concrete, and other abrasive materials can end up losing their fingerprints. So can people who handle a lot of paper or cardboard. These changes are temporary: since only the topmost layer of skin is abraded, it will come back with the same pattern.

A Modern Problem

In the 21st century, people noticed a new form of fingerprint annihilation. A small number of cancer patients being treated with a chemotherapy drug called capecitabine found that they no longer had fingerprints. In the past, people easily went entire lifetimes without ever noticing their fingerprints, but modern banking and security often require them, and fingerprint technology has now shown up in the protection of personal devices like smartphones and laptops. Patients have been questioned in banks, locked out of their own computers, and stopped in security lines—as if being treated for cancer wasn't hard enough.

Researchers aren't sure whether or not these cancer patients' damaged or obliterated fingerprints will eventually grow back. But how could it be ambiguous like this? The answer is in the makeup of our fingerprints, which grow up into the topmost layer of skin *from* the layer beneath. If you trimmed down your rose bushes for the winter, would you expect tulips the next spring? Without doing major damage to your fingertips, you simply can't get rid of your fingerprints.

The Imperfect Solution

Villains of the past have tried a lot of superficial ways to scrub off their identities. The first known and famous case was a gangster in the 1930s named Handsome Jack, who made cuts to disrupt his fingerprints. His case made a huge splash in the press, but his mutilated prints were even easier to spot than they had been. Most documented attempts to change fingerprints fall into this category, where the actions taken to obscure the fingerprints instead create a new, even more identifiable, scarred fingerprint.

To truly distort and alter your fingerprints, you would need to disrupt the connection between the two layers of skin where fingerprints root and grow. As of yet, there's no reasonable way to do this. Criminals and other wannabe disappearers may have more luck in the future.

Brain Power

Mental gymnastics are not the only way to keep your brain sharp. Physical gymnastics—or any form of exercise—are just as important.

* * * *

LOOKING TO ROLL back the cognitive clock? Or maybe just remember where you put your car keys? Cognitive "training"—that is, doing mentally challenging activities such as crossword puzzles and sudoku—has been all the rage for a while now because it's been shown to help preserve brain function. But there's another kind of workout that could be the real ticket to keeping your brain young.

Stay Sharp and in Shape

Studies show that regular, moderate exercise helps our brains stay sharp. Researchers have found that one hour of aerobic exercise three times a week can increase brain volume, which in turn may delay some age-related changes. It takes only a few months to start seeing results, suggesting that it's never too late to start exercising. Other studies done on animals concur that more exercise equals better brain function. Even the dormant neural stem cells in elderly mice "wake up" once the critters hit the running wheel.

In one study that examined older men and women with memory problems, mental workouts (in the form of brain teasers and puzzles) produced encouraging results when combined with physical exercise and a heart-healthy diet. What's good for the cardiovascular system appears to also be good for our gray matter, giving us another reason to lower our blood pressure, weight, and cholesterol levels.

Experts caution not to waste money on dietary supplements and vitamins that claim to have "anti-aging" benefits, and we don't need to buy special "brain fitness" computer programs.

Engaging in a pleasurable activity, such as studying a foreign language, reading, or playing a musical instrument, is a great way to keep us thinking and learning.

For Crying Out Loud!

Every new parent dreams about getting a good night's sleep, but there's a lot of disagreement over the best way to help a baby sleep soundly.

❋ ❋ ❋ ❋

OF COURSE, NO one expects a newborn or young infant to sleep through the night. But according to some experts, how parents put their infants to bed and how they tend to them when they awaken may make a difference in their sleep habits in the long run.

There are two main schools of thought on the subject. Respected medical experts continue to weigh in with differing theories and studies. Well-meaning—and experienced—friends and family add to the confusion by offering their surefire bedtime strategies.

To Soothe or Not to Soothe?

On one side of the debate are proponents of routines that purportedly train a baby to fall asleep by himself. This approach advises parents to let the baby cry for short intervals of time. On the other side are those who don't believe a crying infant should ever be left alone and that it's a parent's responsibility to make the baby feel secure enough to fall asleep.

The controversy over sleep routines escalated in 1985 with the publication of *Solve Your Child's Sleep Problems*, by pediatrician Richard Ferber. He advocates allowing babies to soothe themselves to sleep, starting around four to six months of age. The basic idea of "Ferberizing," as it has come to be known, is to put the baby in the crib while still awake, after a calming bedtime routine. Parents are told to not pick up the baby or feed him, even if he cries. Instead, Ferber advocates letting the baby cry for a few minutes before returning to the room to comfort him. Parents are instructed to gradually increase the amount of time the baby is left alone. Ferber calls this approach "progressive waiting," and he believes it eventually teaches the baby to fall asleep without parental intervention. Successfully Ferberized babies will be able to fall back asleep without crying out for their parents when they awaken during the night.

Critics of the Ferber method (which was somewhat revised in the 2006 edition of his book) argue that letting a baby "cry it out" only teaches an infant that nobody cares about him. Two Harvard researchers have gone so far as to say that leaving crying babies alone in their cribs can be traumatic, leading to emotional problems later in life.

Parental Prerogative

There is much anecdotal evidence from parents to support both of these points of view. Many moms and dads praise Ferber's system, saying that their babies began sleeping through the night after following the routine for a few days. Other parents tell horror stories of letting a baby cry himself to sleep, only to later discover that the child was suffering from, say, a painful ear infection.

Meanwhile, the American Academy of Pediatrics says that it is perfectly normal for babies to wake up during the night. A newborn is hungry every one and a half to three hours and needs nourishment in order to fall back asleep.

Around the age of six months, most babies can sleep a span of six to eight hours without nursing or having a bottle. However, some doctors insist that it is unreasonable to expect an infant to sleep soundly through the night before he turns a year old.

A Filling Idea

A common misconception is that introducing solid food early will help solve a baby's sleep problems. But some pediatricians say babies should be given only breast milk or formula until they are six months of age. Younger babies cannot properly digest solid food and may end up getting a stomachache instead of a good night's sleep.

The Man from S.E.X.: Alfred Kinsey and His Institute for Sex Research

Proper folks never mentioned it. It was best kept under the covers, with the lights off. If there was a problem, it was never solved. "It" was sex.

* * * *

IN THE 20TH century, American society's view on sex was basically the same as it had been in the 19th century . . . and the 18th century, as well. Heavily influenced by religious tenets and conventions, sex was for procreation only, not pleasure. The common belief held that it was rather dirty; something about which civilized people never spoke. It was kept locked tight, packed away, and zipped up. Then along came Alfred Kinsey.

Setting the Scene

The eldest of three, Kinsey was born in Hoboken, New Jersey, in 1894, to devout Methodist parents who forbade any discussion or thoughts of sex. This, combined with a sickly childhood, may have contributed to Kinsey's interest in his future profession. He joined the Boy Scouts and quickly became an Eagle Scout. This exposure to the outdoors eventually led to his fascination with gall wasps—they became the subject of his doctoral thesis. He ultimately amassed a collection of more than five million specimens, which was later donated to the American Museum of Natural History in New York.

Gifted with a keen interest in biology, Kinsey graduated magna cum laude from Bowdoin College in Maine, with degrees in biology and psychology. Continuing at Harvard University, he earned his doctorate in 1919 and joined the faculty of Indiana University as an assistant professor of zoology. He wrote a textbook on the subject, which was widely adopted by high schools across the country.

Foreplay

Clara "Mac" McMillen was a recent graduate of Indiana University when she met Dr. Kinsey. They were married within a year, although the start of their relationship was anything but ideal. Honeymooning by hiking and camping through the White Mountains of New Hampshire led to an unconsummated beginning. There is some question as to the cause—some suggest the less-than-romantic environment of howling winds and a lean-to tent, while others point to some physical issues that made the act inherently difficult. Many believe that this experience was only one of the many factors that led to Kinsey's ultimate research.

The professor quickly realized that many people shared the same difficulties and frustrations about sex as he faced with Mac. What's more, Kinsey was angered at the total lack of scientific information that could explain the phenomenon of sex and all its variations. No one had ever made a serious study of human sexuality as something that wasn't dirty and unmentionable—until Kinsey.

By the late 1930s, Kinsey offered a course on marriage (a not-so-subtle substitution for "sexual behavior") at the university. Working without notes (although he did have explicit slides), Kinsey addressed the anatomy and mechanics of sex, the acts of love and self-satisfaction, and the resulting pleasure from the natural biological activities of humans. The reaction was overwhelmingly positive, yet Kinsey needed more from those who took his work seriously. He had a few questions that he still wanted to ask.

Doing the Deed

Kinsey and his staff began a systematic process of interviewing men—more than 5,000 of them. Using codes to ensure confidentiality, the surveys were clinical and to the point.

They began with 12 basic demographic questions to identify the respondent: age, race, education level, occupation, and the like. The interview then moved to 350 queries based on sexual data—specific questions concerning likes, dislikes, techniques, partners (even for those with nonhuman interests), variety, and frequency of occurrences. One assumption that Kinsey's team made was that "everyone had done everything." Questions never asked, "Have you ever . . . ?" Rather, they began, "When did you first . . . ?" Their results were amazing.

Respondents talked freely about sex—sex with wives, partners (paid or not), lovers, themselves. But Kinsey also broke ground in areas never dreamed of, such as adolescent memories, along with the most unpleasant subjects—sex with children or animals, bodily functions. The mass of information he compiled was beyond anything previously attempted.

Kinsey's research also tackled the subject of homosexuality. Author Gore Vidal once observed that most Americans believed that "homosexuality was a form of mental disease, confined for the most part to interior decorators and ballet dancers." But it wasn't just Americans in general. Until 1973, the American Psychiatric Association classified homosexuality as a mental disorder. Kinsey's inquests refuted that notion, showing that more than a third of males had taken part in a homosexual experience at least once and that 8 percent were "exclusively homosexual" for at least three years at some point in their lives. He even pointed out that, biologically speaking, almost every animal species exhibited some sort of homosexual behavior. Kinsey's interest in the subject may have been due to his own bisexuality.

Climax

Kinsey founded the Institute for Sex Research at Indiana University in 1947 to further his investigations. The next year he published his findings in a landmark book, *Sexual Behavior in the Human Male*. Within six months, more than 150,000 copies had been sold, and the book was translated into French and Italian. If people didn't want to hear about such things, they certainly seemed ready to read about them. Known as the Kinsey Report, the book ruffled the feathers of more than a few: Kinsey found himself accused of being party to a Communist plot on one end of the spectrum and chastised for largely ignoring the emotional elements of sex on the other. Kinsey took no payment from the book sales, instead contributing his profits to the institute for further research.

By 1953, Kinsey and his team had completed interviews with nearly 6,000 females and issued *Sexual Behavior in the Human Female*. A Gallup poll reported that people, by a two-to-one margin, believed the Kinsey Reports were "a good thing" to have available. Of course, church groups responded with rage: Billy Graham denounced the book (without actually reading it), saying it was "impossible to estimate the damage . . . to the deteriorating morals of America."

Much of this criticism was deserved—Kinsey's samples didn't accurately reflect the makeup of the country—and he readily admitted this. Yet, his work clearly changed society's views of human sexuality, laying the foundation for the sexual revolution of the '60s. Researchers Masters and Johnson furthered this work in the '70s, leading to a new posture surrounding "the birds and bees" in America.

The Intelligence on IQ Tests

IQ scores are best known as quantitative representations of a person's intelligence. Yet the original IQ test was intended to predict future scholastic achievement, not intelligence.

* * * *

Relatively Smart?

THE FAMED DUMBBELL Forrest Gump had an IQ of 75, but he did pretty well for himself. He was a military hero, savvy businessman, exceptional table tennis player, and beloved son, husband, and father. An IQ test is supposed to measure one's intelligence, but there is much debate over what an IQ score actually means.

A person's intelligence quotient is calculated according to his or her performance on a standardized test. This means that the score is not derived from how many questions are answered correctly but on how many the person gets right *relative to others who have taken the same test.* That's an important distinction. IQ tests are usually standardized so that 100 is the mean score, and half of the scores lie within 10 points of the mean—so half the population has an IQ between 90 and 110. "IQ test" actually refers to a number of popular tests that are standardized in a similar fashion, such as the Wechsler or Stanford-Binet tests.

The first IQ test was developed in the late 1800s, hand-in-hand with the appearance of special-education programs in schools. Administrators needed a reliable way to identify those who were unable to learn as easily or quickly as others. From the beginning, then, IQ tests were meant to measure one's ability to perform academic tasks; this is not necessarily synonymous with intelligence.

Kinds of Smart

IQ test questions measure such functions as short-term memory, vocabulary, perceptual speed, and visual-spatial reasoning. These are all skills that help a person succeed in a school, work, daily tasks, or even social environment. Not surprisingly, high IQ scores are positively correlated with one's future academic success. They are also correlated, though not as strongly, with the socioeconomic status of one's parents, as well as on future income and future job performance.

Many researchers have pointed out that IQ tests neglect to calculate many types of talent that could also fall under the "intelligence" heading. Psychologist Howard Gardner developed his theory of multiple intelligences, which include linguistic, logical-mathematical, spatial, bodily-kinesthetic, musical, interpersonal, intrapersonal, and naturalist. Many multiple-intelligence tests try to include indicators of "books smarts," "street smarts," and "creativity smarts."

Testing IQ Tests

The reliability of IQ tests as meters of intelligence is also suspect because, on average, African American, Native American, and other minority or immigrant populations tend to score lower than populations of Euro-American descent. And these minority groups tend to come from areas where there is a high dropout rate and limited access to quality education. IQ tests are administered in standard English, which partly accounts for these low scores (especially in the verbal section) among people who speak other dialects of English or speak English as a second language.

Down the Drain

Dismayed at the high prices of traditional funerals? Why not wash the remains down the drain?

✻ ✻ ✻ ✻

The Pressure Cooker

FUNERALS ARE BIG business. A traditional funeral can cost upward of $15,000, while an "inexpensive" funeral runs around $2,000. This can cause big problems—perhaps the bereaved is unable to foot the bill. Or maybe it was the deceased's wishes to have a less expensive, more eco-friendly method of disposing of his or her remains.

Enter an interesting approach. In 1992, scientists came up with a unique way to dispose of animal carcasses used in medical research: alkaline hydrolysis. Today the method is sometimes used on human remains as well. The process uses a combination of a high-alkaline product (usually lye) and a temperature of about 300 degrees Fahrenheit. It also takes a specially designed stainless-steel container that acts much like a large pressure cooker.

After a moderate amount of "cooking," the carcass is reduced to a coffee-color syrup, with the approximate consistency of motor oil. The liquid is sterile and is generally safe enough to be poured down the drain. For those who would like to have something more tangible to remember their loved ones by, bone residue can be captured, dried, and placed in an urn, similar to cremation. The cost would also be akin to that of cremation.

The New Alternative to Burial

The process is actually nothing new. For years, Hollywood movie plots have incorporated similar methods—think of gangster flicks in which some wise guy's body is dissolved in a bathtub full of lye. But there are real-life benefits: Alkaline hydrolysis is gaining favor because of its environmental benefits. Unlike cremation, there are no dangerous emissions like carbon dioxide or the ill effects of the disposal of mercury and silver dental fillings.

Even so, alkaline hydrolysis has been a tough sell. While human burial and cremation have become mainstream methods for disposing human bodies, there's something unsettling about watching grandma circle down the drain. Minnesota and New Hampshire have legalized the process, but only one funeral director has stepped up to offer the service.

Many detractors of the service are uncomfortable with the idea. "We believe this process, which enables a portion of human remains to be flushed down a drain, to be undignified," says Patrick McGee, a spokesperson for the Roman Catholic Diocese of Manchester. A number of detractors are more concerned with the process of disposing human waste.

On the other hand, George Carlson, an Industrial Waste Manager with the New Hampshire Department of Environmental Services, says that substances the public might find more troubling routinely flow into U.S. sewage treatment plants all the time—including blood and embalming fluid from funeral homes.

But others, such as New Hampshire State Representative Barbara French, agree that it might be time for a change toward hydrolysis. "I'm getting near that age and thought about cremation but this is equally as good and less of an environmental problem," she said. "It doesn't bother me any more than being burned up."

The Big Kahuna of Hydrolysis

The chief proponent of the process is Brad Cain, president of BioSafe Engineering. The company manufactures the required steel containers and estimates that as many as 50 facilities (including veterinary schools, universities, pharmaceutical companies, and the U.S. government) use his equipment to dispose of animal carcasses and other types of medical wastes. Cain would like to offer the service to the public as an alternative to cremation.

Currently, there are only two facilities in the country that use the process—the University of Florida in Gainesville and the Mayo Clinic in Rochester, Minnesota. Both are research facilities and neither of them offer the service to the public. Chad Corbin, a Manchester funeral director, was issued a permit to operate a hydrolysis tank in 2007, but the process became mired in delays and red tape. Now he has to begin the approval process all over again. "I don't know how long it will take," he said, "but eventually it will happen."

Unusual Medical Maladies

The human body is able to play some nasty tricks on its owner. While none of them are considered life-threatening, the six syndromes and disorders described below are documented cases of unusual medical maladies.

* * * *

Who Is That?

TAKE THE INTERESTING and perplexing diagnosis of **Capgras Syndrome,** a rare psychological disorder that makes sufferers suspicious of their loved ones or even their own reflections. For a number of reasons, including schizophrenia, epilepsy, and malformed temporal lobes of the brain, Capgras victims have difficulty making physical and emotional connections with the people, places, and things they see, even ones that have been a part of their lives for years. Sufferers see themselves in a mirror or other shiny surfaces and wonder who the stranger is that's peering back at them.

According to Dr. V. S. Ramachandran, director of the Center for Brain and Cognition at the University of California, San Diego, people diagnosed with this particular disorder can also find themselves suspicious of animals or other objects, such as a pair of running shoes. In such cases, they convince themselves that someone has broken into their home and replaced familiar objects with imposters.

Can You Direct Me to the Loo?

Foreign Accent Syndrome is even more rare: a disorder that causes the afflicted to suddenly and unexplainably speak in an unfamiliar dialect. One of the first cases of FAS was discovered in 1941 after a young Norwegian woman sustained a shrapnel injury to her head during a wartime air raid. Although she had never been out of her home country, she suddenly began speaking with a German accent, which resulted in her being shunned by her family and friends. In Indiana, a 57-year-old woman suffered a stroke in 1999 and began speaking with a British accent, including colloquialisms like "bloody" and "loo."

Get Your Hands Off of Me!

If there was ever a malady that a high-school boy might envy, it's **Alien Hand Syndrome,** also known as **Dr. Strangelove Syndrome.** Alien Hand Syndrome is caused by damage to the parietal or occipital lobe of the brain. Those afflicted often find one of their hands operating independently from the rest of their body and sometimes completely against their conscious will. AHS sufferers often report incidences of a "rogue hand" getting involved in disobedient behavior such as undoing buttons or removing clothing. One patient reported a bizarre incident in which her right hand put a cigarette into her mouth. Before she could light it, her left hand yanked the cigarette out and crushed it in an ashtray.

Please Pass the Dirt

At one time or another all kids will experiment by eating an occasional handful of dirt. The good news is that it's a passing phase for the majority of youngsters. The bad news is that if this fascination with eating nonfood items persists longer than a month, your child could be afflicted with **Pica.** Associated with developmental disabilities such as autism or mental retardation, Pica typically affects children younger than 24 months. It can also appear in people with epilepsy and pregnant women.

Pica sufferers find themselves craving and consuming a wide variety of nonfood items such as dirt, sand, hair, glue, buttons, paint chips, plaster, laundry starch, cigarette butts, paper, soap, and even feces. There was even one documented case of "cutlery craving," in which a 47-year-old Englishman underwent more than 30 operations to remove various items from his stomach—including eight dinner forks.

Another form of Pica, called **Geophagia,** is practiced by cultures that eat earth substances such as dirt and clay to relieve nausea, morning sickness, diarrhea, and to remove toxins from their bodies.

Something Smells Fishy Around Here

Bad breath, body odor, and the occasional flatulence—we've all had to deal with them in one way or another. But how would you like to live with someone who constantly smelled of pungent fish? A rare metabolic disorder called **Fish Odor Syndrome** (also known as trimethylaminuria or TMAU) results in the afflicted releasing an enzyme called trimethylamine through their sweat, urine, and breath. This enzyme also happens to give off a strong "fishy" odor.

The condition appears to be more common in women than men, and researchers suspect that female sex hormones such as estrogen or progesterone may be at fault.

While there is no cure for Fish Odor Syndrome, people afflicted can control the disease by avoiding eggs, certain meats, fish, legumes, and foods that contain choline, nitrogen, and sulfur. And, of course, showering regularly.

A Permanent Bad Hair Day

If you've suffered from the occasional bad hair day, consider yourself lucky—you could be afflicted with **Uncombable Hair Syndrome.** UHS is a rare disease that affects boys and girls before puberty. In fact, it's so rare that there have only been 60 cases reported in medical literature between 1973 and 1998.

UHS is an inherited disease with subtle hair changes noted in several preceding generations. It begins with a hair follicle that produces triangular hair shafts with several longitudinal grooves that also has very little pigment and is exceptionally dry and brittle. Because the hair is so dry, it rarely lies down; instead, the hair grows straight out from the scalp.

So what should you do if you are diagnosed with UHS? First, cancel your appointment with your hairdresser. People afflicted with UHS typically experience alopecia, or periodic baldness. The hair that does grow frequently breaks off before it has time to mature. And there is hope: There has been some success with medication, and some cases have recovered spontaneously several years after the first outbreak.

Gotta Dance

What do wild, maniacal dancing, a strange disease, and patron saints have in common with each other? If you guessed the medieval epidemic called "dancing mania," you'd be right!

* * * *

Taking the Fun Out of "Fungus"

KNOWN AS "DANCING mania," this affliction is often associated with several diseases: St. Anthony's Fire, St. Vitus's Dance, and St. John's Dance. All three describe bizarre neurological diseases that hit Europe between the 13th and 18th centuries. The first outbreak of dancing mania affected the majority of the inhabitants of Aachen, Germany, in 1374, and it reached its pinnacle 100 years later in Strasbourg, France.

While there are a number of plausible theories about how the disease took hold of nearly an entire population, the most widely accepted hypothesis involves the ingestion of ergot, a fungus that infects rye with toxic and psychoactive chemicals, including lysergic acid—the same acid that would ultimately be synthesized into LSD.

The symptoms of dancing mania were as peculiar as their origin. Those affected exhibited uncontrolled and painful seizures, diarrhea, paresthesias (a pins-and-needles feeling), itching, foaming at the mouth, maniacal laughter, erratic gyrations, jerking movements, nausea, hallucinations, headaches, and vomiting.

But the most appalling symptoms were involuntary muscular contortions of the face and extremities that appeared to resemble dancing. That's where the bizarre treatments came in.

Take Two Tunes and Call Me in the Morning

Medicine, being as primitive as it was, had no plausible theories or cures for the diseases; the "science" often took as much from medieval witchcraft as it did from the Catholic Church. And because the diseases affected thousands of people, physicians needed to come up with an effective treatment for the masses, and fast. After a number of unsuccessful treatments, the medical community finally agreed on a solution: music.

In the 14th century, music was considered a "magic bullet" for just about everything and was used extensively to drive demons from anyone with an inexplicable malady. So, the town's elders gathered the afflicted and marched them through the center of town to the accompaniment of upbeat music in hopes that it—coupled with copious amounts of sweat generated from the movement—would exorcise the demons.

Being under the spell of the gyrating rhythms (and the ingredients of LSD), many of the afflicted ended up tearing off their clothes and dancing through the streets naked until they fell to the ground, exhausted. The few who weren't immediately cured were eventually hauled off to the nearest cathedral where they offered themselves in prayer—that's where the saints come in.

The Saints Come Marching

St. Vitus was the only son of a well-to-do Sicilian senator. He (the son) went on to perform a number of documented miracles. The Sicilian administrator (think mayor) was fed up with Vitus's antics and sentenced him to martyrdom. Just before losing his head—literally—Vitus prayed to God that those afflicted by the dance mania be cured.

Dancing mania also came to be known as St. John's disease because one of the first major outbreaks arose in St. Johannestanz, Germany.

A continent away, St. Anthony was an Egyptian monk who lived between AD 251 and 356. Withdrawing into an abstemious life at an early age, he finally emerged from years of solitary confinement to establish one of the first monasteries. During his self-imposed confinement, it is said that he battled with the devil, who attacked him with wild beasts and temptations of exotic feasts and naked women. But Anthony's prayers and penitence prevailed over evil. During the 12th century, the Order of Hospitallers of St. Anthony in Grenoble, France, became the destination of those afflicted with dancing mania, which is also called St. Anthony's Fire.

The Appeal of Dear Old Dad

Scientists call it "sexual imprinting"—when a woman falls for a guy who is just like her father.

* * * *

ASK A WOMAN if she wants a man as dependable, kind, and quietly confident as dear old Dad, and the chances are good that she'll say yes. Ask if she wants a man who looks like dear old Dad, and she'll probably say, "Huh?" or "Eww!"

Nonetheless, dear old Dad may be a little hotter than some women are willing to admit. According to a couple of studies by European researchers, many women fall for men who look like their fathers, whether they intend to or not.

Researchers at Hungary's University of Pécs measured proportions of the facial features of members of fifty-two families. Based on this data, they found that the faces of young men and their fathers-in-law had some strong resemblances, most notably around the nose and eyes. Comparable results came out of a similar study conducted jointly by Durham University in England and the University of Wroclaw in Poland.

For the record, the Oedipal implications of the findings extend to men as well: Most guys go gaga for girls who resemble Mom. The studies found that resemblances between young women and their mothers-in-law occurred in the lips and jaw. (Insert mother-in-law-mouth joke here.)

Scientists offered no substantive explanations for why we choose mates who look like our parents. However, it seems that the likelihood of a woman choosing a man who looks like her father depends on her relationship with him—if the relationship tends to be negative, the woman might migrate to other facial types.

Sure, laugh at Pops when he emerges from the bathroom scratching his posterior and belching like a hippo. But to many women, dear old Dad's still got it goin' on.

Getting a Charge Out of Life

It may seem odd to compare the human body to an electric power generator, but rare cases around the world have shown that some people are born with shocking abilities . . . literally. Jacqueline Priestman, a British woman, consistently produces ten times the static electricity of a normal human being.

✻ ✻ ✻ ✻

How to "Conduct" Oneself

PRIESTMAN, WHO IRONICALLY married an electrician before she knew about her strange ability, grew up with no more than the usual mild electromagnetic field that surrounds every human. But when she turned 22, sparks began to fly. Priestman noticed that her mere touch would cause ordinary household appliances to short out and fizzle, while others could use the same appliances with no problem. She could also change the channels on her TV by going near it.

Priestman has had to buy at least 30 new vacuum cleaners in her married life, plus five irons and several washing machines. Michael Shallis, a lecturer at Oxford University and a former astrophysicist, studied Priestman and told a British newspaper in 1985 that she was actually able to transmit tiny bolts of "lightning" that could affect any electrical system nearby. He had no explanation for the phenomenon but did say that most similar cases he had investigated involved women. For example, Pauline Shaw flooded her house every time she tried to do laundry because the washing machine fuses would blow when she touched the dials. The washer's door would then pop open and turn the machine into a fountain.

For more than four years, Shallis studied 600 people with Priestman's condition and, eventually, wrote a book about them called *The Electric Connection.*

SLI-ding Through Life

There is a name for those like Priestman and Shaw. Because people with abnormal amounts of static electricity often cause streetlights to flicker when they pass by, scientists call the strange disorder Street Light Interference, or SLI. People with the condition are called SLI-ders, or Sliders.

An older name for the phenomenon is High Voltage Syndrome, or HVS. Around 1930, one HVS patient, Count John Berenyi of Hungary, was reportedly able to make neon light tubes glow merely by holding them. And according to author Vincent Gaddis, the National Safety Council investigates what he calls "human spark plugs"—people who can start fires with the electrical abundance of their mere presence. One woman made a rather poor vocational choice in the early 1940s when she got a job gluing shoes together with rubber cement, a highly flammable substance. She allegedly started at least five fires in the factory and could ignite a pail of rubber cement merely by standing near it. She had to quit after suffering severe burns in one of the fires.

Even infants can act as superconductors. In 1869, a child born in France was so highly charged that anyone who approached him received a sharp electric shock. The infant was even said to exhibit a faint glow around his hands. The infant died from undetermined causes when he was only nine months old, and, according to witnesses, his entire body radiated light at the time of his passing.

Radiant Blood

The strange baby was not the only human known to glow. Luminous people have been reported in many circumstances, and their abilities are often tied to medical conditions. Anna Monaro, an Italian woman, gained attention in 1934, when her breasts began to spontaneously emit blue phosphorescent light while she was sleeping. The weird condition lasted for weeks and drew many eminent doctors and scientists to study her firsthand. They were even able to capture the glow on film. Many theories were offered, from "electrical and magnetic organisms in the woman's body" to "radiant blood." Eventually, the bizarre condition went away and did not return.

Through No Fault of Her Own

A Welsh woman named Mary Jones set off a religious fervor in 1905, when amazing forms of light appeared to emit from her body. Jones had already gained some notoriety as a local preacher when people began to observe glowing, exploding balls of lightning and electric-blue rectangles hovering near her as she spoke.

The light show lasted for several months and attracted hundreds of believers, along with a cadre of scientific observers. Various explanations were offered for the lights, from a misidentification of the planet Venus to fault lines under the chapel where Jones preached. Scientists speculated that movements of the earth had stressed the bedrock, issuing gases that resulted in geomagnetic anomalies in the air above.

Lightning Reactions

Not everyone with an electric attraction finds the sensation enjoyable. Grace Charlesworth, a woman from the UK, had lived in a house for almost 40 years when, in 1968, she began receiving unexplainable shocks both indoors and out. The weird voltage was strong enough to spin Charlesworth's body in a complete circle, and at times, it would even make her head shake uncontrollably. The voltage was sometimes visible as sparks, and she could escape only by leaving her house or yard, as she was never bothered elsewhere.

Charlesworth blamed her problem on the noise from a compressor in a nearby factory, but fixing the compressor did not stop the mysterious electricity. One possible contributing factor was that the house had been hit by lightning five times.

Some people become so sensitive to electrical currents that they cannot even live in homes with any sort of wiring or appliances. An Irish woman named Margaret Cousins had to move to a cabin with no utilities in 1996 because her condition had become so painful. But two years later she had to move again after two cell phone towers were installed nearby and caused her pain to return.

The Science of Sleep

Insomnia is a serious issue in America—more than a third of all adults suffer from sleepless nights (and not because of active social lives, either). As anyone who's ever tossed and turned till dawn can tell you, shutting down isn't always simple. Here's a quick look at the science of sleep.

* * * *

The Faces of Insomnia

INSOMNIA COMES IN several variations. You can have a tough time falling asleep, problems staying asleep, or trouble sleeping late enough. You can also just have problems getting the kind of sleep that leaves you feeling refreshed. An occasional bad night isn't necessarily a problem; it's when the sandman stays away night after night that insomnia comes into play.

Anyone can suffer from insomnia, although certain groups of people are considered more susceptible: women, people over the age of 60, and those who are depressed. Stress, noise, and an environment that's too hot or too cold can make the condition worse, as well as the overuse of caffeine and alcohol.

People who suffer from insomnia on a long-term basis may have some physiological factor bringing those bags under their eyes every morning. Sleep apnea, narcolepsy, and restless legs can all keep you up, as can more serious conditions such as kidney disease or heart failure.

Finding a Fix

Plenty of pills are available for sleep-starved zombies, both over-the-counter and by prescription. The over-the-counter stuff won't usually do much for true insomniacs (though melatonin is often a viable option for some people), but medications such as Ambien and Lunesta can provide relief (and perhaps unpleasant side effects). Some doctors question whether pills are always needed and instead recommend working on learning to properly relax. Which, of course, would be far less daunting after a good night's sleep.

Gum Makes You Hungry

When hunger pangs strike, does chewing a piece of gum stave them off or stoke them? Scientists are still chewing on the answer, but prevailing research gives some clues.

* * * *

To Chew or Not to Chew?

DIETERS SOMETIMES SHUN gum because they fear it will exacerbate their feelings of deprivation and emptiness. They think—or feel—that chewing gum starts the gastric juices flowing by stimulating saliva in anticipation of some real food. When the juices find nothing to digest, it makes the person feel like devouring something.

Scientists have found evidence to the contrary, however. Stimulating saliva by chewing gum has not been shown to increase hunger. In fact, recent studies indicate that gum can *decrease* one's appetite. A study presented at the 2007 Annual Scientific Meeting of the Obesity Society found that chewing gum before an afternoon snack helped reduce hunger, diminish cravings, and promote fullness among people who were trying to limit their calorie intake.

That Adds Up!

The people who chewed gum before a snack consumed 25 fewer calories from that snack than the non-chewers. The study even touted the benefits of chewing gum for appetite control, saying that it is an easy, practical tool for weight management. If you think 25 calories is insignificant, here's a little weight-loss math: 25 calories times 7 equals 175 calories per week; 25 times 7 times 52 weeks equals 9,100 calories (or 2.6 pounds) per year.

Other research studies have shown that hunger and the desire to eat are significantly suppressed by chewing gum at one-, two-, and three-hour intervals after a meal.

If you're still convinced that chewing gum makes you hungry, consider whether boredom, habit, or stress are responsible for those cravings instead. Until science discovers otherwise, chewing gum is a tasty alternative to an "unnecessary" meal.

Why Do We Sweat?

What's the deal with our built-in sprinkler system?

❋ ❋ ❋ ❋

SWEAT ENABLES US to cool off when the exterior temperature rises (due to changes in the weather) or when our interior temperatures rise (due to exercise, anxiety, or illness). Sweat is one of the mechanisms that our bodies use to keep us at a steady—and healthy—98.6 degrees Fahrenheit.

Here are the basics: Humans have about 2.6 million sweat glands, but not all of these glands produce the same kind of sweat. Sweat has two distinct sources: eccrine and apocrine glands. Eccrine glands exist all over the body and are active from birth. They constantly release a salty, nearly odorless fluid onto the skin, though you probably only notice this sweat when it's really hot or you've been working out really hard.

Apocrine glands, on the other hand, are concentrated in the armpits, on the soles of the feet, in the palms of the hands, and in the groin. They become active during puberty. Yes, puberty and perspiration go hand in hand.

Apocrine glands don't secrete liquid directly onto the skin. Instead, each gland empties into a hair follicle. When a person is under emotional or physical stress, the tiny muscle around the follicle contracts, pushing the liquid onto the skin, where it becomes sweat. Apocrine glands carry lipids and proteins, as well as water and salt. When these substances mix with the sebaceous oils in the hair follicles and then meet the bacteria on the skin, well, that's when you begin to hold your nose.

But before you start thinking of eccrine as "good" sweat and apocrine as "bad," consider this little nugget of information: Apocrine sweat has been found to contain androsterone pheromones, those mysterious musky odors that are responsible for sexual arousal. So sweat can be sexy, too. Just don't take this as an excuse to wear unwashed gym socks on a date—a few pheromones go a long way.

To banish body odor, a little dab of deodorant should do. Deodorants are based on mildly acidic compounds that dry the skin before the odor starts. Antiperspirants, another popular option, actually block sweat with aluminum salts. Some people think that these salts may be unhealthy, but so far, clinical evidence has failed to connect them to any disease.

If you feel that you sweat too much, or too little, see your doctor. Excessive sweating, officially known as hyperhidrosis, and lack of sweat, called anhidrosis, are genuine medical conditions with serious complications. Fortunately, both are treatable.

For most of us, however, dealing with sweat is fairly simple: Take a shower and wear loose and absorbent clothing. For goodness sake, don't sweat about sweat!

Go Ahead—Try to Tickle Yourself

Your brain is expecting your attempts at self-tickling, so they won't work. When someone else tickles you, however, the contact is unexpected, and the shock contributes to the effect.

⁕ ⁕ ⁕ ⁕

WHEN THE NERVES of your skin register a touch, your brain responds differently depending on whether you're responsible for it. MRI scans show that three parts of the brain—the secondary somatosensory cortex, the anterior cingulated cortex, and the cerebellum—react strongly when the touch comes from an external source. Think of it like this: When you see a scary movie for the first time, you jump when the maniac suddenly appears and kills the high school kids as punishment for having teenage sex. The second time you see the movie, it isn't a surprise, so you don't jump. The same goes for tickling: It's the element of surprise that causes the giddy laughter of the ticklish.

Why do we laugh hysterically when other people tickle us? Scientists believe that it's an instinctual defense mechanism—an exaggerated version of the tingle that goes up your spine when an insect is crawling up your leg. This is your body's way of saying, "You may want to make sure whatever is touching you won't kill you." The laughter is a form of panic due to sensory overload.

If you're in desperate need of tickling but have no friends or family willing to help, you can invest in a tickling robot. People do respond to self-initiated remote-control tickling by a specialized robot that was developed by British scientists in 1998. There's a short delay between the command to tickle and the actual tickle, which is enough to make the contact seem like a surprise to the brain and induce fits of laughter.

When "He" Becomes "She," and Vice Versa

Have you ever met a married couple who looked so alike you could have sworn they were brother and sister? Well, scientific research has come up with a few explanations as to why.

* * * *

Compatibility Counts

FOR STARTERS, IT seems that we seek out mates who have features that are similar to our own. Recent studies suggest that we're attracted to those who look like us because they tend to have comparable personalities.

It's often said that women "marry their fathers." Research at the University of Pécs in Hungary supports this notion. Women tend to choose husbands who resemble their natural fathers—even if they're adopted. What science characterizes as "sexual imprinting," is actually known to occur in many animal species. Glenn Weisfeld, a human ethologist at Wayne State University in Detroit, says that there seems to be an advantage to selecting mates who are similar to ourselves: "Fortuitous genetic combinations" are retained in our offspring.

This doesn't give you carte blanche to marry your cute first cousin Betty. When it comes to mating, it's best to avoid people who are members of your family tree. But it doesn't hurt to pick a guy or gal who shares your dark features or toothy smile. Studies show that partners who are genetically similar to each other tend to have happier marriages.

Togetherness

It seems the longer couples stay together, the more their likenesses grow. A study by Robert Zajonc, a psychologist at the University of Michigan, found this to be the case—even among couples who didn't particularly look alike when they first got hitched. In Zajonc's study, people were presented with random photographs of men's and women's faces and asked to match up couples according to resemblance. Half of the photos were individual shots of couples that were taken when they were first married; the other half were individual shots of the same couples after twenty-five years of wedlock.

What do you know? People were able to match up husbands and wives far more often when looking at photographs of the couples when they were older than when they were younger. It seems that with time, the couples' similarities became much more discernible.

Why? Zajonc says husbands and wives start looking like each other because they spend decades sharing the same life experiences and emotions. Spouses often mimic the facial expressions of each other as a sign of empathy and closeness. Think of that the next time you and your spouse exchange smiles, sighs, or looks of contempt. Before you know it, you'll be sharing a life complete with matching facial sagging and wrinkle patterns! Hey, it's better than being told you look like your dog.

Are You Related to Genghis Khan?

Your DNA may carry the stuff you need to conquer the world.

* * * *

From Riches to Rags to Riches

GENGHIS KHAN WAS one of the first self-made men in history. He was born to a tribal chief in 1162, probably at Dadal Sum, in the Hentii region of what is now Mongolia. At age 9, Genghis was sent packing after a rival tribe poisoned his father. For three years, Genghis and the remainder of his family wandered the land living from hand to mouth.

Genghis was down, but not out. After convincing some of his tribesmen to follow him, he eventually became one of the most successful political and military leaders in history, uniting the nomadic Mongol tribes into a vast sphere of influence.

The Mongol Empire lasted from 1206 to 1368 and was the largest contiguous dominion in world history, stretching from the Caspian Sea to the Sea of Japan. At the empire's peak, it encompassed more than 700 tribes and cities.

A Uniter, Not a Divider

Genghis gave his people more than just land. He introduced a writing system that is still in use today, wrote the first laws to govern all Mongols, regulated hunting to make sure everybody had food, and created a judicial system that guaranteed fair trials. His determination to create unity swept old tribal rivalries aside and made everyone feel like a single people, the "Mongols."

Today, Genghis Khan is seen as one of the founding fathers of Mongolia. However, he is not so fondly remembered in Asia, the Middle East, and Europe, where he is regarded as a ruthless and bloodthirsty conqueror.

Who's Your Daddy?

It seems that Genghis was father of more than the Mongol nation. Recently, an international team of geneticists determined that one in every 200 men now living is a relative of the great Mongol ruler. More than 16 million men in central Asia have been identified as carrying the same Y chromosome as Genghis Khan.

A key reason is this: Genghis's sons and other male descendants had many children by many women; one son, Tushi, may have had 40 sons of his own, and one of Genghis's grandsons, Chinese dynastic ruler Kublai Khan, fathered 22 sons with recognized wives and an unknown number with the scores of women he kept as concubines.

Genetically speaking, Genghis continues to "live on" because the male chromosome is passed directly from father to son, with no change other than random mutations (which are typically insignificant). When geneticists identify those mutations, called "markers," they can chart the course of male descendants through centuries. Is the world large enough for 16 million personal empires? Time—and genetics—will reveal the answer.

Living in Fear

Approximately 20 percent of people in the United States have an intense, irrational fear of common things or experiences, such as spiders, heights, or confined spaces. Sometimes they fear something more unusual, such as pine trees or public bathrooms.

* * * *

Acrophobia is the fear of heights. This fear is often very specific. A person may be able to ski the Alps with no problem but be overcome with panic on a fifth-floor balcony.

Astraphobia, also known as tonitrophobia, brontophobia, or keraunophobia, is a paralyzing fear of thunder and lightning. As with other phobias, the reaction often causes a rapid heartbeat or labored breathing. People, and even pets, often seek shelter in confined spaces such as closets and basements.

Cacophobia is the fear of ugliness. Sufferers aren't just repulsed by unattractive people or things; they actually have intense panic attacks around them. When they see someone or something they consider ugly, they often turn away and flee.

Dendrophobia is the fear of trees. A child on a camping trip may be afraid there will be bears wandering among the pines. A fear of bears is not unusual, but the child may subsequently develop a paralyzing fear of pine trees. People with dendrophobia usually have strange stories about why they are terrified of a particular type of tree.

Friggatriskaidekaphobia is the fear of Friday the 13th. This fear is more typical among people who are from England, Poland, Germany, Bulgaria, or Portugal—countries in which the number 13 is traditionally deemed unlucky.

Nyctophobia is an irrational fear of nighttime or the dark. Rationally, an adult (the fear is common among children) may understand that there is nothing to be afraid of, but he or she may still experience heightened anxiety when the lights go out.

Trypanophobia is an exaggerated fear of injection with a hypodermic needle. This phobia has a history in genetic memory. Thousands of years ago, the people who avoided being stabbed in general were the most likely to survive.

Further Phobias

Arachibutyrophobia—a fear of peanut butter sticking to the roof of the mouth

Basophobia—fear of standing, walking, or falling over

Catagelophobia—fear of being ridiculed

Chorophobia—fear of dancing

Didaskaleinophobia—fear of going to school

Doxophobia—fear of expressing opinions or of receiving praise

Elurophobia—fear of cats

Ergophobia—fear of work

Gamophobia—fear of marriage

Hexakosioihexekontahexaphobia—fear of the number 666

Iatrophobia—fear of doctors or going to the doctor

Linonophobia—fear of string

Ochlophobia—fear of crowds or mobs

Paralipophobia—fear of neglecting duty or responsibility

Phengophobia—fear of daylight or sunshine

Soceraphobia—fear of parents-in-law

Xanthophobia—fear of the color yellow or the word "yellow"

✳ Chapter 8

Fringe Theories and Wild Speculations

The Philadelphia Experiment

In 1943, the Navy destroyer USS Eldridge *reportedly vanished, teleported from a dock in Pennsylvania to one in Virginia, and then rematerialized—all as part of a top-secret military experiment. Is there any fact to this fiction?*

✳ ✳ ✳ ✳

The Genesis of a Myth

THE STORY OF the Philadelphia Experiment began with the scribbled annotations of a crazed genius, Carlos Allende, who in 1956 read *The Case for the UFO*, by science enthusiast Morris K. Jessup. Allende wrote chaotic annotations in his copy of the book, claiming, among other things, to know the answers to all the scientific and mathematical questions that Jessup's book touched upon. Jessup's interests included the possible military applications of electromagnetism, antigravity, and Einstein's Unified Field Theory.

Allende wrote two letters to Jessup, warning him that the government had already put Einstein's ideas to dangerous use. According to Allende, at some unspecified date in October 1943, he was serving aboard a merchant ship when he witnessed a disturbing naval experiment. The USS *Eldridge* disappeared, teleported from Philadelphia, Pennsylvania, to Norfolk,

Virginia, and then reappeared in a matter of minutes. The men onboard the ship allegedly phased in and out of visibility or lost their minds and jumped overboard, and a few of them disappeared forever. This strange activity was part of an apparently successful military experiment to render ships invisible.

The Navy Gets Involved

Allende could not provide Jessup with any evidence for these claims, so Jessup stopped the correspondence. But in 1956, Jessup was summoned to Washington, D.C., by the Office of Naval Research, which had received Allende's annotated copy of Jessup's book and wanted to know about Allende's claims and written comments. Shortly thereafter, Varo Corporation, a private group that does research for the military, published the annotated book, along with the letters Allende had sent to Jessup. The Navy has consistently denied Allende's claims about teleporting ships, and the impetus for publishing Allende's annotations is unclear. Morris Jessup committed suicide in 1959, leading some to claim that the government had him murdered for knowing too much about the experiments.

The Fact in the Fiction

It is not certain when Allende's story was deemed the "Philadelphia Experiment," but over time, books and movies have touted it as such. The date of the ship's disappearance is usually cited as October 28, though Allende himself cannot verify the date nor identify any other witnesses. However, the inspiration behind Allende's claims is not a complete mystery.

In 1943, the Navy was in fact conducting experiments, some of which were surely top secret, and sometimes they involved research into the applications of some of Einstein's theories. The Navy had no idea how to make ships invisible, but it *did* want to make ships "invisible"—i.e., undetectable—to enemy magnetic torpedoes. Experiments such as these involved wrapping large cables around Navy vessels and pumping them with electricity in order to descramble their magnetic signatures.

An Underground Mystery: The Hollow Earth Theory

For centuries, people have believed the Earth is hollow. They claim that civilizations may live inside Earth's core or that it might be a landing base for alien spaceships. This sounds like fantasy, but believers point to startling evidence, including explorers' reports and modern photos taken from space.

* * * *

A Prize Inside?

HOLLOW EARTH BELIEVERS agree that our planet is a shell between 500 and 800 miles thick, and inside that shell is another world. It may be a gaseous realm, an alien outpost, or home to a utopian society.

Some believers add a spiritual spin. Calling the interior world Agartha or Shambhala, they use concepts from Eastern religions and point to ancient legends supporting these ideas.

Many Hollow Earth enthusiasts are certain that people from the outer and inner worlds can visit each other by traveling through openings in the outer shell. One such entrance is a hole in the ocean near the North Pole. A November 1968 photo by the ESSA-7 satellite showed a dark, circular area at the North Pole that was surrounded by ice fields.

Another hole supposedly exists in Antarctica. Some Hollow Earth enthusiasts say Hitler believed that Antarctica held the true opening to Earth's core. Leading Hollow Earth researchers such as Dennis Crenshaw suggest that President Roosevelt ordered the 1939 South Pole expedition to find the entrance before the Germans did.

The poles may not hold the only entrances to a world hidden beneath our feet. Jules Verne's novel *Journey to the Center of the Earth* supported another theory about passage between the worlds. In his story, there were many access points, including waterfalls and inactive volcanoes. Edgar Allan Poe and Edgar Rice Burroughs also wrote about worlds inside Earth. Their ideas were based on science as well as fantasy.

Scientists Take Note

Many scientists have taken the Hollow Earth theory seriously. One of the most noted was English astronomer Edmund Halley, of Halley's Comet fame. In 1692, he declared that our planet is hollow, and as evidence, he pointed to global shifts in Earth's magnetic fields, which frequently cause compass anomalies. According to Halley, those shifts could be explained by the movement of rotating worlds inside Earth. In addition, he claimed that the source of gravity—still debated in the 21st century—could be an interior world.

In Halley's opinion, Earth is made of three separate layers or shells, each rotating independently around a solid core. We live on the outer shell, but the inner worlds might be inhabited, too.

Halley also suggested that Earth's interior atmospheres are luminous. We supposedly see them as gas leaking out of Earth's fissures. At the poles, that gas creates the *aurora borealis*.

Scientists Look Deeper

Hollow Earth researchers claim that the groundwork for their theories was laid by some of the most notable scientific minds of the 17th and 18th centuries. Although their beliefs remain controversial and largely unsubstantiated, they are still widely discussed and have a network of enthusiasts.

Some researchers claim that Leonhard Euler (1707–1783), one of the greatest mathematicians of all time, believed that Earth's interior includes a glowing core that illuminates life for a well-developed civilization, much like the sun lights our world. Another mathematician, Sir John Leslie (1766–1832), suggested that Earth has a thin crust and also believed the interior cavity was filled with light.

In 1818, a popular lecturer named John Cleves Symmes, Jr., proposed an expedition to prove the Hollow Earth theory. He believed that he could sail to the North Pole, and upon reaching the opening to Earth's core, he could steer his ship over the lip of the entrance, which he believed resembled a waterfall. Then he would continue sailing on waters inside the planet. In 1822 and 1823, Symmes petitioned Congress to fund the expedition, but he was turned down. He died in 1829, and his gravestone in Hamilton, Ohio, is decorated with his model of the Hollow Earth.

Proof Gets Woolly and Weird

In 1846, a remarkably well-preserved—and long extinct—woolly mammoth was found frozen in Siberia. Most woolly mammoths died out about 12,000 years ago, so researchers were baffled by its pristine condition.

Hollow Earth enthusiasts say there is only one explanation: The mammoth lived inside Earth, where those beasts are not extinct. The beast had probably become lost, emerged into our world, and froze to death shortly before the 1846 discovery.

Eyewitnesses at the North Pole

Several respected scientists and explorers have visited the poles and returned with stories that suggest a hollow Earth.

At the start of the 20th century, Arctic explorers Dr. Frederick A. Cook and Rear Admiral Robert E. Peary sighted land—not just an icy wasteland—at the North Pole. Peary first described it as "the white summits of a distant land." A 1913 Arctic expedition also reported seeing "hills, valleys, and snow-capped peaks." All of these claims were dismissed as mirages but would later be echoed by the research of Admiral Richard E. Byrd, the first man to fly over the North Pole. Hollow Earth believers suggest that Byrd actually flew into the interior world and then out again, without realizing it. They cite Byrd's notes as evidence, as he describes his navigational instruments and compasses spinning out of control.

Unidentified Submerged Objects

Support for the Hollow Earth theory has also come from UFO enthusiasts. People who study UFOs have also been documenting USOs, or unidentified submerged objects. These mysterious vehicles have been spotted—mostly at sea—since the 19th century. USOs look like "flying saucers," but instead of vanishing into the skies, they plunge beneath the surface of the ocean. Some are luminous and fly upward from the sea at a fantastic speed . . . and without making a sound.

UFO enthusiasts believe that these spaceships are visiting worlds beneath the sea. Some are certain that these are actually underwater alien bases. Other UFO researchers think that the ocean conceals entries to a hollow Earth, where the aliens maintain outposts.

The Search Continues

Scientists have determined that the most likely location for a northern opening to Earth's interior is at 84.4 N Latitude, 141 E Longitude. It's a spot near Siberia, about 600 miles from the North Pole. Photos taken by *Apollo 8* in 1968 and *Apollo 16* in 1972 show dark, circular areas confirming the location.

Some scientists are studying seismic tomography, which uses natural and human-made explosions as well as earthquakes and other seismic waves to chart Earth's interior masses. So far, they have stated that Earth is comprised of three separate layers. And late 20th-century images supposedly a mountain range at Earth's core.

What may seem like fantasy from a Jules Verne novel continues to hold fascination for the fringe few. Hollow Earth societies around the world continue to look for proof of this centuries-old legend . . . and who knows what they might find?

Franz Mesmer Transfixes Europe

The Age of Enlightenment saw the explosion of new ideas. One of these was the possibility of tapping into a person's subconscious, causing them to enter a dreamlike state where they might find relief from various ailments, whether through actual effect or merely by the power of a hypnotist's suggestion. One early practitioner of this technique became so famous that his very name became synonymous with the ability to send his patients into a trance—the art of mesmerism.

* * * *

FRANZ ANTON MESMER was a late bloomer. Born in Germany in 1734, Mesmer had difficulty finding a direction in life. He first studied for the priesthood, then drifted into astronomy and law before finally graduating at age 32 from the University of Vienna with a degree in medicine. He set up practice in Vienna and married a well-to-do widow, becoming a doctor to the rich and famous and using his connections to cater to an upper-crust clientele. He lived comfortably on a Viennese estate and counted among his friends Wolfgang Amadeus Mozart, who wrote a piece for Mesmer to play on the glass harmonica, an instrument lately arrived from America.

At first, Mesmer's medical prescriptions were unremarkable; bleeding and purgatives were the order of the day, and Mesmer followed accepted medical convention. But Mesmer's attention was also drawn to the practice of using magnets to induce responses in patients, a technique much in vogue at the time. Mesmer experimented with magnets to some effect and came to believe that he was successfully manipulating tides, or energy flows, within the human body. He theorized that illness was caused by the disruption of these flows, and health could be restored by a practitioner who could put them back in order. He also decided that the magnets themselves were an unnecessary prop and that he was performing the manipulation of the tides himself, because of what he termed his animal magnetism—the word "animal" merely stemming from the Latin term for "breath" or "soul." He would stir the tides by sitting in a chair opposite a patient, knees touching, gazing unblinkingly into their eyes, making passes with his hands, and massaging the areas of complaint, often continuing the treatment for hours until the patient felt the magnetic flows moving inside their body.

Europe Becomes Mesmerized

Mesmer gained notoriety as a healer, his fame growing to the point where he was invited to give his opinion in other famous cases of the day. He investigated claims of unusual cures and traveled around Switzerland and Germany, holding demonstrations at which he was able to induce symptoms and their subsequent cures by merely pointing at people, much to the amazement of his audience. He also took on more challenging cases as a doctor, but a scandal involving his treatment of a blind piano player—he temporarily restored her sight, only to have her lose her audiences because the novelty of watching her play was now gone—caused Mesmer to decide that 1777 was an opportune year to move to Paris.

France would prove to be a fertile ground for Mesmer. He resumed seeing patients, while at the same time seeking approval from the scientific community of Paris for his techniques. The respect and acknowledgment he felt he deserved from his peers was never to come, but his popular reputation soared; Marie Antoinette herself wrote Mesmer and begged him to reconsider when he once announced that he intended to give up his practice.

His services were in such demand that he could no longer treat patients individually; he resorted to treating groups of patients with a device he called a baquet, a wooden tub bristling with iron rods around which patients would hold hands and collectively seek to manipulate their magnetic tides. Mesmer himself would stride back and forth through the incense-laden room, reaching out and tapping patients with a staff or finger. For a complete cure, Mesmer believed the patients needed to undergo a convulsive crisis—literally an experience wherein they would enter a trancelike state, shake and moan uncontrollably, and be carried to a special padded chamber until they had come back to their senses. The treatment proved particularly popular with women, who outnumbered men 8–1 as patients of Mesmer. This statistic did not go unnoticed by the monitors of public decency, who drew the obvious conclusion that something immoral was taking place, though they were unable to produce much more than innuendo in support of their accusations.

When I Snap My Fingers . . .

Unfortunately, Mesmer's incredible popularity also made him an easy target for detractors. Mesmerism became such a fad that the wealthy even set up baquets in their own homes. But, as with many trends, once over they are held up for popular ridicule. As a result, Mesmer saw his client base decline and even found himself mocked in popular theater.

Copycats emerged to the extent that in 1784, the king set up a commission—including representatives from both the Faculty of Medicine and the Royal Academy of Science—to investigate all claims of healing involving animal magnetism. Benjamin Franklin, in Paris as an ambassador at the time, was one of the investigators. In the end, the commission determined that any treatment benefits derived from Mesmerism were imagined. This rejection by the scientific community combined with the erosion of his medical practice drove Mesmer from Paris in 1785. He kept an understandably low profile after that, spending some time in Switzerland, where he wrote and kept in touch with a few patients. He died in 1815.

Mesmer's legacy remains unresolved. Some still view him as a charlatan of the first order. Others see in his techniques the foundation of modern hypnotherapy, which has become a well-recognized practice in modern psychiatry. Regardless, it is indisputable that Franz Anton Mesmer's personal animal magnetism continues to capture our imagination even today.

Eating After 8:00 P.M. Makes You Gain Weight?

It seems logical enough: You eat a huge cheesesteak and fries, hit the hay a few hours later, and lay there like a lump, burning off only negligible calories when you roll over or unconsciously pass wind. Logically, one would expect all of those calories to turn into fat by sunrise.

* * * *

But medical degrees aren't handed out to people who buy into a theory because it seems logical. Scientists with actual credentials have looked into this issue—specifically, in a study at Oregon Health and Science University in 2006—and proved once again that logic doesn't always equal medical fact.

To conduct their study, researchers fed a group of rhesus monkeys (whose DNA is a 93 percent match to that of humans) identical high-fat diets. They monitored the times that the monkeys ate and found a wide range of results. Some monkeys ate more than half of their daily calories at night, while others consumed as little as 6 percent after dark—the monkeys' eating habits were essentially in step with those of people. The monkeys all had the same weight gain, no matter when they ate. The study concluded that the timing of food intake isn't a factor in weight gain.

The issue, however, doesn't end there. Some doctors suggest that the habit of eating at night is linked to poorer dietary choices (ice cream sandwiches, anyone?) that can result in weight gain. For example, some people might be too busy to eat before 8:00 p.m. and end up hitting a fast-food drive-through at night. Or they might wait so long to have dinner that they become ravenous and eat enormous portions to compensate. Or at the end of a long day, they lose the will to stick to a strict diet and eat whatever they want.

The bottom line: Eat as late at night as you want—just don't stay at the table too long.

Curse of the (Polish) Mummy

In 1973, a group of research scientists entered the tomb of King Casimir IV, a member of the Jagiellon dynasty that once ruled throughout central Europe. Within weeks of entering the tomb, only two scientists remained alive.

* * * *

The Jagiellon Curse

INDIANA JONES DIDN'T have it easy, but as archaeologist work hazards go, there are actually worse fates than snake pits and big rolling boulders. For example, there are strains of mold fungi that eat your body from the inside out. This was the inauspicious fate of several scientists who opened a tomb that had been shut for centuries, thereby unleashing a powerful mummy's curse—or, more realistically and less fantastically—powerful microorganisms.

The tomb of King Casimir IV of Poland and his wife, Elizabeth of Habsburg, is located in the chapel of Wawel Castle in Krakow, Poland. Casimir served as king for more than 40 years in the 13th century. He left behind 13 children, many of whom went on to positions of great power. In 1973, Cardinal Wojtyla (who later went on to become Pope John Paul II) gave a group of scientists permission to open King Casimir's tomb and examine its contents. Within the tomb, the unlucky group found a heavily rotted wooden coffin—not so surprising, given the box had been decaying for nearly 500 years. However, within a few days, four of the twelve researchers were dead; six more died soon after.

Killer Fungi

While sensationalists blamed the tragedy on a mummy's curse, the scientific-minded questioned whether the sudden deaths were related to the icky molds, fungi, and parasites that would linger in a room that had been sealed off for centuries. This was precisely the suspicion of Dr. Boleslaw Smyk, one of the two surviving scientists. He set out to discover what exactly had killed his colleagues, and he came up with three species of fungi mold that had lingered in King Casimir's tomb: *Aspergillus flavus, Penicillim rubrum,* and *Penicillim rugulosum.*

Not a Mummy, but No Less Scary

These are not the kindest of specimens. *Aspergillus flavus* is toxic to the liver, while *Penicillim rubrum* causes, among a host of other afflictions, pulmonary emphysema. These toxins grow on decaying wood and lime mortar, both of which were in Casimir's tomb. The toxins remained in the tomb in the form of mold spores, which can survive for thousands of years in closed environments. It is likely the researchers breathed in the spores immediately upon entering the coffin, since the sudden flow of fresh air into a closed tomb would blow the spores about. Toxic spores that are inhaled in this fashion can lead to organ failure and death in a very short time.

It's therefore unsurprising that whisperings of a "mummy's curse" abound. The more famous legend came from the 1922 Egyptian excavation of Pharaoh Tutankhamun's tomb. Lord Carnarvon, one of the main financiers of the King Tut excavation, died a few months after he entered the fungi-laden tomb—the same fungi spores that were identified in King Casimir's tomb were also present in King Tut's. Stories of a mummy's curse followed, although it's unclear whether Carnarvon's death actually was related to his archaeological pursuits: Carnarvon had a cut on his cheek that became infected weeks after the excavation. He fell ill and eventually died of pneumonia and septicemia from the cut.

Whether or not Carnarvon died of natural causes, rumors of the supernatural took on a life of their own. After news of his death spread, fantastical stories grew regarding the grisly deaths of anyone who had entered King Tut's tomb. Today, even modern archaeologists are warned of their potential exposure to the dreaded Mummy's Curse.

Luna Ticks

A full moon holds mysterious attractions, prompting love at first sight, criminal malfeasance, and boosted birthrates. Does the gleaming globe really have magical powers, or is it just our state of mind?

* * * *

For centuries, there have been reports of abnormal human behavior under the whole of the moon. Full moons have been linked to fluctuating rates of birth, death, crime, suicide, mental illness, natural and spiritual disasters, accidents of every description, fertility, and all kinds of indiscriminate howling. People with too much spare cash and not enough common sense have been known to buy and sell stocks according to phases of the moon. The word *lunatic* was coined to describe irrational and maniacal individuals whose conduct is seemingly influenced by the moon; their desolate domicile is dubbed the "loony bin."

Don't Blame It on the Moon

So is there a scientific relationship between the moon and human behavior? In 1996, researchers examined more than 100 studies that looked into the effects of the moon—full or otherwise—on an assortment of everyday events and anomalies, including births and deaths, kidnappings and car-jackings, casino payouts and lottery paydays, aggression exhibited by athletes, assaults and assassinations, suicides and murders, traffic accidents and aircraft crashes.

Dr. Ivan Kelly, a professor of educational psychology and human behavior, found that the "phases of the moon accounted for no more than 3/100 of 1 percent of the variability in activities usually termed lunacy." This represents a percentage so close to zero that it can't be considered to have any theoretical, practical, or statistical interest. Because there was no significant correlation between the aforementioned occurrences and the periods and phases of the moon, it's safe to assume that the only moonshine that's causing trouble is the kind that's brewed in the Ozarks.

Can Coca-Cola Burn a Hole in Your Stomach?

The world's most famous soft drink has been the subject of seemingly countless urban legends.

* * * *

ONE LEGEND SUGGESTS that Coca-Cola can cause death from carbon dioxide poisoning, another says that it dissolves teeth, and still another persistent theory posits that it makes an effective spermicide.

The topic here is whether Coke can burn a hole in your stomach. The answer is, quite simply, no. Your stomach is designed to withstand punishment—it's the Rocky Balboa of internal organs—and it can handle a lot worse than what little old Coca-Cola throws at it.

Your stomach takes every culinary delight that you consume and prepares it for the body to use as fuel. It breaks down food using hydrochloric acid—a substance that, in its industrial form, is used to process steel and leather, make household cleaning products, and even aid in oil drilling in the North Sea. Since this acid is highly corrosive, a mucus is secreted to protect the stomach lining.

The strength of an acid is measured on a pH scale that ranges from zero to fourteen. A pH level of seven is considered neutral; any substance with a pH level of less than seven is acidic. Where does your stomach's hydrochloric acid fall on the pH scale? Its pH level is one, meaning that it is among the most potent acids in existence. Coca-Cola contains phosphoric acid, a substance with a pH level of about 2.5. Phosphoric acid, then, is less potent than what is already inside you. In other words, Coca-Cola isn't going to burn a hole in your stomach.

Still, there are some reasons to hesitate before you take the pause that refreshes. Coca-Cola contains the stimulant caffeine. (There is a caffeine-free Coca-Cola, but we're talking about the original version.) The stomach reacts to stimulants by creating more acid, which isn't an issue when the stomach is working well. But when the stomach contains ulcer-causing bacteria called *Helicobacter pylori*, the production of extra acid can exacerbate the problem. Further, people with gastroesophageal reflux disease (GERD) should avoid caffeinated drinks. And finally, phosphoric acid has been linked to osteoporosis.

But under ordinary circumstances, a big swig of Coca-Cola isn't going to harm your stomach, or any other part of your body. Sip and enjoy.

Eat Worms, Lose Weight!

No matter what anyone tells you, the only way to successfully lose weight is to eat less and exercise more. Yet this common-sense knowledge hasn't stopped millions from trying anything to make the road to weight loss a little smoother—including purposely ingesting parasites. Wouldn't it be easier to just go on a walk and skip dessert?

* * * *

Those Wacky Early 1900s

THERE WAS A time when cocaine was the cure for a sore throat and smoking was considered a healthy habit. So not many feathers were ruffled when ads showed up advertising a tapeworm pill for ladies looking to slim down. The ads, which first appeared between 1900 and 1920, claimed that by ingesting a pill containing tapeworm larvae, you could give a hungry worm a happy home and lose that pesky weight. You could eat all you wanted, content in the knowledge that your new friend would be eating up most of the calories you consumed, thus allowing you to lose weight without thinking twice about it.

No one can prove that the pills advertised back then actually contained worm larvae. The pills could've been placebos, and for the foolish folks who tried the diet fad, we can only hope that's what they were.

The Worm Is Back!

The weight loss via tapeworm idea died down for many years (obesity was not as much of an issue during the Great Depression and both World Wars), but talk of it resurfaced in the 1960s. Rumors that a new appetite suppressant candy introduced to the market contained worm eggs started getting around, though of course, this was entirely false.

After a remarkable weight loss of an estimated 65 pounds, acclaimed opera star Maria Callas endured heavy gossip that she had purposely acquired a tapeworm to do it. Though the singer indeed was diagnosed with a tapeworm, her doctor suggested it was due to her fondness for eating beef tartare. Other celebrities are rumored to have swallowed tapeworm pills to whittle down their figures, including model Claudia Schiffer, though this was never confirmed.

An Internet search these days reveals companies that advertise "sterile tapeworms" for a variety of medicinal uses (whether they're selling a real product or scamming the public is another article). The fine print is lengthy, however, as using tapeworms to treat any condition has not been approved by the USDA. To get your worms, you'll likely have to go to Mexico. These stowaways will get you in big trouble if you try to bring them back across the border.

Tapeworms: Not a Good Pet

A lot of time and attention is spent around the world trying to keep worms from getting into the human body via water, food, or skin. Simply put: Having a tapeworm is not a good thing. In the case of the fish tapeworm, especially, the essential vitamin B12 is sucked out of the host's body and depletes the vital ingredient for making red blood cells.

Adult tapeworms can grow up to 50 feet long and live up to 20 years. Depending on the worm, a host's symptoms range from epileptic seizures, diarrhea, nausea, fatigue, a swollen belly (oh, the irony), and even death. While it's likely a person with a worm will lose weight, they'll also suffer from malnutrition—B12 isn't the only nutrient eaten by the parasite. And tapeworm eggs are an inevitable byproduct of a tapeworm. The fish tapeworm can produce a million eggs in a single day, and the larvae tend to burrow out of the intestines and find homes elsewhere in the body, like the brain, for example. Worms also have the habit of popping out of various orifices without warning, too.

Still interested in tapeworms as a form of weight loss? Then perhaps it's your head, and not your pants size, that's the issue!

Hard as Nails

You may be soft at heart, but you want your nails to be as hard as . . . well, nails.

* * * *

GELATIN CAN TURN a liquid into a solid—think Jell-O after it sets. So perhaps it makes sense to think that ingesting it or soaking your fingertips in it would strengthen your nails and make them more resistant to chipping and cracking.

The logic may be flawed, but millions of people fell for a marketing scheme that connected gelatin with strong nails. And it's a tribute to that wildly successful advertising campaign that millions of people still believe it today. In 1890, Charles Knox developed gelatin, a product made from slaughterhouse waste. He sold the public on it by touting its nail-enhancing benefits. The animals (cows and pigs) used to produce gelatin had strong hooves, Knox reasoned, so eating their by-products would give people nails just as strong. Consumers fell for it, lock, livestock, and barrel.

But no matter how many people swear by it, there is no scientific proof that the kind of gelatin found in Jell-O hardens your nails. Gelatin is made from the skin, connective tissue, and bones of cattle and pigs, and so it is full of protein and collagen. The protein is not, however, in a form that's usable by humans. And unless your diet is deficient in protein, which is unlikely, eating more protein is not going to solve your nail problems. The best thing you can do for your nails is to pick up some petroleum jelly and use it to moisturize them.

Science vs. Séance

In which we examine the curious nature of ectoplasm and the story of famed medium and ectoplasm producer Margery Crandon's examination by a scientific committee.

⁂ ⁂ ⁂ ⁂

The Essence of Ectoplasm

TODAY MOST PEOPLE (of a certain age, perhaps) associate the term "ectoplasm" with the film *Ghostbusters* in which Bill Murray's character gets slimed by a slovenly ghost. To spiritualists of the late 18th and early 19th centuries, however, ectoplasm was an essential substance produced by mediums during a séance.

The ectoplasm was typically white and often luminescent, and it appeared to have the consistency of cheesecloth (which, skeptics wryly observed, was due to it actually being cheesecloth). While many believers regarded ectoplasm as a physical manifestation of the spiritual realm, to others it was part of the 19th-century spiritualism craze.

The first use of the word *ectoplasm* is credited to French scientist and 1913 Nobel Prize winner Charles Richet, who used it to describe the substance produced by a European medium during a séance in 1894. That a highly regarded scientist such as Richet should have a deep and abiding interest in spiritualism did not strike his contemporaries as a contradiction of interests. In fact, the idea that the spiritual world could transfer messages and substances into the world of the living was an appealing notion to many different Victorians. It was, after all, an era that bridged widespread Victorian beliefs in the metaphysical with the emerging fields of quantum theory and technological experimentation.

Enter Margery, Mistress of Ectoplasm

In 1924, Mina "Margery" Crandon, the wife of a Boston society surgeon, was on her way to becoming the world's most celebrated medium. With the alleged cooperation of Walter, her long-dead brother and spirit-world contact, Crandon had been able to levitate objects, manifest writing, and produce auditory emanations. She had traveled to Europe and submitted to scientific tests in Paris and London to prove her supernatural talents.

Upon Crandon's return to the United States, she perfected the art of producing copious amounts of ectoplasm in the form of glowing strands, hands, rods and, in one instance, the fully formed figure of a tiny girl. At different times the ectoplasm streamed from all of her orifices, including mouth, nose, ears, and vagina.

Put to the Test

In 1923, *Scientific American* magazine offered $2,500 to any medium that could conduct a successful séance while under professional scientific scrutiny. Several amateur mediums tried and were proven frauds. Finally, the magazine's associate editor, J. Malcolm Bird, convinced Crandon to sit for the committee.

The individuals who assembled to observe Crandon in Boston in the spring and summer of 1924 included Dr. William McDougall, professor of psychology at Harvard; Dr. Daniel Comstock, former Massachusetts Institute of Technology professor; professional magician and spiritualist skeptic Harry Houdini; Dr. Walter Prince, researcher at the American Society of Psychical Research; and amateur magician and author Hereward Carrington.

Although Crandon was able to provide many examples of her intimacy with the spirit world, including the production of ectoplasm, Houdini became convinced that Crandon was a fraud. He delivered a series of lectures denouncing her and accusing Bird of incompetence and collaboration.

Mired in mistrust, the committee members published their disparate findings separately in November 1924. The committee's inability to disprove Crandon's abilities, however, catapulted Margery Crandon into national prominence. Her career peaked several years later when she manifested her deceased brother's fingerprint in dental wax. Subsequent examination, however, showed that the fingerprint was that of her very much still-living dentist.

This episode, coupled with Bird's admission that Crandon's husband had asked him to collaborate in fooling the *Scientific American* committee, thoroughly debunked her abilities.

A Fiery Debate: Spontaneous Human Combustion

Proponents contend that the phenomenon—in which a person suddenly bursts into flames—is very real. Skeptics, however, are quick to explain it away.

* * * *

The Curious Case of Helen Conway

A PHOTO DOCUMENTS THE gruesome death of Helen Conway. Visible in the black-and-white image—taken in 1964 in Delaware County, Pennsylvania—is an oily smear that was her torso and, behind, an ashen specter of the upholstered bedroom chair she occupied. The picture's most haunting feature might be her legs, thin and ghostly pale, clearly intact and seemingly unscathed by whatever it was that consumed the rest of her.

What consumed her, say proponents of a theory that people can catch fire without an external source of ignition, was spontaneous human combustion. It's a classic case, believers assert: Conway was immolated by an intense, precisely localized source of heat that damaged little else in the room. Adding to the mystery, the investigating fire marshal said that it took just twenty-one minutes for her to burn away and that he could not identify an outside accelerant. If Conway's body ignited from within and burned so quickly she had no time to rise and seek help, hers wouldn't be the first or last death to fit the pattern of spontaneous human combustion.

The phenomenon was documented as early as 1763 by Frenchman Jonas Dupont in his collection of accounts, published as *De Incendis Corporis Humani Spontaneis*. Charles Dickens's 1852 novel *Bleak House* sensationalized the issue with the spontaneous-combustion death of a character named Krook. That humans have been reduced to ashes with little damage to their surroundings is not the stuff of fiction, however. Many documented cases exist. The question is, did these people combust spontaneously?

How Does It Happen?

Theories advancing the concept abound. Early hypotheses held that victims, such as Dickens's Krook, were likely alcoholics so besotted that their very flesh became flammable. Later conjecture blamed the influence of geomagnetism. A 1996 book by John Heymer, *The Entrancing Flame*, maintained emotional distress could lead to explosions of defective mitochondria. These outbursts cause cellular releases of hydrogen and oxygen and trigger crematory reactions in the body. That same year, Larry E. Arnold—publicity material calls him a parascientist—published *Ablaze! The Mysterious Fires of Spontaneous Human Combustion*. Arnold claimed sufferers were struck by a subatomic particle he had discovered and named the "pyrotron."

Perhaps somewhat more credible reasoning came out of Brooklyn, New York, where the eponymous founder of Robin Beach Engineers Associated (described as a scientific detective agency) linked the theory of spontaneous human combustion with proven instances of individuals whose biology caused them to retain intense concentrations of static electricity.

A Controversy Is, Er, Sparked

Skeptics are legion. They suspect that accounts are often embellished or important facts are ignored. That the unfortunate Helen Conway was overweight and a heavy smoker, for instance, likely played a key role in her demise.

Indeed, Conway's case is considered by some to be evidence of the wick effect, which might be today's most forensically respected explanation for spontaneous human combustion. It holds that an external source, such as a dropped cigarette, ignites bedding, clothing, or furnishings. This material acts like an absorbing wick, while the body's fat takes on the fueling role of candle wax. The burning fat liquefies, saturating the bedding, clothing, or furnishings, and keeps the heat localized.

The result is a long, slow immolation that burns away fatty tissues, organs, and associated bone, leaving leaner areas, such as legs, untouched. Experiments on pig carcasses show it can take five or more hours, with the body's water boiling off ahead of the spreading fire.

Under the wick theory, victims are likely to already be unconscious when the fire gets going. They're in closed spaces with little moving air, so the flames are allowed to smolder undisturbed, doing their work without disrupting the surroundings or alerting any passersby.

Nevertheless, even the wick effect theory, like all other explanations of spontaneous human combustion, has scientific weaknesses. The fact remains, according to the mainstream science community, that evidence of spontaneous human combustion is entirely circumstantial, and that not a single proven eyewitness account exists to substantiate anyone's claims of "Poof—the body just went up in flames!"

How Murphy Got His Own Law

Murphy's Law holds that if anything can go wrong, it will. Not surprisingly, the origin of the law involves a guy named Murphy.

* * * *

Project M3981

IN 1949, CAPTAIN Edward A. Murphy, an engineer at Edwards Air Force Base, was working on Project M3981. The objective was to determine the level of sudden deceleration a pilot could withstand in the event of a crash. It involved sending a dummy or a human subject (possibly also a dummy) on a high-speed sled ride that came to a sudden stop and measuring the effects. George E. Nichols, a civilian engineer with Northrop Aircraft, was the manager of the project. Nichols compiled a list of "laws" that presented themselves during the course of the team's work. For example, Nichols's Fourth Law is, "Avoid any action with an unacceptable outcome."

"If There Is Any Way to Do It Wrong . . ."

These sled runs were repeated at increasing speeds, often with Dr. John Paul Stapp, an officer, in the passenger seat. After one otherwise-flawless run, Murphy discovered that one of his technicians had miswired the sled's transducer, so no data had been recorded. Cursing his subordinate, Murphy remarked, "If there is any way to do it wrong, he'll find it." Nichols added this gem to his list, dubbing it Murphy's Law. Not long after, Stapp endured a run that subjected him to forty Gs of force during deceleration without injury. Prior to Project M3981, the established acceptable standard had been 18 Gs, so the achievement merited a news conference. Asked how the project had maintained such an impeccable safety record, Stapp cited the team's belief in Murphy's Law and its efforts to circumvent it. The law, which had been revised to its current language before the conference, was quoted in a variety of aerospace articles and advertisements, and gradually found its way into pop culture.

Beyond Murphy

It should be noted that "laws" very similar to Murphy's—buttered bread always lands face down; anything that can go wrong at sea will go wrong, at some point—had been in circulation prior to Project M3981. But even if Edward Murphy didn't break new ground when he cursed a technician in 1949, it's his "law" we quote when things go wrong, and that's all right.

Don't Blame Beer for Your Beer Belly

You see them at the corner bar, and maybe even in the recliner in your TV room: gargantuan guts, worn by dudes who have devoted countless hours to quaffing. But is it really the beer that's responsible for those whopping waistlines?

* * * *

THE CZECH REPUBLIC is the world champion in per capita beer consumption, and in 2007 a team of Czech researchers studied 2,000 male and female beer drinkers. They found no direct link between obesity and the amount of beer one consumes. That's not to say beer can't make you fat—it can. Each glass of your favorite malt contains plenty of gravity-enhancing calories. But beer on its own is apparently not the culprit. Swiss physiologists in 1992 determined that alcohol in the bloodstream can slow the body's ability to burn fat by about 30 percent. That means high-fat foods become even more potent when combined with alcohol. And it doesn't take a scientific survey to determine that in a room full of beer drinkers, a plate of celery and carrots will go unmolested while bowls of potato chips disappear faster than you can say "myocardial infarction."

Further, results of a study in Italy suggest that some men are genetically predisposed to develop a big midsection, regardless of what they eat and drink. So beer can play a role in the development and maintenance of a beer belly—but it's not required.

✳ Chapter 9

Animal Oddites

Why Don't Animals Need Glasses?

Humans are so quick to jump to conclusions. Just because you've never sat next to an orangutan at the optometrist's office or seen a cat adjust its contact lenses, you assume that animals don't need corrective eyewear?

✳ ✳ ✳ ✳

ANIMALS DO DEVELOP myopia (nearsightedness), though it seems less widespread in nature than among humans. For one thing, nearsighted animals—especially carnivores—would have an extremely difficult time hunting in the wild. As dictated by the rules of natural selection, animals carrying the myopia gene would die out and, thus, wouldn't pass on the defective gene.

For years, nearsightedness was thought to be mainly hereditary, but relatively recent studies have shown that other factors may also contribute to the development of myopia. Some researchers have suggested that myopia is rare in illiterate societies and that it increases as societies become more educated. This doesn't mean that education causes nearsightedness, but some scientists have speculated that reading and other "close work" can play a role in the development of the condition. In accordance with this theory, a study of the Inuit in Barrow, Alaska, conducted in the 1960s found that myopia was much more common in younger people than in older generations, perhaps coinciding with the introduction of schooling and mass literacy in Inuit culture that had recently occurred. But schooling was just one component of a larger shift—from the harsh, traditional lifestyle of hunting and fishing at the edge of the world to a more modern, Western lifestyle. Some scientists believe that the increase of myopia was actually due to other changes that went along with this shift, such as the switch from eating primarily fish and seal meat to a more Western diet. This diet is heavier on processed grains, which, some experts believe, can have a bad influence on eye development.

And this brings us back to animals. Your beloved Fido subsists on ready-made kibble that's heavy on processed grains, but its ancient ancestors ate raw flesh. If this switch to processed grains might have a negative effect on the eyesight of humans, why not in animals, too?

Unfortunately, there's not much we can do for a nearsighted animal. Corrective lenses are impractical, glasses would fall off, and laser surgery is just too darn expensive. Sorry, Fido!

How Do Carrier Pigeons Know Where to Go?

No family vacation would be complete without at least one episode of Dad grimly staring straight ahead, gripping the steering wheel, and declaring that he is not lost as Mom insists on stopping for directions.

* * * *

MEANWHILE, THE KIDS are tired, night is falling, and nobody's eaten anything except a handful of Cheetos for the past six hours. But Dad is not lost. He will not stop.

It's well known that men believe they have some sort of innate directional ability—and why not? If a creature as dull and dim-witted as a carrier pigeon can find its way home without any maps or directions from gas-station attendants, a healthy human male should certainly be able to do the same.

Little does Dad know that the carrier pigeon has a secret weapon. It's called magnetite, and its recent discovery in the beaks of carrier pigeons may help solve the centuries-old mystery of just how carrier pigeons know their way home.

Since the fifth century BC, when they were used for communication between Syria and Persia, carrier pigeons have been prized for their ability to find their way home, sometimes over distances of more than five hundred miles. In World War I and World War II, Allied forces made heavy use of carrier pigeons, sending messages with them from base to base to avoid having their radio signals intercepted or if the terrain did not allow for a clear signal. In fact, several carrier pigeons were honored with war medals.

For a long time, there was no solid evidence to explain how these birds were able to find their way anywhere, despite theories that ranged from an uncanny astronomical sense to a heightened olfactory ability to an exceptional sense of hearing. Recently, though, scientists made an important discovery: bits of magnetic crystal, called magnetite, embedded in the beaks of carrier pigeons. This has led some researchers to believe that carrier pigeons have magneto reception—the ability to detect changes in the earth's magnetic fields—which is a sort of built-in compass that guides these birds to their destinations.

Scientists verified the important role of magnetite through a study that examined the effects of magnetic fields on the birds' homing ability. When the scientists blocked the birds' magnetic ability by attaching small magnets to their beaks, the pigeons' ability to orient themselves plummeted by almost 50 percent. There was no report, however, on whether this handicap stopped male pigeons from plunging blindly forward. We'd guess not.

Cockroaches: Nuke-Proof, up to a Point

We've all heard that cockroaches would be the only creatures to survive a nuclear war. But unless being exceptionally gross is a prerequisite for withstanding such an event, are cockroaches really that resilient?

⁕ ⁕ ⁕ ⁕

COCKROACHES ARE INDEED that resilient. For one thing, they've spent millions of years surviving every calamity the earth could throw at them. Fossil records indicate that the cockroach is at least three hundred million years old. That means cockroaches survived unscathed whatever event wiped out the dinosaurs, be it an ice age or a giant meteor's collision with Earth.

The cockroach's chief advantage—at least where nuclear annihilation is concerned—is the amount of radiation it can safely absorb. During the Cold War, a number of researchers performed tests on how much radiation various organisms could withstand before dying. Humans, as you might imagine, tapped out fairly early. Five hundred Radiation Absorbed Doses (or rads, the accepted measurement for radiation exposure) are fatal to humans. Cockroaches, on the other hand, scored exceptionally well, withstanding up to 6,400 rads.

Such hardiness doesn't mean that cockroaches will be the sole rulers of the planet if nuclear war breaks out. The parasitoid wasp can take more than 100,000 rads and still sting the heck out of you. Some forms of bacteria can shrug off more than one million rads and keep doing whatever it is that bacteria like to do. Clearly, the cockroach would have neighbors.

Not all cockroaches would survive, however—definitely not the ones that lived within two miles of the explosion's ground zero. Regardless of the amount of radiation a creature could withstand, the intense heat from the detonation would liquefy it. Still, the entire cockroach race wouldn't be living at or near ground zero—so, yes, at least some would likely survive.

Forget the Swiss—the Mouse Wants a Dorito

Who moved your cheese? Chances are, if you are a mouse, you won't care all that much. But relocating your peanut butter, corn chips, or chocolate? That's another matter entirely.

* * * *

Earth-Shaking News

We've all seen the cute cartoons of mice nibbling away on a big, luscious hunk of cheese. In 2006, however, Dr. David Holmes, an animal behaviorist at Britain's Manchester Metropolitan University, stunned the world when he announced: No, mice really don't like cheese.

"Mice Hate Cheese!" the venerable *Manchester Guardian* declared. Other news sources mourned the break-up of the old mice-cheese love team. Not surprisingly, Holmes and his colleagues were bombarded with messages from irate cooks, telling them that mice certainly do eat cheese, along with fried chicken, salami sandwiches, and anything else they can get their thieving paws on, including the plastic coating that insulates copper wires. (Mice have been to blame for more than one short circuit in the kitchen.)

To defuse the tense situation, Holmes explained that his research was intended to identify those foods preferred by a wide range of animals under optimal conditions. Yes, mice eat many things, he stated, but they evolved as vegetarians. That means their ideal meal consists of grains, nuts, seeds, beans, fruits, and other substances high in carbohydrates and sugar—which explains the little rodents' predilection for chocolate.

The Origins of a Myth

Where did the myth that mice love cheese come from? No one knows, exactly. One thought is that mice were known to be stowaways on ships, hiding themselves in the holes of Swiss cheese. When sailors found mice in the cheese, they assumed it was because mice loved cheese.

Before refrigeration, cheese was stored in the pantry. Because making cheese is a labor-intensive process and uses a lot of precious milk, people were probably angrier then usual when they discovered that they were sharing their cheddar with furry invaders. Or perhaps the old folktale about the city mouse that impressed its country cousin with a gourmet spread of cosmopolitan treats created the impression that mice like rich, fatty foods. In reality, researchers discovered that high concentrations of fat can give mice indigestion, which puts cheese fairly far down on their list of preferred snacks. So you can keep your Brie, Stilton, and Camembert. If you really want to catch a mouse, try a dab of peanut butter, a piece of potato, or a few raisins—the chocolate-covered kind should do nicely.

Forever Slim: Finally, Something to Admire about Insects

It's not surprising that you don't see any obese insects ambling around your yard. If you existed on a diet of leaves, garbage, and rotting corpses, would you overeat? But as it turns out, insects can't get fat even if they want to—their bodies won't allow it.

⁕ ⁕ ⁕ ⁕

The Skinny on Exoskeletons

INSECTS (AS WELL as other arthropods, such as spiders, scorpions, and crustaceans) have exoskeletons—rigid outer body parts that are made of chitin and other material—instead of internal skeletons like humans have. The hard stuff is all on the outside, while the fat and other squishy stuff is all on the inside. The only way for an insect to get bigger is to molt, which involves forming a new exoskeleton underneath the old one and casting off the old material.

Some species start off as smaller versions of full-grown adults and go through progressive molts until they reach their full sizes. Others start off in larval stages, grow steadily, and then enter pupal stages so that they can metamorphose into adults. (For example, a caterpillar forms a cocoon and turns into a butterfly or moth.) If an insect eats a lot while it's still growing, it will simply molt sooner rather than get chunky.

Were an insect to overeat after reaching full maturity, the fat wouldn't have anywhere to go because the exoskeleton is rigid. The results would be catastrophic. Researchers learned this by severing the stomach nerves of flies, so that the flies couldn't sense that they had had their fill. The flies kept feeding until they burst open.

The Perfect Metabolism

Bugs have an innate sense of exactly how much sustenance they need. There's evidence that insects adapt their metabolisms over multiple generations, depending on how much food is in their environments. A study published in 2006 showed that diamondback moth caterpillars that lived in carbohydrate-rich environments going back eight generations could load up on more carbs without adding fat than could caterpillars that had evolved in a more carbohydrate-poor environment over the same time period.

The study could be a sign that other animals, including humans, will evolve metabolic adaptations based on the food in their environments. This doesn't help us much today, but if we start pounding Whoppers, nachos, and Big Macs now, perhaps our descendants will be able to scarf them down without gaining a single pound.

The Narwhal: Mother Nature's 35 Million-Year-Old Joke

Perhaps if Herman Melville had made Moby Dick a narwhal instead of a boring old white whale, we'd all have a little more love for this bizarre-looking (but lovable) aquatic animal.

⁕ ⁕ ⁕ ⁕

- A typical narwhal averages somewhere between 11.5 and 16.4 feet, and weighs in at around 3,500 pounds.
- Narwhals swim upside down! Researchers are still trying to figure out exactly why narwhals spend 80 percent of their time inverted, but they think it's because the animals send sonar signals underwater to detect prey. Swimming upside down may direct the sonar beam downwards, where lunch is likely to be most abundant.
- Narwhal enemies include polar bears and killer whales, but human poachers pose the biggest threat.
- The most distinguishing characteristic of a narwhal is its long, unicornlike tusk. The tusk points slightly downward, which is another reason narwhals may swim on their backs—while looking for food on the sea floor, they don't want to bust their horns on a rock.
- Over 10 million tiny nerve endings are found on the surface of a narwhal tusk, making it an essential tool for sensory perception.
- Narwhals lack a dorsal fin, which is unusual in underwater creatures of their kind. Scientists believe the narwhal evolved without a dorsal fin as an adaptation to navigate beneath ice-covered waters.

* Newborn narwhals are called "calves." Calves are weaned after a year or so, and then they move on to a regular diet of fish, squid, and shrimp.
* If you see a group of narwhals above water, rubbing their tusks together, they're "tusking." This activity helps the narwhals clean their tusks—kind of like when humans brush their teeth.

Tugging on a Loose Thread

Spider silk is not stronger than steel. But it has plenty of interesting potential uses.

❋ ❋ ❋ ❋

Spinning Yarns

THIS MYTH REALLY has legs—eight legs, in fact. (Sorry to all the arachnophobes who've just closed this book and thrown it across the room in disgust.) Why is that? Well, there are many possible factors. Something appeals to us on a basic level in the idea that a totally natural product like spider silk could be stronger than steel. It feels like a David and Goliath story, although not a David and Goliath birdeater spider (*Theraphosa blondi*) story.

Spiders are also one of the creatures that seems most alien to humans, from their strange-seeming number of legs to their array of differently sized, opaque, black eyeballs. Severe fears that qualify as clinical phobias are more studied, but researchers find that up to 75% of people are casually afraid of spiders, with an emphasis on women over men. There are competing explanations for the gender gap, but the overall fear of spiders comes from their unpredictability and seeming strangeness. Very few people are harmed by spiders, and most spider bites are hardly worse than a mosquito bite.

Think of the related myth that an average person "eats" eight spiders per year while sleeping. There's no rational reason to believe it! Why would a spider want to crawl into the moist, windy cave of your snoring open mouth? There isn't even anything for a spider to eat in there, hopefully. But our shared pessimistic imagination around spiders seems to have no limits.

Many Threads of Study

Spider silk is definitely strong, and scientists do study its strength. But other qualities about spider silk are more interesting from a scientific and technical standpoint. Manmade strong and light materials like Kevlar are made with chemicals that can be very costly, and these processes may create caustic or otherwise dangerous waste products. Spider silk is made by the same general family of body processes that make human hair or nails, and we don't generate caustic byproducts. Proteins like spider silk are generated in living cells and have no byproducts other than water.

Steel and other strong alloys need to be superheated in almost sterile settings in order to perform their best as manufactured products. This requires a great deal of energy, both to generate the heat in the first place and to absorb and neutralize the heat after it has been used. Alloys require mined raw materials in addition to simple, abundant elements like carbon, but spider silk uses only the abundant elements to make remarkable results at room temperature.

These factors give scientists different ways to approach and play with spider silk. They can study the chemical makeup of the strands and mimic this using synthetic fibers. They can try to create biological factories to string out their own strong natural proteins. They can even experiment with combinations of abundant elements to make new kinds of strands altogether.

Immortality in the Ocean

By repeating its life cycle, a certain ocean creature can reverse the aging process.

⁕ ⁕ ⁕ ⁕

A Handy Talent

IN THE 1990S, scientists noticed something peculiar about *Turritopsis nutricula*, a certain hydrozoan related to the jellyfish: It seemed to be able to live forever. The tiny creature, seen bobbing around blue Caribbean oceans, is capable of returning to its juvenile polyp stage, even after reaching sexual maturity and mating.

The hydrozoan evolved this ability through a cell development process called "transdifferentiation." Certain other animals can do this in order to regenerate an appendage—say, an arm or a leg—but this creature is the only one capable of regenerating its whole body. First, it turns into a bloblike shape, which then develops into a polyp colony. *Discover* magazine described this process as similar to a "butterfly turning back into a caterpillar."

Actually, the creatures seem to only revert to this early stage when in trouble, (e.g., they are starving or are in danger of getting picked on by other sea creatures). Crazily enough, they can repeat this process over and over again, whenever it is necessary. In this way, despite the ongoing threat of hunger and bullies, they do not have to die.

Panama or Bust

More recently, scientists have noticed that the *Turritopsis* are traveling great distances. Although they originated in the Caribbean, certain types of the hydrozoan have been turning up in the waters near Spain, Italy, Japan, and Panama. They're being extremely resourceful in finding a ride across the sea: According to *Discover*, "Researchers believe the creatures are criss-crossing the oceans by hitchhiking in the ballast tanks of large ships." In any case, one thing's for sure: the *Turritopsis*—with its ability to turn back the clock—is bound to become the envy of everyone in Hollywood.

How Do Cats Always Find Their Way Home?

You can count on two things from your local television news during sweeps week: a story about a household appliance that is a death trap and another about a cat that was lost but somehow trekked thirty miles through a forest, across a river, and over an eight-lane highway to find its way home.

⁕ ⁕ ⁕ ⁕

YOU THINK, "NO, I don't think my electric mixer is going to give me cancer, but, oh, that kitty . . ."

What's the deal with felines? How do they always seem to be able to make it home, regardless of how far away home might be? No one knows for sure, but researchers have their theories. One study speculates that cats use the position of the sun as a navigational aid. Another posits that cats have a sort of built-in compass; this is based on magnetic particles that scientists have discovered on the "wrists" of their paws. While these are merely hypotheses, scientists know that cats have an advanced ability to store mental maps of their environments.

Exhibit A is Sooty, one of the felines chronicled on the PBS program *Extraordinary Cats*. Sooty traveled more than a hundred miles in England to return to his original home after his family moved. Sooty's feat, however, was nothing compared to that of Ninja, another cat featured on the program. A year after disappearing following his family's move, Ninja showed up at his old house, 850 miles away in a different state; he went from Utah back to Washington.

But there are limits to what a cat can do—that's why odysseys of felines like Sooty and Ninja are extraordinary. In other words, the odds aren't good that Snowball will reach your loving arms in Boston if you leave her in Pittsburgh.

Alex Was No Birdbrain

Alex was an African Grey parrot who could talk, count, and follow orders from his trainers. Alex died in 2007, but the knowledge he lent to science about animal cognition ensures his legacy.

* * * *

The Avian Learning Experiment

There was once a time when Dr. Irene Pepperberg was not particularly interested in parrots. Sure, she had a parakeet as a child, but that was about it. In 1973, she was working toward a PhD in theoretical chemistry when she saw a television documentary about animal communication and intelligence. Inspired, Pepperberg continued working toward her doctorate while studying birds on the side, in an attempt to understand how animals think.

Pepperberg decided to test animal intelligence by working closely with a single animal from a young age. She picked an African Grey parrot because they live a long time (upwards of 70 years), they are known for their intelligence, and they possess the anatomy required to imitate human sounds and syllables. Pepperberg went to a pet shop and chose a one-year-old parrot. She named him Alex, an acronym for her project, the Avian Learning Experiment.

A Nutty Way of Learning

The scientific community questioned whether animals learn through simple operant conditioning, or whether some animals are capable of a more nuanced associative learning. The former involves simple input and output. If you want to teach a bird how to say "hello," you teach them through repetition—

a person's entrance will coincide with a positive stimulus, but only if the word "hello" is also spoken. Previous researchers had failed to teach parrots a high vocabulary by training them in this manner.

Pepperberg suspected the problem was that researchers assumed birds are incapable of associative learning. In the wild, an associative learning scenario might go something like this: A parrot encounters a large variety of berries. Over time, the parrot learns that some berries are like or different than other berries, and should be eaten at different times of the year and in different quantities. This leads to mental categorization. From this comes *representation,* or the ability to represent objects within these categories through social interaction. One bird pecking another bird's head might mean "macadamia nuts." This is linked to functionality, so that interaction might mean, "let's go find macadamia nuts." This implies the ability to have abstract thoughts about desire, intention, and future activities.

We may never know how birds think—the abstract thoughts and representations in avian brains may work in ways we can never fully conceive. But one path to understanding is to teach them *our* mode of representational communication: words. How parrots use these words might indicate whether they have abstract thought.

Alex's Abilities

Scientists commonly believe that intelligence and learning is rooted in social interaction, so Pepperberg taught Alex in a social manner. Alex competed with his trainers for rewards, and learned words in their social context so that the word was not disjointed from its meaning. Alex learned quickly; soon he knew over 150 words and could accurately label objects according to their color, shape, and material. He also answered questions about relative size and quantity. According to reports, when he was tired of testing, he said, "I'm going to go away." If a trainer got annoyed, Alex replied, "I'm sorry."

Some argue that Alex was not really capable of representational thought. When he says, "I want nut," he might not know what the words "want" and "nut" mean. However, many instances suggested he understood what the words communicated. For example, Alex was once shown an apple, for which he did not have a word. He did, however, know the words "banana" and "cherry." Alex spontaneously said "banerry," which suggests he understood the words represented categories and items. When he was taught a new color, he immediately labeled objects accurately according to his new word. Unfortunately, Pepperberg's experiment was cut short by Alex's early death; another parrot, Griffin, is his successor.

Elephant Graveyard

Do dying elephants actually separate themselves from their herd to meet their maker among the bones of their predecessors?

* * * *

JUST AS SEARCHING for the Holy Grail was a popular pastime for crusading medieval knights, 19th-century adventurers felt the call to seek out a mythical elephant graveyard. According to legend, when elephants sense their impending deaths, they leave their herds and travel to a barren, bone-filled wasteland. Although explorers have spent centuries searching for proof of these elephant ossuaries, not one has ever been found, and the elephant graveyard has been relegated to the realm of metaphor and legend.

Elephants Never Forget

Unlike most mammals, elephants have a special relationship with their dead. Researchers from the United Kingdom and Kenya have revealed that elephants show marked emotion—from actual crying to profound agitation—when they encounter the remains of other elephants, particularly the skulls and tusks. They treat the bones with unusual tenderness and will cradle and carry them for long periods of time and over great distances. When they come across the bones of other animals, they show no interest whatsoever. Not only can elephants distinguish the bones of other elephants from those of rhinoceroses or buffalo, but they also appear to recognize the bones of elephants they were once familiar with. An elephant graveyard, though a good way to ensure that surviving elephants wouldn't be upset by walking among their dead every day, does not fit with the elephants' seeming sentiment toward their ancestors.

Honor Your Elders

The biggest argument against an elephant burial ground can be found in elephants' treatment of their elders. An elephant would not want to separate itself from the comfort and protection of its herd during illness or infirmity, nor would a herd allow such behavior. Elephants accord great respect to older members of a herd, turning to them as guiding leaders. They usually refuse to leave sick or dying older elephants alone, even if it means risking their own health and safety.

But What about the Bones?

Although there is no foundation for the idea that the elephant graveyard is a preordained site that animals voluntarily enter, the legend likely began as a way to explain the occasional discovery of large groupings of elephant carcasses. These have been found near water sources, where older and sickly elephants live and die in close proximity. Elephants are also quite susceptible to fatal malnutrition, which progresses quickly from extreme lethargy to death. When an entire herd is wiped out by drought or disease, the remaining bones are often found en masse at the herd's final watering hole.

There are other explanations for large collections of elephant bones. Pits of quicksand or bogs can trap a number of elephants; flash floods often wash all debris (not just elephant bones) from the valley floor into a common area; and poachers have been known to slay entire herds of elephants for their ivory, leaving the carcasses behind.

In parts of East Africa, however, groups of elephant corpses are thought to be the work of the *mazuku*, the Swahili word for "evil wind." Scientists have found volcanic vents in the earth's crust that emit carbon monoxide and other toxic gases. The noxious air released from these vents is forceful enough to blow out a candle's flame, and the remains of small mammals and birds are frequently found nearby. Although these vents have not proved to be powerful enough to kill groups of elephants, tales of the *mazuku* persist.

The Term Trudges On

Although no longer considered a destination for elephants, the elephant graveyard still exists as a geologic term and as a figure of speech that refers to a repository of useless or outdated items. Given how prominent the legend remains in popular culture, it will be a long time before the elephant graveyard joins other such myths in a burial ground of its own.

The Cat Toss

Cats are curious creatures: Many people believe that a dropped kitty will right itself and land safely on it feet, only to step away aloof and unaffected.

* * * *

A BELGIAN LEGEND HAS it that in AD 962, Baldwin III, Count of Ypres, threw several cats from a tower. It must have been a slow news year, because the residents of Ypres named the last day of their annual town fair "Cat Wednesday" and commemorated it by having the village jester throw live cats from a belfry tower—a height of almost 230 feet. But there's no need to call PETA: The last time live cats were used for this ceremony was in 1817, and since then stuffed animals have been thrown in their place.

As cruel as this custom was, it is unclear whether the cat toss was meant to kill cats or to demonstrate their resilience. After that last live toss in 1817, the village record keeper wrote the following: "In spite of the height of the fall, the animal ran off quickly so that it might never be caught again in a similar ceremony." How could the cat have survived such a tumble?

Twist and Meow

Cats have an uncanny knack for righting themselves in midair. Even if a cat starts falling head first, it almost always hits the ground on its paws. The people of Ypres weren't the only ones amazed and amused by this feline feat. In 1894, French physiologist Etienne-Jules Marey decided to get to the bottom of the mechanics of cat-righting by taking a series of rapid photographs of a cat in midfall. Marey held a cat upside down by its paws and then dropped it several feet onto a cushion.

The resulting 60 sequential photos demonstrated that as the cat fell, it initiated a complex maneuver, rotating the front of its body clockwise and then the rear part counterclockwise. This motion conserved energy and prevented the cat from spinning in the air. It then pulled in its legs, reversed the twist again, and extended its legs slightly to land with minimal impact.

High-Rise Syndrome

The story gets even more interesting. In 1987, two New York City veterinarians examined 132 cases involving cats that had fallen out of the windows of high-rise buildings (the average fall was five and a half stories). Ninety percent of the cats survived, though some sustained serious injuries. When the vets analyzed the data, they found that, predictably, the cats suffered progressively greater injuries as the height from which they fell increased. But this pattern continued only up to seven stories; above that, the farther the cat fell, the greater chance it had of surviving relatively unharmed.

The researchers named this peculiar phenomenon High-Rise Syndrome and explained it this way: A cat that fell about five stories reached its terminal velocity—that is, maximum downward speed—of 60 miles per hour. If it fell any distance beyond that, it had the time not only to right itself in midair but also to relax and spread itself out to slow down its fall, much like a flying squirrel or a parachute.

Nature's Nerds: The World's Brainiest Animals

It's notoriously difficult to gauge intelligence, both in humans and animals. Comparing animal IQs is especially tricky, since different species may be wired in completely different ways. But when you look broadly at problem-solving and learning ability, several animal brainiacs do stand out from the crowd.

* * * *

Great Apes Scientists generally agree that after humans, the smartest animals are our closest relatives: chimpanzees, gorillas, orangutans, and bonobos (close cousins to the common chimpanzee). All of the great apes can solve puzzles, communicate using sign language and keyboards, and use tools. Chimpanzees even make their own sharpened spears for hunting bush babies, and orangutans can craft hats and roofs out of leaves. One bonobo named Kanzi has developed the language skills of a three-year-old child—and with very little training. Using a computer system, Kanzi can "speak" around 250 words and can understand 3,000 more.

Dolphins and Whales Dolphins are right up there with apes on the intelligence scale. They come up with clever solutions to complex problems, follow detailed instructions, and learn new information quickly—even by watching television. They also seem to talk to each other, though we don't understand their language. Scientists believe some species use individual "names"—a unique whistle to represent an individual—and that they even refer to other dolphins in "conversation" with each other. Researchers have also observed dolphins using tools. Bottlenose dolphins off the coast of Australia will slip their snouts into sponges to protect themselves from stinging animals and abrasion while foraging for food on the ocean floor. Marine biologists believe whales exhibit similar intelligence levels as well as rich emotional lives.

Elephants In addition to their famous long memories, elephants appear to establish deep relationships, form detailed mental maps of where their herd members are, and communicate extensively over long distances through low-frequency noises. They also make simple tools, fashioning fans from branches to shoo away flies. Researchers have observed that elephants in a Kenyan national park can even distinguish between local tribes based on smell and clothing. The elephants are fine with one tribe but wary of the other, and for good reason: That tribe sometimes spears elephants.

Monkeys They're not as smart as apes, but monkeys are no intellectual slouches. For example, macaque monkeys can understand basic math and will come up with specific cooing noises to refer to individual objects. Scientists have also trained them to learn new skills by imitating human actions, including using tools to accomplish specific tasks. They have a knack for politics, too, expertly establishing and navigating complex monkey societies.

Dogs If you're looking for animal brilliance, you might find it right next to you on the couch. Dogs are good at learning tricks, and they also demonstrate incredible problem-solving abilities, an understanding of basic arithmetic, and mastery of navigating complex social relationships. A 2009 study found that the average dog can learn 165 words, which is on par with a two-year-old child. And dogs in the top 20 percent of intelligence can learn 250 words. Border collies are generally considered the smartest breed, followed by poodles and German shepherds. One border collie, named Rico, actually knew the names of 200 different toys and objects. When his owners asked for a toy by name, he would go to the next room and retrieve it for them.

The Purpose of Purring

Felines are forever mysterious to mere humans. One of our favorite ponderings is the act of purring.

* * * *

Don't cats purr because they're content? When cats purr, people are happy. But cats aren't always happy when they purr. There are actually several reasons why cats purr, and happiness is only one of them. Purring begins at birth and is a vital form of communication between mother and kitten. The kitten purrs to let its mother know it's getting enough milk, and the mother cat purrs back to reassure her kitten.

What's the significance of purring? Cats purr throughout their lives and often at times you wouldn't expect. Cats purr when they are frightened, ill, or injured, and they even purr while giving birth. Animal behaviorists believe that cats purr under stressful conditions to comfort themselves and to signal their feelings to other cats. A frightened cat may purr to indicate that it is being submissive or non-threatening, and an aggressive cat may purr to let other cats know that it will not attack. Some cats purr even when they're dying. Interestingly, domestic cats aren't the only felines that purr. Some of the big cats—lions, cougars, and cheetahs—also exhibit this endearing behavior. But it's much more soothing—not to mention safer—to stroke a purring pet cat than the king of the jungle.

The Exact Pace of a Snail

The word "slow" hardly begins to cover it. These animals make all others look like Speedy Gonzales. Next to the snail, tortoises look like hares, and hares look like bolts of furry brown lightning.

❋ ❋ ❋ ❋

ALL OF THIS snail talk brings to mind a bad joke: What did the snail riding on the tortoise's back say? *Whee!* But enough of the jokes—you really want to know just how fast a snail travels.

Garden snails have a top speed of about 0.03 mile per hour, according to *The World Almanac and Book of Facts*. However, snails observed in a championship race in London took the 13-inch course at a much slower rate—presumably because snails lack ambition when it comes to competition. To really get a snail moving, one would have to make the snail think its life was in jeopardy. Maybe the racing snails' owners should be hovering behind the starting line wearing feathered wings and pointed beaks, cawing instead of cheering.

The current record holder of the London race, the Guinness Gastropod Championship, is a snail named Archie, who made the trek in two minutes, 20 seconds in 1995. This calculates to 0.0053 mile per hour. At that rate, a snail might cover a yard in 6.4 minutes. If he kept going, he might make a mile in a little less than eight days.

In the time it takes you to watch a movie, your pet snail might travel about 56 feet. You could watch a complete trilogy, and your snail might not even make it out of the house. Put your pet snail on the ground and forget about him—he'll be right around where you left him when you get back. So long as no one steps on him, that is.

Hummingbirds

When early Spanish explorers first encountered hummingbirds in the New World, they called them joyas voladoras*—or "flying jewels." But the hummingbird is more than just beautiful: Its physical capabilities put the toughest human being to shame.*

⁕ ⁕ ⁕ ⁕

⁕ The ruby-throated hummingbird—the only hummingbird species east of Mississippi—migrates at least 2,000 miles from its breeding grounds to its wintering grounds. On the way, it crosses the Gulf of Mexico—that's 500 miles without rest. Not bad for a creature that weighs just an eighth of an ounce and is barely three inches long.

⁕ A hovering hummingbird has an energy output per unit weight about ten times that of a person running nine miles per hour. If a person were to do the same amount of work per unit weight, he or she would expend 40 horsepower.

⁕ A man's daily energy output is about 3,500 calories. If one were to recalculate the daily energy output of a hummingbird—eating, hovering, flying, perching, and sleeping—for a 170-pound man, it would total about 155,000 calories.

⁕ The ruby-throated hummingbird can increase its weight by 50 percent—all of it fat—just before its winter migration. This provides extra fuel for the long, nonstop flight across the Gulf of Mexico. In comparison, a 170-pound man would have to pack on enough fat to increase his weight to 255 pounds in just a few weeks.

⁕ The wing muscles of a hummingbird account for 25 to 30 percent of its total body weight, making it well adapted to flight. However, the hummingbird has poorly developed feet and cannot walk.

* Due to their small body size and lack of insulation, hummingbirds lose body heat rapidly. To meet their energy demands, they enter torpor (a state similar to hibernation), during which they lower their metabolic rate by about 95 percent. During torpor, the hummingbird drops its body temperature by 30° F to 40° F, and it lowers its heart rate from more than 1,200 beats per minute to as few as 50.
* Hummingbirds have the highest metabolic rate of any animal on Earth. To provide energy for flying, they must consume up to three times their body weight in food each day.
* Unlike other birds, a hummingbird can rotate its wings in a circle. It can also hover in one spot; fly up, down, sideways, and even upside down (for short distances); and is the only bird that can fly backward.
* The smallest bird on Earth is the bee hummingbird (*Calypte helenae*), native to Cuba. With a length of only two inches, the bee hummingbird can comfortably perch on the eraser of a pencil.
* The most common types of hummingbirds include the Allen's, Anna's, berylline, black-chinned, blue-throated, broad-billed, broad-tailed, buff-bellied, Costa's, Lucifer, Magnificent, ruby-throated, Rufous, violet-crowned, and white-eared.
* Like bees, hummingbirds carry pollen from one plant to another while they are feeding, thus playing an important role in plant pollination. Each bird can visit between 1,000 and 2,000 blossoms every day.
* There are about 330 different species of hummingbirds. Most of them live and remain in Central and South America, never venturing any farther north. Only 16 species of hummingbirds actually breed in North America.

The Secrets of Hibernation

They binge, then go comatose. Does this sound like something that's happened to you after you polished off a quart of Häagen-Dazs? Can you imagine that food coma lasting all winter?

* * * *

Preparation Is the Key

ANIMALS THAT HIBERNATE have triggers that warn them to glut themselves for the winter ahead. As the days get shorter and colder, the critters' internal clocks—which mark time through fluctuations of hormones, neurotransmitters, and amino acids—tell them to fill up and shut down. Bingeing is important; if these creatures don't build up enough fat, they won't survive. The fat that they store for hibernation is brown (rather than white, like human body fat) and collects near the brain, heart, and lungs.

Animals have a number of reasons for hibernating. Cold-blooded creatures such as snakes and turtles adjust their body temperatures according to the weather; in winter, their blood runs so cold that many of their bodily functions essentially stop. Warm-blooded rodents can more easily survive the extreme chills of winter, but they have a different problem: finding food. They most likely developed their ability to hibernate as a way of surviving winter's dearth of munchies.

After an animal has heeded the biological call to pig out, its metabolism starts to slow down. As it hibernates, some bodily functions—digestion, the immune system—shut down altogether. Its heartbeat slows to ten or fewer beats per minute, and its senses stop registering sounds and smells. The animal's body consumes much less fuel than normal—its metabolism can be as low as 1 percent of its normal rate. The stored fat, then, is enough to satisfy the minimal demands of the animal's body, provided the creature found enough to eat in the fall and is otherwise healthy.

Waking Up

It can take hours or even days for the animal's body temperature to rise back to normal after it awakens from hibernation. But time is of the essence—the beast desperately needs water, and thirst drives it out of its nest. However, the animal is groggy and slow of foot—it walks like a drunk—so it can be easy prey if it doesn't hydrate quickly.

Which animals hibernate? Small ones, mostly—cold-blooded and warm-blooded critters alike. The first category includes snakes, lizards, frogs, and tortoises; the second includes dormice, hedgehogs, skunks, and bats.

But what about the bear, the animal that is most closely associated with hibernation? Here's a shocker: Bears don't hibernate. They slow down, sleep a lot, and lose weight during winter, but they don't truly hibernate. So if you're ever taking a peaceful nature walk on a sunny winter morn, beware. A bear might be out there.

Toad-ally High

Bored with all of the traditional ways of getting high: marijuana, cocaine, and ecstasy? Looking for a new disgusting and unhygienic way to tune out for a while? Look no further than the banks of the Colorado River, dude.

* * * *

HOPPING ALONG THE river's shores in southern Arizona, California, and northern New Mexico, the *Bufo alvarius* (also called the "Cane Toad" and "Colorado River Toad") would normally be in danger of being the main course for a wolf or Gila monster. That is, it would if it weren't for a highly toxic venom that this carnivorous toad produces whenever it gets agitated: the same venom that can get you high as a kite if properly ingested.

The toad's venom is a concentrated chemical called bufotenine that also happens to contain the powerful hallucinogen 5-MeO-DMT (or 5-methoxy-dimethyltryptamine). Ingested directly from the toad's skin in toxic doses (such as licking its skin), bufotenine is powerful enough to kill dogs and other small animals. However, when ingested in other ways—such as smoking the toad's venom—the toxic bufotenine burns off, leaving only the 5-MeO-DMT chemicals. Those can produce an intense, albeit, short-lived rush that has been described as 100 times more powerful than LSD or magic mushrooms, even if it takes a lot more work to get it.

As one of the few animals that excrete 5-MeO-DMT, *Bufo alvarius* are leathery, greenish-gray or brown critters that can grow up to seven inches long. They have four large glands that are located above the ear membranes and where their hind legs meet their bodies. Toad-smokers first milk the venom from the amphibian by rubbing its glands, which causes it to excrete the bufotenine. Then they catch the milky white liquid in a glass dish or other container. After the bufotenine has evaporated into a crystalline substance, it is collected using a razor blade or other sharp instrument and put in a glass-smoking pipe, and then lit and inhaled. Sounds, uh, fun!

Sea-Monkey See, Sea-Monkey Do

In the past 50 years, millions of people have ordered Sea-Monkeys. Alas, these people were disappointed to find their microscopic pets don't wear tiaras or even live very long. Here's a look at the PR machine that turned a homely crustacean into a generation-spanning fad.

* * * *

The Man Behind the Monkey

IN 1957, MAIL-ORDER marketer Harold von Braunhut had already given the world Invisible Goldfish and X-Ray Spex. Upon encountering brine shrimp, he saw potential for his next great venture, which he called "Instant Life." Brine shrimp, or *artemia salina,* are the perfect pet for someone who doesn't have a lot of space and doesn't mind if their pet has zero personality. Fully grown, they are less than an inch long. The official website says the name stems from their monkey-like "funny behavior and long tail," but any simian resemblance is pretty questionable. What appealed to von Braunhut was their cost-effectiveness: Brine shrimp eggs can exist out of water for years, dormant and seemingly lifeless. But just add water and the eggs hatch, making them perfect for warehouse storage and mail-order shipping.

A Shrimp by Any Other Name . . .

Alas, initial sales were lackluster. In 1962, von Braunhut renamed his product "Sea-Monkeys," marketing them through colorful ads in comic books that depicted them as a family of smiling, playful merfolk. Despite fine-print warnings that these were not accurate representations, people were smitten. Von Braunhut began selling other products that people could buy to show their Sea-Monkeys affection, such as special desserts and aphrodisiac elixirs. Soon, competing toy companies began carrying Sea-Monkey accessories, including racetracks, ski lodges, and elaborately themed aquariums. Sea-Monkey ads are still ubiquitous. The first generations of Sea-Monkey owners may have since wised up, but each year another generation begs to order them.

Canines Are Anti-Perspirant

Do dogs "sweat" through their mouths by panting?

✻ ✻ ✻ ✻

NO, THEY DON'T. This myth just fundamentally misunderstands the way sweating works. Humans, who are relatively hairless, have a large and smooth surface area from which moisture can evaporate. The evaporation itself is the cooling mechanism, and this action doesn't translate to the small, wet surface area inside your mouth. Try it: Open your mouth and imitate a panting dog. Odds are good that the only change you feel is that your mouth is more dry, and the air being exhaled from your lungs is warm, not cool. Then run water over your forearm and blow gently to evaporate the water. You'll feel cooler or even cold almost right away. When dogs pant, the goal is to bring fresh air into their lungs and cool their bodies from the inside out, not to cool the actual surface area inside of their mouths. It's apples and oranges—both refrigerated, of course.

Sponge Party

Deep-sea sponges + fiber optics = undersea rave?

* * * *

Deep-Sea Darkness

NO ONE LIKES to be left in the dark—including giant sea sponges. Yet the bottom of the ocean, where these sponges are found, is a very dark place indeed. To deal with this problem, some massive deep-sea sponges have evolved fiber optic exoskeletons by which they illuminate their surroundings.

These aren't sponges of the squishy, ideal-for-washing-dishes variety; they're sturdy sponges that produce reinforced glass tubes known as *spicules*. Although sponges are some of the world's most primitive creatures, these spicules are astoundingly intricate, growing up to about three feet in length and rivaling some of humanity's spiffier architectural designs.

Light Show

How does a sponge grow such an intricate exoskeleton? First, it spins silica from the ocean into microscopic bits of glass. It then adheres together these thin pieces of glass to form little fibers. The sponge arranges these fibers into a complex lattice, which it reinforces with a kind of glass cement. According to a report on NPR's *All Things Considered*, this lattice design resembles techniques used for building skyscrapers, such as the Swiss

Tower in London or the Eiffel Tower in Paris. And just like these manufactured structures, the sponges light up.

The undersea light show occurs because the spicules behave like fiber-optic rods, enabling the sponge to transport light throughout its tissue. Many tiny life-forms live inside the sponge—such as green algae and glass shrimp—because these organisms need its light to survive. According to scientists, the spicules conduct light in such a way that they're more sophisticated than many human-made fiber-optic cables. Plus, they don't require the high temperatures needed for manufactured fibers. So if you ever find yourself on the ocean floor and in need of a light source, look for the sponges with spicules. It's where the little shrimp like to party!

Laika—First Casualty of the Space Race

The first living being to be launched into orbit was a three-year-old mongrel named Laika, an unassuming stray who became an unwilling pawn in the space race. Unfortunately, contingency plans to bring the dog back to Earth were never developed, and Laika also became the first animal to die in space.

* * * *

THE SPACE RACE was launched on October 4, 1957, when the Soviet Union successfully sent Sputnik 1 into orbit. And with that, the mad dash to be the first country to get a man into space began.

Russian Premier Nikita Khrushchev ordered members of his space program to launch a second spacecraft into orbit on the 40th anniversary of the Russian Revolution. The primary mission of Sputnik 2 was to deliver a living passenger into orbit. The spacecraft—which was designed and built in just four short weeks—blasted into space on November 3, 1957.

Initially, three dogs were trained for the Sputnik 2 test flight. Laika, a 13-pound mixed-breed rescued from a Moscow shelter, was ultimately chosen for the historic mission. Though the Soviets affectionately nicknamed the dog Little Curly, the Western press referred to her by a more derogatory nickname—Mutt-nik.

3, 2, 1 . . . Blast Off!

Sputnik 29s capsule was equipped with a complete life-support system, padded walls, and a harness. During liftoff, Laika's respiration rocketed to four times its prelaunch rate, and her heartbeat tripled. Sensors indicated the cabin's temperature soared to 104 degrees.

The Soviets waited until after the spacecraft was launched to reveal that Laika would never return to Earth. The mission had been so rushed there hadn't been time to plan for the dog's return. She had only enough food and oxygen to last her about ten days.

The Soviets issued conflicting information regarding Laika's cause of death. Initial reports indicated that Laika survived four days in orbit—evidence that a living being could tolerate space. Her eventual death was attributed to either oxygen deprivation or euthanization via poisoned dog food—depending on the source. Laika's remains were destroyed when Sputnik 2 fell out of orbit and burned up reentering the earth's atmosphere.

Russians named postage stamps, cigarettes, and chocolate in Laika's honor, treating the dog as a national hero. In the meantime, there was outrage in the West over what many perceived to be the cruel and inhumane treatment of the animal.

The Aftermath

In October 2002, a Moscow scientist revealed that Laika had actually died hours into the mission due to overheating and stress. In March 2005, a patch of soil on Mars was unofficially named in Laika's honor. Dozens of other animals—including monkeys, frogs, mice, worms, spiders, fish, and rats—have since successfully made the journey into space.

Guano Guano Everywhere

Bird and bat droppings were once a significant part of the world's economy.

✻ ✻ ✻ ✻

In our mysterious world, people's "nonsense detectors" operate in a donut shape. We tend to believe mild-sounding myths and rumors because they can seem too small and harmless for anyone to bother lying or exaggerating about. And we tend to believe *the most outrageous* things we hear because they're too strange not to be true. In the middle are a lot of things we disbelieve more or less correctly.

Guano, the local and industry term for bat and bird droppings, fall into the category of too ludicrous to disbelieve. The enormous value of such a gross-sounding product lies in its chemical makeup, and to explain why, we need to briefly dive into some agricultural backstory.

Why Do We Farm?

Imagine the first person, or the first several people who were all in different areas, deciding to try to plant a seed in order to grow food. (This thought exercise works with almost everything we eat. Imagine the first person to eat the wedge of old milk that had solidified and begun to smell a little like feet. Imagine the first person to bring an oyster into the house and say, "Nah, I don't think we need to cook it first.")

This event likely took place in the Fertile Crescent, the collective term for an area that includes much of the Arabian Peninsula, the Nile Delta, and parts of the Mediterranean coast. About 12,000 years ago, these ancient and ingenious people dropped wild barley and other seeds into the ground. Maybe it happened by accident in a sort of Isaac Newton apple-on-the-head moment of serendipity and someone noticed that there were tiny plants growing out of some discarded old food. But it's hard to imagine food being thrown away by hunter-gatherer groups.

They also began keeping animals around, eventually domesticating them. These weren't the first domesticated animals *period*, because dogs have been found buried alongside their people at sites that predate the advent of agriculture. But having meat and dairy animals and a more reliable source of food crops changed the course of human history.

The ancient (but way less ancient) Egyptians and others experimented with soil, ash, manure, and other substances they thought might improve their crops. But it wasn't until the explosion of industrial farming that the commercial fertilizer industry took off. In the meantime, ideas like crop rotation also helped to boost productivity on farms around the world.

The Big 3

In the 1800s, scientists began experimenting in earnest to find why some substances helped plants to grow better while others did not. They used the scientific method, designing experiments and using both control and experimental groups to measure their results accurately.

One by one, these pioneering scientists uncovered three ingredients that are vital to fertilizing plants: nitrogen, phosphate, and potassium. Plants leech these three nutrients from soil in larger quantities than any other nutrients, and scientists realized that stripped nutrients were one reason why crop rotation had improved outcomes for farmers. Different crops require different nutrient profiles from the soil.

The Bat (and Bird) Signal

The element phosphorus doesn't show up anywhere in nature in its pure form. It readily combines with other elements to form phosphates, meaning compounds with one phosphorus atom and four oxygen atoms. In turn, phosphates form crystals with other molecules, and these are known as phosphate minerals. Turquoise is an example of a phosphate mineral.

One major source of phosphates is bird and bat droppings, or guano. Over many thousands of years, these droppings form phosphate rock. But even as fresh droppings, guano contains all three major fertilizer elements. Sea birds make more valuable guano but bat guano is also valued. By the time imperialist European explorers discovered the value of guano, it had been used for a millennium by the Quechua people whose language gave it its name.

Nauru'd Awakening

At the turn of the nineteenth century, the first foreign explorer found his way to the tiny island nation of Nauru. He gave it an English name but that didn't last long, and the island has been Nauru for two hundred years. The Nauruan people were and remain a mix of descendants from other Pacific island groups, and beginning in the 1800s, some Europeans joined the genealogy.

This tiny island has just eight square miles of surface area, most of which was taken up by huge deposits of phosphate rock: many thousands of years worth of being the local hangout for Pacific sea birds. Nauru and a couple of fellow tiny Pacific islands almost singlehandedly floated the phosphate, and therefore fertilizer, industry for decades.

But brutal, aggressive strip mining has left Nauru bare. The phosphate is gone, leaving barren, spiky expanses of rock behind. Runoff from the mines has killed much of the surrounding marine life. Nauru has almost no arable land and can't support a seaport. The same birds who once thrived in their own paradise aren't interested in returning to a bleak expanse of rock. Mining operations abandoned Nauru to its fate and the island became independent in the 1960s.

The Industrial Aftermath

Companies began researching and manufacturing fertilizers long before Nauru was strip mined, but the discovery of these rich natural sources of readymade fertilizer derailed the synthetic fertilizer industry for quite a while. Eventually, when the Pacific island supply ran out, these fertilizer companies reclaimed their market.

Today, the chemical makeup of guano is still the same basic profile used by major fertilizers. Scientists have worked to craft seeds that use fewer nutrients and fertilizers that have added benefits beyond the Big 3 nutrients. But as with the guano boom of the early twentieth century, several gigantic companies dominate the world fertilizer market.

Agribusiness and agricultural research science also contributed to the development of new strains and varieties of world staple crops, making them part of a portfolio of cutting-edge technologies to help every development level of farming to produce better, stronger, more productive crops.

What Ever Happened to Nauru?

Newly independent Nauru became a democracy without political parties, probably because there are only about ten thousand people in the entire nation. It has scrambled to attract any kind of industry, and mistakes have been made in that scramble. In 2003, *This American Life* ran a long feature by Jack Hitt about Nauru's brush with tax-haven greatness and other attempts to create a new economic base, including a failed musical about Leonardo da Vinci.

Global climate change will affect low-lying islands like Nauru more than almost anyplace else, because a matter of just several feet of water could cover almost all the inhabitable land on Nauru and others like Kiribati. Imperialist powers destroyed most of their island, and climate change will likely destroy the rest—all because of the saga of guano.

Flying Fish

They fly through the air with the greatest of ease—to escape predators such as swordfish, tuna, and dolphins.

⁕ ⁕ ⁕ ⁕

- A flying fish can glide through the air for 10 to 20 feet—farther if it has a decent tailwind. It holds its outstretched pectoral fins steady and "sails" through the air, using much the same action as a flying squirrel.
- If you'd like to catch sight of a member of the *Exocoetidae* family, you'll have to travel to the Atlantic, Indian, or Pacific oceans, where there are more than 50 species of flying fish.
- Whiskers are not an indication that a flying fish is up there in years; actually, it's the opposite. Young flying fish have long whiskers that sprout from the bottom jaw. These whiskers are often longer than the fish itself and will disappear by early adulthood.
- To get a taste of flying fish, stop by a Japanese restaurant. The eggs of flying fish, called *tobiko,* are often used in sushi.
- Attached to the eggs of the Atlantic flying fish are long, adhesive filaments that enable the eggs to affix to clumps of floating seaweed or debris for the gestation period. Without these filaments, the eggs (which are more dense than water) would sink.
- Beyond their useful pectoral fins, all flying fish have unevenly forked tails, with the lower lobe longer than the upper lobe. Many species also have enlarged pelvic fins and are known as four-winged flying fish.
- Flying fish can soar high enough that sailors often find them on the decks of their ships.

Freaky Facts: Bedbugs

Good night, sleep tight. Don't let the bedbugs bite! For decades, bedbugs were all but extinct in many places. But recently, increased travel and less-toxic modern pesticides have allowed a resurgence of these creepy crawlies.

* * * *

* When full of blood, bedbugs can swell as large as three times their normal size.
* It takes a bedbug five minutes to drink its fill of blood.
* A female bedbug can lay up to five eggs per day and 200–500 eggs in her lifetime.
* When bedbugs bite, proteins in their saliva prevent the wound from closing.
* Bedbugs can consume as much as six times their weight in blood at one feeding.
* Normally, a bedbug is brown. After eating, however, its body appears dark red.
* Bedbugs are nocturnal—they become active when humans are sleeping.
* Bedbugs only eat once every seven to ten days.
* Other names for bedbugs include "mahogany flats," "redcoats," and "chinches."
* Bedbugs molt five times before reaching adulthood.
* Bedbugs migrated from Europe to the United States during the 1600s.
* Contrary to popular belief, the presence of bedbugs does not indicate a dirty house.

Rats! They're . . . Everywhere?

Urban folklore would have us believe that we're never farther than a few feet from a rat. The thought is enough to make your skin crawl, but are there really that many rats around us?

* * * *

Why are rats so reviled? Not everyone hates rats. The Jainist religious sect of south Asia honors all life—even that of a rat. People love their pet rats. Your weird friend (you know which one) even likes wild rats. Of course, that might change when he contracts bubonic plague.

Beyond that, the only creatures that like rats are rat predators. The same animal lover who would feed and care for stray dogs would likely pay an exterminator good money to dispose of stray rats. Wild rats carry diseases and filth, eat unspeakable things, are very difficult to kill, can grow to an enormous size, and run in large packs that could overwhelm any human. To the majority of people, rats are the stuff of nightmares, as Winston Smith finds in George Orwell's *1984*.

So just how close to us are they? Do you spend a lot of time in the alleys of a large city's slums, cuddled up next to a garbage can, drinking in the smell of fermenting everything? Do you often seek shelter in the cool, tranquil comfort of your favorite sewer pipe? Do you spend idle afternoons sifting through that landfill you love so well in search of rare treasures?

If you answered yes to any of these questions, you've been in real close proximity to rats. Then again, if these are your preferred haunts, you know that already. For your safety, you might want to peruse the February 13, 1998, *Morbidity and Mortality Weekly Report*. In it you'll find an article that describes a couple of bona-fide cases of rat-bite fever.

Or we could just avoid those places. Unfortunately, rats aren't picky about where they live. Some estimates say there is one rat per U.S. resident, which is hard to confirm because vermin don't answer the Census. But suppose there are that many. They'd be concentrated in big cities where there's also a lot of food and places to hide and scurry. Any poorly secured storage of food, either fresh or discarded, will attract them. People living in immaculate suburban mansions probably don't have a homey woodpile or trash heap in their backyards, but that's not to say rats don't roam idyllic family neighborhoods.

What do we do if we encounter a rat? The number of rats reported to health officials in the suburbs has been steadily increasing, and it's now common for municipalities to offer some sort of "rat patrol" to assist citizens in the fight against these critters. Have you heard the horror stories about rats that get into residential toilets after swimming up through sewer pipes? We'd like to say that those are also urban folklore—but they're not.

Fallacies & Facts: Snakes

How well do you know your serpents?

✻ ✻ ✻ ✻

Fallacy: You can identify poisonous snakes by their triangular heads.

Fact: Many non-poisonous snakes have triangular heads, and many poisonous snakes don't. You would not enjoy testing this theory on a coral snake, whose head is not triangular. It's not the kind of test you can retake if you flunk. Boa constrictors and some species of water snakes have triangular heads, but they aren't poisonous.

Fallacy: A coral snake is too little to cause much harm.

Fact: Coral snakes are indeed small and lack long viper fangs, but their mouths can open wider than you might imagine—wide enough to grab an ankle or wrist. If they get ahold of you, they can inject an extremely potent venom.

Fallacy: A snake will not cross a hemp rope.

Fact: Snakes couldn't care less about a rope or its formative material, and they will readily cross not only ropes but live electrical wires.

Fallacy: Some snakes, including the common garter snake, protect their young by swallowing them temporarily in the face of danger.

Fact: The maternal instinct just isn't that strong for a mother snake. If a snake has another snake in its mouth, the former is the diner and the latter is dinner.

Fallacy: When threatened, a hoop snake will grab its tail with its mouth, form a "hoop" with its body, and roll away. In another version of the myth, the snake forms a hoop in order to chase prey and people!

Fact: There's actually no such thing as a hoop snake. But even if there were, unless the supposed snake were rolling itself downhill, it wouldn't necessarily go any faster than it would with its usual slither.

Fallacy: A snake must be coiled in order to strike.

Fact: A snake can strike at half its length from any stable position. It can also swivel swiftly to bite anything that grabs it—even, on occasion, professional snake handlers. Anyone born with a "must grab snake" gene should consider the dangers.

Fallacy: The puff adder can kill you with its venom-laced breath.

Fact: "Puff adder" refers to a number of snakes, from a common and dangerous African variety to the less aggressive hog-nosed snakes of North America. You can't defeat any of them with a breath mint, because they aren't in the habit of breathing on people, nor is their breath poisonous.

Fallacy: Snakes do more harm than good.

Fact: How fond are you of rats and mice? Anyone who despises such varmints should love snakes, which dine on rodents and keep their numbers down.

Fallacy: Snakes travel in pairs to protect each other.

Fact: Most snakes are solitary except during breeding season, when (go figure) male snakes follow potential mates closely. Otherwise, snakes aren't particularly social and are clueless about the buddy system.

How Well Do You Understand Sharks?

They look frightening and strike fear into the hearts of nearly everyone who dips their toes into the ocean. Unfortunately, sharks are still one of the most misunderstood creatures on Earth. Do you know which rumors are true and which aren't?

* * * *

Sharks are vicious man-eaters. False! People are not even on their preferred-food list. Every hunt poses a risk of injury to sharks, so they need to make every meal count. That's why they go for animals with a lot of high-calorie fat and blubber—they get more energy for less effort. Humans are usually too lean and bony to be worth the risk.

Sharks are loners. That depends on the shark. Some species, such as the great white, are rarely seen in the company of other sharks. However, many other species aren't so antisocial. Blacktip reef sharks hunt in packs, working together to drive fish out of coral beds so every shark in the group gets a meal. Near Cocos Island off Costa Rica, hammerheads have been filmed cruising around in schools that consist of hundreds of sharks.

You have a greater chance of being killed by a falling coconut than by a shark. True! Falling coconuts kill close to 150 people every year. In comparison, sharks kill only five people per year on average. The International Shark Attack File estimates that the odds of a person being killed by a shark are approximately 1 in 264 million.

Sharks have poor vision, and most attacks are cases of mistaken identity. As popular as this belief is, it's wrong. Scientists have observed that sharks' behavior when they are hunting differs significantly from what most people report when a bite occurs. Sharks are extremely curious creatures and, since they don't have hands, they frequently explore their environment with the only things they have available—their mouths. Unfortunately for humans, a curious shark can do a lot of damage with a "test" nibble, especially if it's a big shark.

Most shark attacks are not fatal. True! There are approxi mately 60 shark attacks around the world each year, and, on average, just 1 percent of those are fatal. What we usually call "attacks" are really only bites. Scientists report that an inquisitive shark that bites a surfboard (or an unlucky swimmer) shows far less aggression than when it is on the hunt and attacks fiercely and repeatedly.

Shark cartilage is an effective treatment for cancer. False! Anyone toting the benefits of shark cartilage as a nutritional supplement to cure cancer is selling snake oil. Multiple studies by Johns Hopkins University and other institutions have shown that shark cartilage has no benefit. This myth got started with the popular but incorrect notion that sharks don't get cancer.

Most shark attacks occur in water less than six feet deep. True! And the reason is obvious—that's where the majority of people are. It makes sense that most of the interactions between humans and sharks happen where the concentration of people is the greatest.

Sharks have to swim constantly or they drown. There are a few species that need to keep moving, but most sharks can still get oxygen when they're "motionless." They just open their mouths to draw water in and over their gills.

Most sharks present no threat to humans. True! There are more than 400 species of sharks, and approximately 80 percent of them are completely harmless to people. In fact, only four species are responsible for nearly 85 percent of unprovoked attacks: bull sharks, great white sharks, tiger sharks, and great hammerhead sharks.

The Impossible 'Possum

Let's face it: Opossums are weird. But here are some interesting facts that might change your mind about the unique opossum.

* * * *

- Opossums are the only marsupial (i.e., a mammal that carries its young in a pouch) found in North America.
- The word "opossum" comes from the Algonquin word *apasum,* meaning "white animal." Captain John Smith used "opossum" around 1612, when he described it as a cross between a pig, a cat, and a rat.
- Although it is often colloquially called the "'possum," the opossum is completely different from a possum, an Australian marsupial.
- The opossum's nickname is "the living fossil," as it dates back to the dinosaurs and the Cretaceous Period, 70 million years ago. It is the oldest surviving mammal family on Earth.
- When cornered, opossums vocalize ferociously and show all 50 of their teeth, which they have more of than any other mammal. But they are lovers, not fighters, and prefer to run from danger.

* If trapped, they will "play 'possum," an involuntary response in which their bodies go rigid, and they fall to the ground in a state of shock. Their breathing slows, they drool a bit, and they release smelly green liquid from their anal sacs. This is enough to convince most predators the opossum is already dead, leaving it alone.
* Despite their predilection for eating anything—including rotting flesh—opossums are fastidious about hygiene. They bathe themselves frequently, including several times during each meal.
* Opossums are extremely resistant to most forms of disease and toxins, including rabies and snake venom, the latter probably due to their low metabolism.
* The idea that opossums mate through the female's nostrils is a myth. Although the male opossum has a forked (bifid) penis, he mates with the female in the normal manner. Conveniently, she has two uteri, so he deposits sperm into both of them at once.

Are Some Bookworms Really Worms?

Librarians hate bookworms. They'd like to banish them from the stacks forever. The literature lovers in question aren't people—they're tiny winged creatures known as book lice or barklice. They resemble flies, not worms, and they don't even like paper. They feed off mold and usually are found in damp, mildewed tomes.

OTHER BUGS, INCLUDING silverfish and cockroaches, feast on things such as the flour and cornstarch found in old library paste. Wood-boring beetles eat wood, naturally, but will consume paper made of wood pulp. Though beetles technically are not worms, they probably inspired the term "bookworm."

Nineteenth-century French book dealer and bibliophile Étienne-Gabriel Peignot reported finding a bookworm that had burrowed clear through a set of twenty-seven volumes, leaving a single hole, like the track of a bullet, in its wake. How far did it go? Given that many old tomes are at least three inches thick, the critter might have traveled nearly seven feet.

The best way to keep bugs out of the books is to stop them before they get in. Contemporary librarians strive to maintain clean, dry buildings. Replacing wood shelving with metal discourages beetles. And those signs that say, "Please do not eat in the stacks"? Heed them. A few moldy cookie crumbs can be a bonanza to hungry book lice. Once insects settle in, it's difficult to get rid of them. Sometimes the only effective method is professional fumigation.

Modern construction and bookbinding methods have done a lot to make libraries free of bookworms—free of the six-legged kind, that is. If you're a human bookworm, come right in. Just check your lunch at the door.

Do Birds Get Tired in Flight?

Flight—especially migration—can be an exhausting experience for any bird. Reducing the amount of energy that is spent in the air is the primary purpose of a bird's body structure and flight patterns. Even so, migrating thousands of miles twice a year takes its toll on a bird's body, causing some to lose up to 25 percent of their body weight. How do they keep on truckin'?

* * * *

LARGE BIRDS CUT energy costs by soaring on thermal air currents that serve to both propel them and keep them aloft, which minimizes the number of times the beasts have to flap their wings. The concept is similar to a moving walkway at an airport: The movement of the current aids birds in making a long voyage faster while expending less energy.

Smaller birds lack the wingspans to take advantage of these currents, but there are other ways for them to avoid fatigue. The thrush, for instance, has thin, pointed wings that are designed to take it great distances while cutting down on the energy expended by flapping. Such small birds also have hollow bone structures that keep their body weights relatively low.

If you've ever seen a gaggle of migrating geese, you likely noticed the distinctive V-formation that they take in flight. They do this to save energy. The foremost goose takes the brunt of the wind resistance, while the geese behind it in the lines travel in the comparatively calm air of the leader's wake. Over the course of a migration, these birds rotate in and out of the leader position, thereby dispersing the stress and exhaustion.

While large birds routinely migrate across oceans, smaller birds tend to keep their flight paths over land—they avoid large bodies of water, mountain ranges, and deserts. This enables them to make the occasional pit stop.

Perhaps the most amazing avian adaptation is the ability to take short in-flight naps. A bird accomplishes this by means of unilateral eye closure, which allows it to rest half of its brain while the other half remains conscious. In 2006, a study of Swainson's Thrush—a species native to Canada and some parts of the United States—showed that the birds took hundreds of in-flight naps a day. Each snooze lasted no more than a few seconds, but in total, they provided the necessary rest.

How sweet is that? After all, who among us wouldn't want to take naps at work while still appearing productive?

No Cats Were Harmed

What's the story with so-called "catgut," the material once used in tennis and musical-instrument strings? It is really made of guts, but the guts don't come from cats.

* * * *

Intestinal Fortitude

THE INTESTINES OF cattle and other livestock are cleaned, stripped of fat, and prepared with chemicals before they can be made into catgut string. Historically, preparation of string from animal tissue dates back thousands of years in recorded history and likely much longer ago in reality. Intestines are uniquely suited because of their combination of strength and elasticity, even in comparison to other pretty robust naturally occurring strings like horse hair or silk. This makes sense intuitively because of the role our intestines play in our bodies, but let's not think too hard about that. One downside is that the prepared gut fibers are still absorbent—enough so that even atmospheric humidity can warp them out of shape.

In Stitches

Historical humans realized surprisingly early that gut string was a good way to sew up wounds, and in this case the absorbency is a bonus. These humans weren't aware of issues like infection, so their ingenuity with gut string was a matter of simple craftsmanship: you should sew with the strongest material you can find, whether for clothing or shelter or for a wound. Thousands of years later, scientists experimented and realized that gut string could dissolve in the human body. With the eventual rise of germ theory, doctors were able to create and use sterilized dissolving sutures that would be recognizable to the ancient Egyptians who first documented their sewing of wounds. In fact, most dissolvable stitches are still made with prepared animal fibers or with synthetics that were designed to mimic animal fibers in the body.

High Strung

For musical instruments, gut strings also date back thousands of years. In both Latin and Greek, the terms for strings and bowstrings (and our modern words chord and cord) were from the original Greek term meaning guts. Musicians found that gut strings made the best sound, but the strings warped, frayed, and broke quite quickly because of the effects of moisture in the air and from musicians' touch. Modern musicians can use strings with a core of gut that's surrounded by a snug winding of very fine metal. In a fine example of art imitating nature, this structure mimics the way our flexible gut fibers are arranged in our intestines, with both lengthwise fibers and circular bands.

The Gut Racket

Gut strings are still considered the gold standard by many high-level tennis players and manufacturers. In tennis, the absorbency of gut strings is counteracted with topical wax that seals the strings. Choosing and making gut strings for tennis rackets is still an artisan craft, and some cattle—most if not all tennis gut strings come from cattle—apparently produce finer quality gut strings than others, creating a Wagyu-beef-like hierarchy among cattle ranchers.

Digesting the Information

Humans have shown remarkable ingenuity since we first diverged from our most recent ancestor, but even by human standards, it's unusual to have a found material that works as both a durable tool and a nourishing food—depending on how you prepare it. Whether you're preparing for your Wimbledon debut or a period-correct Baroque chamber orchestra, consider the millennia-old tradition of the gut string.

White Buffalo Miracle

When Dave and Valerie Heider decided to raise buffalo on their hobby farm in the late 1990s, they had no idea what fate had in store for them.

* * * *

Follow the Star

HOPING TO EARN a little extra money for their retirement years, the Heiders raised buffalo in their spare time and kept their day jobs in Janesville, Wisconsin. Then in August 1994, one of the buffalo cows prepared to give birth and her white calf was not just a surprise—she was a miracle. And that's how she got her name.

The Heider family considered Miracle's appearance to be a bit unusual, but they were totally unprepared for the attention that soon descended on their little farm. It turns out that white buffalo are very significant in Native American mythology. The Associated Press picked up the story, and soon people from all over the country wanted to see the baby calf that was causing such a stir.

In fact, the day after the story hit the presses, the first Native Americans had arrived in Janesville. To the Native American tribes of the Midwest, a white buffalo is just about akin to the second coming of Christ, according to a Lakota medicine man who saw the calf. They wanted merely to see her, pray, and leave an offering.

What's It All Mean?

Buffalo, of course, were very important to certain Native American peoples. They relied on the beasts for food, clothing, and shelter, as well as tools and utensils. But even more important for this story, in light of everything buffalo added to their lives, Native Americans forged a spiritual relationship with the buffalo that did not exist with other animals.

The Legend of the White Buffalo varies a little depending on who tells it, but the most important points remain the same. Long ago, when the Sioux inhabited the Great Plains, a group of hunters went out in search of game. Some say the hunters saw a beautiful woman, while others say they saw a white buffalo that then turned into a woman.

The woman instructed the hunters to return to their village and tell people that she would be coming. They did and she came, bringing a sacred pipe. Before leaving, the White Buffalo Calf Woman promised she would return as a white female buffalo calf. This event would symbolize a new harmony among people of all colors.

Along Came a Miracle

So imagine the excitement among Native Americans when a white calf was born. While some said she should have been born to one of the Native American nations, others thought the fact that she was born on a farm owned by a "white man" was significant—and was possibly a necessity for restoring peace among nations.

Whatever the reason that Miracle was born on the Heider farm, it seemed to be a stroke of luck. The Heiders had never heard the white buffalo tale before her birth, and Dave Heider himself admits he saw dollar signs when she first appeared. But once the family saw the religious and cultural significance that Miracle held, they changed their minds. They considered Miracle a special gift and went so far as to open their farm to visitors. Onlookers were not allowed within the gates—and in fact, the Heiders eventually needed to install a sturdy fence to protect the buffalo and to hold the numerous gifts and offerings that visitors brought with them and draped over the fence to honor her.

And people did come—thousands of them. They came from all over to pay their respects, satisfy their curiosity, and to meditate. One man even came all the way from Ireland.

The family had opportunities to sell the calf but turned down each offer. Ted Turner, who owns a large private buffalo herd, made an offer. So did Ted Nugent, who wrote a song about a white buffalo. Circuses and carnivals came calling, as well, but in each case, the answer was no.

Miracle was never about money to the Heiders. They didn't sell posters or mugs. They didn't charge admission. They finally sold photos of the calf for a dollar, if only to discourage visitors from taking their own pictures and selling them for profit.

One in a Million . . . or So

The reason for all the uproar is that a white buffalo, as you may have guessed, doesn't come along every day. In fact, the last documented white buffalo died in 1959. And just for the record, Miracle was not an albino, a genetic oddity; she had brown eyes. One source says the odds of the birth of a true white female buffalo are as low as one in a million. Other sources maintain that the chance is considerably less than that. In any case, your odds of winning the lottery are quite possibly higher. Miracle lived a relatively short life (she died of natural causes at age ten), considering many buffalo live as long as 40 years. During her short life, she didn't remain white but changed color four times. This too was part of the ancient prophecy, to unify the four peoples of red, white, black, and yellow.

Native Americans believe Miracle lived up to her name as the return of the White Buffalo Calf Woman. No matter what, she was surely a symbol of hope. Today, you'll find a statue of the White Buffalo Calf Woman erected in her honor on the Heider's farm in Janesville.

✳ Chapter 10

Discoveries and Inventions

The Theory of Natural Selection

The Galapagos Islands are an archipelago of volcanic islands that lie in the Pacific Ocean 563 miles west of Ecuador. These isolated islands are home to a vast collection of unusual animal species found nowhere else in the world, including the Galapagos giant tortoise, the flightless cormorant, and the Galapagos penguin. The unique flora and fauna of the area has made the Galapagos a favorite destination of biologists, naturalists, birdwatchers, and nature-loving tourists. These little islands have played a big role in our discovery of the vast complexity of life on Earth.

✳ ✳ ✳ ✳

PHILOSOPHERS AND SCIENTISTS had spent much time debating where so many different animals came from, and even how our own human characteristics came about. The Greek philosopher Empedocles believed that creatures arose accidentally due to environmental factors like cold and heat, producing animals with no particular rhyme or reason. But the great philosopher Aristotle criticized this idea, saying that the form of each species had a purpose, which was obvious from the way the characteristics of each animal were passed on to the next generation.

For centuries, very little changed beyond what the early philosophers debated, but by the 19th century, some major new discoveries were made. First, geologists were finding that the

Earth was old—very old—much older than can be explained by a literal interpretation of biblical creation, which many in the Western world believed. Second, fossil discoveries were proving that at one time, animals that looked nothing like any living animal in the world roamed the Earth, and these fossils became less similar as the geological strata became deeper. And finally, some scientists, like French naturalist Jean-Baptiste Lamarck, began to consider the idea of an evolutionary relationship within species.

Lamarck theorized that the environment surrounding an animal would influence its characteristics, suggesting, for instance, that giraffes have long necks because their environment requires them to stretch upward for food. Although Lamarck was not the first scientist to consider an evolutionary theory, he was the first to develop a comprehensive idea and published several works in the early 1800s describing his hypothesis.

Darwin's Dilemma

Then, in 1831, a young Charles Darwin graduated from Christ's College in Cambridge, England, where he had studied to become a country parson. But his true love was the natural world: Darwin learned taxidermy as a teenager, loved collecting beetles, and joined the Plinian Society—a group of students interested in natural history—while he was in college. So when Robert FitzRoy, the captain of the HMS *Beagle*, invited Darwin to serve as a naturalist on a voyage to chart the coast of South America, the young nature-lover jumped at the chance to see the world before settling into his new career as a parson.

During the five-year voyage, Darwin spent as much time on land as he could, retrieving geological samples and taking notes on zoology and natural history. After charting the coast of South America, the *Beagle* headed for the Galapagos Islands. For someone like Darwin, who was so interested in the natural world, the islands provided some fascinating finds. He observed that the tortoises on each island, while similar, exhibited slight differences. The same was true for a group of mockingbirds and finches, later dubbed "Darwin's finches," which each displayed slight differences from island to island.

Puzzled by the characteristics exhibited by the fauna on the Galapagos Islands, Darwin began to research his finds once he returned home from the sea voyage. After presenting the Zoological Society of London with bird specimens he had collected from the islands, it was discovered that the birds were not simply different varieties of the same bird, but rather separate species. This revelation began to influence Darwin's thinking and ideas, and in 1838 he started forming his theory of natural selection and evolution, pondering the idea that "one species does change into another" to explain the different species of animals living on each of the Galapagos Islands.

Darwin's theory stated that species change over time and space, with populations in different geographic regions differing in form or behavior. But despite the differences, Darwin also posited that different organisms share common ancestors with other organisms, which explained certain similarities within organisms that were classified together. The naturalist also looked to the fossil record to show that changes in evolution are slow and gradual, occurring over millions of years. And the mechanism that produces these changes was outlined in Darwin's theory of natural selection.

Natural selection includes four components: variations in appearance and behavior; inheritance of traits from parent to offspring; a high rate of population growth, which leads to a struggle for resources and mortality; and differential survival and reproduction, whereby individuals possessing traits suitable for the struggle for resources will produce more offspring. This struggle for resources is what leads to certain traits appearing more frequently within a population. These "adaptations" must be heritable, and they must give individuals an advantage in the fight for survival; this is natural selection.

One famous example of natural selection concerns the peppered moth, or *Biston betularia,* which is common in England. The moths are generally light in color, with black speckles across their wings, giving them their spicy name. Their coloring allows them to blend in with the lichen on tree trunks, providing them with camouflage against hungry birds and other predators. But there is a genetic mutation in these moths that can result in some moths having all-black wings, making them much more obvious when they rest on tree trunks, and therefore much less common than their lighter counterparts.

But during the Industrial Revolution, coal from fires spat out black soot, covering buildings and tree trunks and killing off the lichen. As a result, the mutated moths were given an advantage, and suddenly birds were much more likely to notice the light-colored moths resting against the dark tree trunks. As dark-colored moths survived and began to reproduce, they became the predominant type of *Biston betularia* found in cities. By 1895, 99 percent of peppered moths in the city of Manchester were dark. But by the mid-1900s, a reduction in air pollution reversed the phenomenon, and once again light-colored moths became the norm.

Darwin spent decades reviewing research, speaking to other scientists, and publishing essays and books about his voyage with the *Beagle* before finally fleshing out his theory of evolution. He published *On the Origin of Species* in 1859, more than 20 years after setting off on the voyage that would ultimately change the way we look at ourselves and the natural world around us.

Theory of Relativity

The greatest theory of all time was a group effort.

* * * *

THERE IS AN oft-repeated story about the 17th century astronomer Isaac Newton that describes how he was sitting under an apple tree one day, when a piece of fruit fell from the tree and hit him on the head. That knock on the noggin suddenly prompted Newton to come up with the law of gravity, inspired by the apple's painful fall toward the earth. This story is not quite true, however, having been embellished over the centuries to make the real story slightly more dramatic.

Newton did propose his theory after spending time in an apple orchard, but he was simply observing the fact that apples occasionally fell to the ground. Newton published his law of gravity in 1687, which stated that every body in the universe is attracted to every other body with a force directly proportional to the product of their masses and inversely proportional to the square of the distance between them.

Although it was a groundbreaking hypothesis at the time, even Newton himself was never quite sure that it was complete. While most of the time his ideas worked perfectly throughout the solar system, there were anomalies that never fit in with his theory. It took centuries before scientists began to come up with new ideas about the forces at work in the universe.

Some of these scientists included Dutch physicist Hendrik Lorentz, French mathematician Henri Poincare, and German theoretical physicist Max Planck, all of whom helped to change the way scientists looked at the universe. Two other scientists, Americans Albert A. Michelson and Edward W. Morley, would make a significant discovery in 1887 that at first seemed to be a failed experiment, but would soon open the door to one of the most famous theories in science.

Moving Through the Aether

Michelson and Morley set out to find the presence and measure the properties of a substance called "aether," which, at the time, was believed to fill empty space. The assumption was that since on Earth, waves in water must move in water and sound waves must move through air, wavelengths of light must also require a substance in which to move. Scientists named this theoretical substance "aether," after the Greek god of light.

So, using a device called an interferometer, which measures the interference properties of light waves, Michelson and Morley set out to prove that aether existed. They theorized that if it existed, it would seem like a moving substance to those on Earth, the same way air feels like a moving substance if you hold your hand out of a moving car's window. The Michelson-Morley experiment aimed to detect this "aether wind" by measuring the speed of light in different directions. A mirror would split a beam of light into two directions, and then recombine them, after which the scientists would measure changes in speed that could be attributed to aether. But to their surprise, Michelson and Morley found no significant differences whatsoever, and their experiment was considered a failure.

A Special Theory

But just because an experiment fails doesn't mean you don't keep researching. After the Michelson-Morley experiment, scientists continued looking for answers about what makes up the universe. And by 1905, Albert Einstein debuted his theory of special relativity, which was based on two statements: One, that the laws of physics appear the same to all observers, and two, that the speed of light, which clocks in at 186,000 miles per second, is unchanging, contrary to what Michelson and Morley originally believed. However, the theory of special relativity only applied when there was an absence of gravity; gravity itself was another puzzle to be solved.

Einstein theorized that the speed of light was the absolute boundary for motion, meaning that nothing can travel faster than the speed of light. According to Newton's law, if the Sun were to start wobbling in space, the Earth would immediately wobble as well, instantaneously affected by the gravity. Einstein didn't accept this idea, since the influence of the sun's gravity would have to travel much faster than the speed of light in order to instantaneously affect the Earth. But Einstein also realized that the Sun does somehow have the ability to reach out across millions of miles of space and pull on the Earth with its gravity. So after he formed his theory of special relativity, he kept searching, knowing there was still more to the story.

A Strange Conundrum

It's easy to see gravity at work here on Earth, where knocking a book off a desk will cause it to fall to the floor. But the Sun somehow exerts a pull of gravity across empty space. Perhaps, Einstein thought, empty space itself was the answer. By 1912, the scientist was working on the question full-time, and in 1915, after a decade of what he considered some of the most intellectually challenging years of his life, Einstein published his theory of general relativity, which expanded on his original theory of special relativity.

Matter, Einstein theorized, such as the Sun and planets, causes the space around it to warp and curve, which then influences how other matter reacts. Gravity, then, is not a force, but rather a curved field that is created by the presence of mass. But that's not all: even time, the scientist believed, could be warped, with time moving more slowly near massive bodies like the Earth. Acceleration causes time to move more slowly, as well, in a phenomenon called time dilation. It may sound like science fiction (and the idea has, in fact, been used in plenty of science fiction books and films), but it's actually science fact!

General relativity predicted that the light from distant stars would warp as it passed by the sun on its way to Earth. But the only way scientists could observe such a phenomenon would be during a total eclipse, as the sun's light is much too bright otherwise. So on the date of the next solar eclipse, May 29, 1919, astronomers gathered in Sobral, Brazil, and on the island of Principe off the west coast of Africa, where the eclipse would be most noticeable. They took photographs as the moon passed in front of the Sun, and then spent the next several months analyzing them.

On September 22, 1919, Einstein got the news that the photographs had confirmed his theory. Einstein had guessed that the light deflection of the starlight near the Sun would be twice what would be expected from Newton's laws, and the photographs proved him correct. The theory of relativity was also used to predict the rate at which two neutron stars orbiting each other will move toward one another. When Einstein's theory was applied to this phenomenon, it was accurate to within a trillionth of a percent, once again confirming his hypothesis.

In fact, the theory of relativity is one of the most oft-tested and subsequently confirmed theories in science, but there is still one test researchers would love to carry out. If Einstein was correct in his theory that space warps and bends due to the mass of objects, then if two objects collide in space, they must cause a sort of "ripple effect" in the fabric of space, not unlike two boats colliding on a lake. So far, scientists have been unable to observe such a phenomenon, as by the time a ripple from any distant colliding stars or black holes reached Earth, the wave would be infinitesimally small and extremely difficult to detect. But new sensors and devices are being developed that can sense the slightest difference in gravitational waves, so it may not be long before one more piece of the relativity puzzle falls into place. In the meantime, those of us who aren't scientists can continue enjoying science fiction films, but perhaps with a new appreciation. Some of that fiction isn't so far from fact, after all.

Radioactivity

In 1895, German physicist William Röntgen discovered a type of radiation he called "X-rays" while experimenting with cathode ray tubes, and immediately the science world was fascinated.

* * * *

AFTER ALL, THESE mysterious rays could pass through all kinds of materials, including human bodies, resulting in ghoulish photos of bones and teeth. At first, people were so amazed by the new technology that public X-ray booths were available, where curiosity seekers could ogle at the bones in their hands or X-ray their feet to purchase better-fitting shoes. However, the public's captivation with this type of radiation would be short-lived, as side effects like burns and hair loss soon became apparent. But that didn't stop scientists from wanting to know more.

Not long after Röntgen's discovery, a French scientist named Henri Becquerel was experimenting with the newly found X-rays using fluorescent minerals. He thought perhaps X-rays and fluorescent minerals, which glow when exposed to sunlight, might be connected in some way. So he devised an experiment to test this theory: Becquerel planned to expose a fluorescent mineral to sunlight, then place it and a metal object on a photographic plate. If the resulting photo showed the metal object, then, he assumed, he could conclude that fluorescent minerals exude X-rays.

But on the day he planned to perform this experiment, the skies of Paris were cloudy and overcast. With no sunlight to make the minerals glow, Becquerel wrapped up the photographic plate, on which he'd already placed a copper Maltese cross and a fluorescent uranium compound called potassium uranyl sulfate, and placed it in a drawer to wait for a sunny day.

A Strange Discovery

A few days later, Becquerel removed the plate from the drawer and unwrapped it, surprised to see that the image of the Maltese cross was visible on the photographic plate, even though the minerals had not been exposed to sunlight and had been wrapped and kept in a dark drawer. He had to surmise that the minerals themselves were emitting radiation, even without the aid of the sun. In March of 1896, Becquerel presented his findings to the Académie des Sciences, and this radiation became known as "Becquerel rays."

Becquerel's discovery particularly interested a student named Marie Curie. Curie was working towards a PhD and thought the subject of these unusual uranium rays could be an interesting topic for her thesis, so she dove into research. Using an electrometer that her husband, Pierre, and his brother had developed, Curie discovered that the air around a sample of uranium became electrically charged, and that this phenomenon changed according to the amount of uranium present. She concluded that the radiation emitted by the uranium was not the result of any sort of reaction between the element and another molecule, but rather was an intrinsic property of the uranium itself.

Continuing her research, Curie began studying the uranium minerals pitchblende (now known as uraninite) and chalcolite with the electrometer, and discovered that they were even more active than uranium itself. She believed that these minerals must contain substances that emitted a greater amount of radiation than uranium, and started to look for what they might be, soon discovering that the element thorium demonstrated this property. Curie's finds were so interesting that Pierre soon dropped his own work and joined her in her radiation research.

By 1898, the search for other elements that emitted radiation led the Curies to the discovery of polonium, named after Marie's homeland of Poland, and radium, named after the

Latin word for "ray." The couple also coined the term "radioactivity" to describe the characteristics of these elements.

Ernest Rutherford

One scientist who was particularly interested in the research done by the Curies was physicist Ernest Rutherford, who spent much of his career studying atoms and would later go on to discover protons. Rutherford, with the help of a young chemist named Frederick Soddy, would finally unravel the mystery of radioactive elements. While Curie herself had theorized that radioactivity involved some sort of process within atoms, at the time it was still widely believed that atoms were indivisible and indestructible. What Rutherford and Soddy discovered turned that notion on its head. Radioactivity, they found, was caused by a decay within the nucleus of an atom. This random, spontaneous disintegration caused the emission of different types of radiation, which Rutherford classified according to their ability to penetrate matter: he called them alpha rays, beta rays, and gamma rays.

By the 1920s, when it was finally understood that atoms were made up of subatomic particles, the process of radioactivity made even more sense. Alpha decay occurs when the nucleus of an atom emits an alpha particle, which consists of two protons and two neutrons, and beta decay occurs when an electron is emitted by an atom's nucleus. When these decay processes occur, the original element eventually transmutes into a new element, due to the loss of protons or electrons. Uranium becomes thorium, and radium becomes radon gas, for instance. Gamma decay usually occurs after an alpha or beta decay has already taken place, as a high-energy nucleus decays into a lower energy form, by emitting gamma rays.

Unfortunately, the early researchers of radioactivity were unaware of its dangers, not realizing that all of these forms of radioactivity can cause damage. Beta rays can moderately penetrate living tissue and cause spontaneous DNA mutations, and gamma rays can cause cell damage and possibly result in cancer. Alpha rays, while usually not a threat because they can be stopped merely by skin, are surprisingly dangerous if ingested or inhaled, causing radiation poisoning and lung cancer. But when it was first discovered, radioactivity was thought to be an effective therapy for illness, with water containing radium sold as a health tonic and enthusiasts soaking in "curative" radioactive baths. By the 1930s, incidents of bone necrosis and the death of many radiation proponents led to the end of these practices.

But this was too late for Curie, who would often carry test tubes of radioactive elements in her pockets or store them in her desk. She died of aplastic anemia in 1934, an illness brought about by her many years of exposure to radiation. Even today, Curie's notes and papers from her time in her lab are considered too radioactive to be handled, and are kept in lead-lined boxes. But for their work, Marie, Pierre, and Becquerel were awarded the 1903 Nobel Prize in Physics. The first woman to be awarded a Nobel Prize, Curie was honored again in 1911 with the Nobel Prize in Chemistry, making her the only person to be awarded the prize in two different sciences.

The Discovery of Oxygen

We tend to take oxygen for granted. In fact, we didn't even know it existed until the 18th century.

❋ ❋ ❋ ❋

THIS ELEMENT IS absolutely necessary for almost all animal life on Earth, with the exception of some anaerobic microorganisms and tiny creatures that live at the bottom of the sea. Oxygen is what feeds our cells and provides us with energy. And breathing is so automatic that we rarely think about it, unless it becomes difficult. But there may be a good reason we hardly ever think about oxygen; after all, it is the third most abundant element in the universe, so it's never hard to find.

Surprisingly, however, oxygen only makes up about 21 percent of the air we breathe. The rest is mostly nitrogen, at around 78 percent, with trace amounts of carbon dioxide, neon, and helium thrown in. The combination of nitrogen and oxygen is interesting, because while our bodies need oxygen, they can't process too much of the gas: breathing pure oxygen for too long would cause irreversible lung damage and eventually death. But the presence of nitrogen, which is a non-reactive, inert gas, allows us to breathe in just the right amount of oxygen. Oxygen, on the other hand, is a very reactive gas, which is exactly what we need to provide our cells with energy. But too much oxygen in the atmosphere would result in a very combustible planet.

Combustible Air

In fact, the combustion of air was one of the first characteristics noticed by early scientists, including Leonardo da Vinci, who noted that a portion of air is consumed not only during respiration, but during combustion, as well. By the late 17th century, scientists like Robert Boyle and John Mayow proved that air was necessary for fire to burn. But around the same time, a theory was developed by German alchemist Johann Becher called the "phlogiston theory." This theory stated that matter contained a substance called "phlogiston" which was released when the matter was burned. This "dephlogisticating" released the phlogiston into the air, where it was absorbed by plants, which then became highly combustible. Though scientists had noticed that fire eventually burns out in an enclosed space, they took this as proof that air was only able to absorb a certain amount of phlogiston. This mysterious phlogiston, which could never seem to be observed or contained, nevertheless was assumed to permeate all things.

Because of the phlogiston theory, even scientists who managed to produce oxygen, including Robert Hooke, Ole Borch, Mikhail Lomonosov, and Pierre Bayen, didn't understand the significance of their findings. It wasn't until the 1770s that two scientists, Swedish pharmacist Carl Wilhelm Scheele and British chemist Joseph Priestley, independently and definitively discovered oxygen.

In 1771, Scheele began experimenting with mercuric oxide, silver carbonate, and magnesium nitrate, heating them to obtain oxygen, which he called "fire air." Familiar with the phlogiston theory, he suggested that this "fire air" combined with phlogiston when materials were burned. But Scheele made another important discovery about air: not only was it composed of "fire air," but something he called "foul air," as well. Scheele's description of fire air and foul air, now known to be oxygen and nitrogen, debunked the long-held assumption that air was just a singular element.

Around the same time, Priestley was running his own experiments, heating up mercuric oxide until it produced oxygen. He made note of the gas's ability to feed bright flames, and called it "dephlogisticated air. He even trapped a mouse in a small container filled with the gas, and was amazed to find that not only did it live four times longer than it would have if it had been breathing regular air, but it seemed to have extra energy once it was released.

Although Scheele made his observations first, his findings weren't published until 1777; Priestley, however, published a paper in 1775 titled "An Account of Further Discoveries in Air," so he is most often credited with the discovery. There was one more scientist, French chemist Antoine Lavoisier, who made contributions in the discovery of oxygen. Lavoisier discredited the phlogiston theory, proving the gas discovered by Scheele and Priestley was a chemical element and being the first to correctly explain how combustion works. He also made a lasting contribution by naming the gas *oxygène*, from the Greek "oxys," meaning acid, and "genes" meaning producer, because he believed the gas was found in all acids. While this was eventually determined to not be the case, the name oxygen stuck.

The discoveries that these scientists made helped to finally bring an end to the ancient Greek theories about "classical elements," which suggested that earth, water, fire, and air were elements themselves. This belief persisted for centuries, which Priestley noted when he said that there are few ideas that "have laid firmer hold upon the mind." Fortunately, there are also few things scientists love more than to expand the knowledge we have of the world around us.

The Periodic Table of Elements

Harry Potter fans know that in the United Kingdom, the first book in the series is titled Harry Potter and the Philosopher's Stone. *In the book, the philosopher's stone is a magical item said to grant its user immortality and the ability to change metal into gold. But surprisingly, the idea of a philosopher's stone isn't merely a literary device: throughout history, it has been believed that creation of such a stone is possible, if the correct elements are combined. What is even more surprising is that an attempt to produce a philosopher's stone was the first step toward the creation of the periodic table of elements.*

⁂ ⁂ ⁂ ⁂

ANYONE WHO HAS taken a high school chemistry class is familiar with the periodic table. It arranges chemical elements according to atomic number (the number of protons in the nucleus of an atom), electron configuration, and chemical properties. Although it was first published by Russian chemist Dmitri Mendeleev in 1869, its creation would not have been possible without contributions that began centuries earlier.

The thought of assembling such a table probably never occurred to early scientists, who were aware of only a few basic elements. Gold, silver, tin, copper, lead, and mercury had been known since antiquity, since they are easily found in their native forms; but this handful of elements hardly warranted any kind of classification system. So before such a table could be amassed, it was, of course, necessary to discover new elements. And this is where the philosopher's stone lent aid: In the mid-1600s, a German merchant by the name of Hennig Brand attempted to create a philosopher's stone.

Chamber Pot Chemistry

After finding a strange recipe which called for a combination of several salts and concentrated urine, which were added to a base metal to create silver (a recipe which does not work, by the way), Brand began experimenting with boiled urine. At one point, the glass vessel he used to boil the urine became red hot and filled with glowing fumes, and a liquid began to drip out. Brand caught the liquid in a jar and covered it, and to his surprise the substance continued to give off a pale-green glow. What Brand had accidentally discovered was the element phosphorus—the first new element to ever be discovered.

Brand's finding helped to answer the question of what it meant for a substance to be an element. Specifically, chemist Robert Boyle stated in 1661 that an element was "those primitive and simple Bodies of which the mixt ones are said to be composed, and into which they are ultimately resolved." The search was on, and over the next two hundred years, dozens of elements would be discovered.

By 1789, French chemist Antoine-Laurent de Lavoisier had written the first textbook about chemistry, *Traité Élémentaire de Chimie* (*Elementary Treatise of Chemistry*). Lavoisier's definition of an element was any substance that could not be broken down into another substance by chemical reaction, and included oxygen, nitrogen, hydrogen, phosphorus, zinc, and sulfur, along with other known elements. Lavoisier was the first scientist to begin organizing the elements, dividing them into basic groups of metals and non-metals.

This attempt to organize the elements was further expanded in 1817 when Johann Wolfgang Döbereiner, a German chemist, began to arrange certain elements with related properties into groups of three that he called "triads." Döbereiner noticed that the atomic weight of strontium fell about midway between the weights of calcium and barium, all elements with similar properties. Using atomic weights as a guide, he began to organize other elements by triads: chlorine, bromine, and iodine; sulfur, selenium, and tellurium; and lithium, sodium, and potassium.

These relationships between elements were studied over the next few decades, until a French geologist, Alexandre-Emile Beguyer de Chancourtois, noticed the periodicity of elements in 1862. He realized that similar elements occurred at regular intervals when they were organized according to atomic weight. He created an early version of a periodic table where the elements were arranged in a spiral pattern in order of decreasing atomic weight. With this arrangement, de Chancourtois noticed that elements of similar properties lined up vertically. Although he was the first to demonstrate that similar elements appear at periodic atomic weights, the geologist's findings garnered little attention until after Mendeleev's table was published several years later.

The Law of Octaves

In 1864, British chemist John Newlands grouped all of the elements that had been discovered, 56 at that time, by similar physical properties. Looking to Döbereiner's ideas, Newlands wanted to look beyond the small groups of triads the German chemist had proposed and instead relate all of the elements to each other. When he arranged them in groups of 11, he noticed that many sets of elements differed by some multiple of eight in atomic weight. He published his discovery in a journal, calling it the "law of octaves" in relation to the notes of a musical scale.

It was Mendeleev's table in 1869 that became the basis for what we know today. To assemble his table, Mendeleev created cards for each of the 63 known elements. Each one listed the element's symbol, atomic weight, and its chemical and physical properties. He then arranged the cards so that elements of similar properties were grouped together, in order of ascending atomic weight, creating his periodic table. This table was superior to previous period tables because it clearly showed the relationships between the elements, and also left spaces for elements yet to be discovered.

The periodic table has changed a bit since Mendeleev's design. Noble gasses were added after English physicist Lord Rayliegh and Scottish chemist William Ramsay discovered argon in the late 1800s. And in 1913, Henry Moseley, an English physicist, discovered a connection between the X-ray wavelengths of elements and their atomic numbers, enabling him to reorganize the table by nuclear charge rather than atomic weight. Today, a total of 118 elements have been discovered, and the table first proposed by Mendeleev has become a familiar sight in schools, universities, and scientific labs around the world.

Nuclear Fission

The ancient Greeks, even with their limited knowledge of the natural world, hypothesized that all things were made up of tiny particles. They named these particles atomos, *meaning "uncuttable," "indivisible," or "that which cannot be divided." And for centuries, the idea that the atom was the smallest possible particle in the universe was a widely-accepted theory. But in the early 20th century, new discoveries about the atom would prompt scientists to rework their notions about the makeup of the universe, and eventually lead to a new source of power—and one of the most destructive forces on Earth.*

* * * *

AFTER ENGLISH PHYSICIST J. J. Thompson discovered electrons, the particles that orbit the nucleus of an atom, in 1897, scientists suddenly realized that atoms were not the smallest particles in the universe. Amazingly, as the observation of electrons proved, atoms were composed of even tinier particles. Their discovery fascinated physicists, who set out to discover exactly what made atoms tick. One of Thompson's students, Ernest Rutherford, suggested in 1911 that the electrons in an atom might surround a small, dense, nucleus. Danish physicist Niels Bohr improved upon Rutherford's concepts in 1913, creating an atomic model where electrons orbit the nucleus like a mini solar system. And in 1919, Rutherford discovered protons, the positively charged particles within the nucleus of an atom.

Now knowing that atoms were composed of an inner structure, scientists began to believe that this inner structure might contain bound up energy. Looking to Albert Einstein's famous 1905 $E=mc^2$ (energy equals mass times the speed of light squared) theory, which basically says that energy and mass are interchangeable, scientists realized that if the energy inside atoms could be released, the power could be immense.

By the 1930s, scientists were learning more about the atom thanks to particle accelerators like Ernest Lawrence's cyclotron, which bombarded atoms with subatomic particles. But there was still a question they were unable to answer: why did the atoms of some elements, such as carbon, weigh different amounts? This question was answered in 1932 when English physicist James Chadwick discovered the neutron. Neutrons, which have no charge, share an atom's nucleus with protons. While the number of protons and electrons is always the same for a particular element, neutrons can differ in number. For instance, carbon always has six protons and six electrons, and in its most abundant form also has six neutrons. This is known as carbon-12. But about one percent of the carbon on earth contains an extra neutron, and is known as carbon-13. The different weights of atoms of the same element are called isotopes.

Hoping to reach the energy within atoms, researchers began attempting to split the nucleus of atoms in particle accelerators. They tried to accomplish this by bombarding atoms with beams of protons, but there was a problem: the nucleus of atoms already contains positively charged protons. So when the same positively charged particles were aimed at the atoms, the two positive charges repelled each other, no matter how fast the beam of protons sped toward the atom. But in 1934, Italian physicist Enrico Fermi had an idea: why not use neutrons, which have no charge, to bombard atoms?

Enrico Fermi

Fermi began using neutrons in particle accelerator experiments, bombarding uranium, the heaviest natural element on earth, with the neutral particles. Around this same time, scientists had also discovered nuclear transmutation, a process in which one element transforms into another element due to a change in the number of protons and neutrons in the nucleus. This can be a natural process, such as in the nuclear fusion in stars, which transmutes light elements into heavier elements, or it can be induced artificially in machines like particle accelerators. When Fermi ran his experiments, he noticed that some kind of transmutation was occurring, but he was unsure of whether it had resulted in new elements, the creation of known elements, or something altogether different.

After publishing the results of his experiments, a German chemist named Ida Noddack suggested it was possible that Fermi hadn't created any new elements, but rather had split up the nucleus of the uranium into fragments.

Atoms Behaving Badly

Fermi's results had sparked curiosity throughout the scientific community, and others began to run similar particle accelerator experiments, including German chemists Otto Hahn and Fritz Strassmann. Hahn and Strassmann had noticed something strange when they performed an experiment with uranium atoms: they seemed to have created isotopes of barium. And while scientists had made note of transmutations before, the resulting elements were always similar, such as the transmutation of nitrogen (atomic number 7) into oxygen (atomic number 8). So with an atomic number of 56, it was extremely unusual that barium would result from uranium, with an atomic number of 92. This was a mass that was 40 percent less, something Hahn and Strassmann couldn't explain.

So in December 1938, Hahn sent a letter to his good friend, physicist Lise Meitner, an Austrian of Jewish ancestry who had fled to Sweden to escape the Nazi occupation of her country. Hahn described his experiment and the results, and asked for Meitner's thoughts. Over the Christmas holiday that year, Meitner and her nephew, physicist Otto Frisch, discussed Hahn's letter as they strolled through the snowy city of Stockholm. Perhaps inspired by the falling flakes, Meitner suggested they think of the uranium nucleus as a drop of liquid. Frisch then sketched out how the nucleus, when hit with a neutron, might begin to elongate like a teardrop of water, until it pinched in the middle and split into two drops.

What's more, Meitner determined that the resulting nuclei, together, would have less mass than the original uranium nucleus, because of the energy expended by the split. When Meitner and Frisch worked out the energy that would result from such a split, it fit perfectly into Einstein's $E=mc^2$ equation. This process was previously undiscovered to the physics world, and Frisch dubbed it "fission," borrowing from a biological term for cell division.

The news of the discovery made by these four scientists quickly spread, with Bohr carrying the findings to America in January 1939. One of the realizations that had been made through the discovery of fission was that it caused the emission of neutrons. Uranium is an unusual element in that it contains a very high number of neutrons in its nucleus, with the most common isotope, uranium-238, having 146 neutrons. Since it was neutrons that caused the fission of the uranium atom in the first place, scientists realized that given the right circumstances, the neutrons that were emitted when the nucleus split could, in turn, split other nearby atoms, which could split other atoms, and so on. This "chain reaction" held the possibility of creating massive amounts of energy, which, if harnessed in a controlled environment, could provide a new heat and power source.

Of course, there was another implication as well: the energy released by fission could also be used to create an extremely powerful bomb.

By the end of 1939, Hitler had invaded Poland and Europe was mired in war. Many feared that the Germans, who had discovered fission, could use it to create a bomb. Einstein himself wrote a letter to President Roosevelt warning of the possibility of Germany creating weapons with uranium. The United States immediately began scrambling to research uranium, experimenting with the element's different isotopes to determine which would be most likely to create a chain reaction.

On December 7, 1941, the Japanese attacked Pearl Harbor, and the United States was officially at war; by March of 1942, the United States had also begun the Manhattan Project. Led by famed physicist J. Robert Oppenheimer, the highly secretive project would ultimately result in the first nuclear weapons, and the bombings of Hiroshima and Nagasaki, Japan, in August of 1945.

While the creation of atomic weapons was a frightening result of the discovery of fission, the process has also enabled some of the cleanest energy possible. Nuclear power results in emissions equal to or even lower than renewable sources of energy like wind and solar, and has prevented approximately 64 billion tons of carbon dioxide from entering our atmosphere. It provides about 10 percent of electricity generation worldwide, originating from 449 fission reactors. And while nuclear accidents such as the Chernobyl disaster in 1986 and the Fukushima disaster in 2011 draw scary headlines, other sources of energy, like coal, petroleum, and natural gas, have caused far more fatalities per unit of energy overall.

It is a bit mind-boggling to consider the energy contained within the tiny confines of the particles that make up our universe. But perhaps it's even more amazing to learn of the contributions of humans—from Einstein and Bohr to Hahn and Meitner—who used their insatiable curiosity to literally unlock the secrets contained within the atom.

Polyethylene

Sometimes, the greatest inventions happen by accident. Play-Doh was created as a wallpaper cleaner, but when children started using it to make arts and crafts projects, the company added fun colors and began marketing it as a toy; cyanoacrylate, now known as Super Glue, was accidentally invented and thought to be completely useless—until it was found to be an incredibly strong bonding agent; and even Popsicles were a mistake, first created when a boy accidentally left a drink and a stir stick out on his cold back porch all night. And accidental discoveries can sometimes change the world, as was the case with polyethylene.

⁂ ⁂ ⁂ ⁂

IN FACT, POLYETHYLENE, now the most common type of plastic in use, was accidentally discovered more than once. The first time a scientist took note of this substance was in 1898, when German chemist Hans von Pechmann created a white, waxy substance when he was researching the chemical compound diazomethane. But at the time, neither von Pechmann nor his colleagues could see any use for the strange substance, and it was all but forgotten.

In 1933, scientists Eric Fawcett and Reginald Gibson, working at Imperial Chemical Industries in Northwich, England, accidentally created the same kind of white, waxy substance when they were experimenting with ethylene and benzaldehyde. They had heated the chemicals to 338 degrees Fahrenheit under extremely high pressure, and at first were unaware that oxygen had contaminated their equipment. But this oxygen provided just the right conditions for polyethylene to form. Since Fawcett and Gibson were unaware of the extra oxygen, replicating the results of their original experiment proved difficult at first. It took another two years before another chemist with Imperial Chemical Industries, Michael Perrin, was able to reproduce the right conditions, and by 1938, scientists were able to create the substance using lower temperatures and pressures. But when World War II broke out, production of the material for commercial use was halted, as it was found to have electrical insulation properties. Great Britain kept the method for creating polyethylene secret, while using it to insulate airborne radar equipment. Because the material was lighter and thinner than the bulky materials the Germans had to use to insulate their equipment, this gave the British a distinct advantage during the war.

Towards the end of the war, Imperial Chemical Industries licensed production of polyethylene to their United States allies, and by 1944, the Bakelite Corporation in Sabine, Texas, and DuPont in Charleston, West Virginia, began large-scale manufacturing of polyethylene products. But the material wasn't always seen as practical, as it was very soft and had a low melting point. But in 1951, chemists Robert Banks and J. Paul Hogan, who worked for Philips Petroleum, developed a new polymerization process that resulted in a sturdier plastic; German chemist Karl Ziegler developed a similar process in 1953. Both processes are still commonly used for industrial polyethylene production today.

A Quirk of Carbon

While it sounds like a complicated material, polyethylene is simple in its chemical makeup. The material is basically just a long chain of carbon molecules, with two hydrogen molecules attached to each carbon atom. The material can also be produced as a "branched" or "low-density" polyethylene, known by the abbreviation LDPE. When these long chains are not present, it is called "linear" or "high-density" polyethylene, also known as HDPE. LDPE is easier to make, but HDPE is stronger. But there is an exception to this rule: when LDPE is created with high molecular weights between three and six million, it is known as "ultra-high molecular weight polyethylene," and is so strong it can be used to make bulletproof vests.

It Began with the Hula Hoop

But it was a much more humble product—the hula hoop—that took polyethylene from a wartime advantage to a household staple. When the plastic, circular toy became popular in the late 1950s, suddenly, homes all over the country were introduced to polyethylene. And the material just took off from there. By 1967, even Hollywood knew that polyethylene was here to stay, when in the film *The Graduate,* Mr. McGuire, played by actor Walter Brooke, tells Dustin Hoffman's Ben Braddock that "There's a great future in plastics."

Today, polyethylene is used in everything from containers and pipes to toys and household electronics. It is the most commonly used plastic in the world, with more than 100 tons being produced annually, accounting for more than a third of the market. Although many worry that the material is not readily biodegradable and builds up in landfills, researchers have been making some amazing discoveries about bacteria and animals that are able to consume the plastic. In 2008, a 16-year-old Canadian named Daniel Burd discovered that a combination of the bacteria *Pseudomonas fluorescens* and *Sphingomonas* can reduce the amount of plastic by 40 percent within three months. And other bacteria, including *Brevibacillus borstelensis* and *Acinetobacter* have similar effects. And more recently, it has been discovered that both Indian mealmoth larvae and the greater wax moth caterpillar can eat plastics.

These bacteria and caterpillars not only help to reduce waste, but make recycling easier. And like polyethylene itself, some of these were discovered accidentally. Which just goes to prove that sometimes, accidents can change the world for the better.

Vulcanized Rubber

Fans of Star Trek *know that the stoic Spock comes from the planet Vulcan. But Vulcan is also the god of fire, volcanoes, and metalworking. He was often honored in ancient Rome by lighting candles and building bonfires. So when we hear the term "vulcanized rubber," we can assume one of two things: either it originated in outer space, or it has something to do with fire. (Sorry,* Star Trek *fans; it's not from outer space.)*

* * * *

THE STORY OF vulcanized rubber begins not in outer space, but rather much closer to home. Thousands of years ago, indigenous peoples in Central and South America would make cuts in the bark of the *Hevea* tree, now commonly known as the Para rubber tree, and would collect the sticky, milky liquid inside. Calling it *caoutchouc*, they would use the substance to make balls for playing games, food containers, and to create waterproof textiles. In the early 1700s, a French explorer named Charles Marie de La Condamine introduced the bouncy, stretchy substance to Europe, and in 1770, English chemist and theologian Joseph Priestley noted that this caoutchouc worked well to rub pencil marks off paper, leading to its now-common name, rubber.

Waterproof, stretchy, and able to be molded into many different shapes, rubber soon made its way to America, where it was all the rage in the early 1800s. Rubber factories popped up in many cities to meet the demand for rubber products, and retail stores sold goods like boots, balls, and life preservers. But in the middle of summer 1834, one of these stores in New York City, owned by America's oldest rubber manufacturer, the Roxbury India Rubber Company, was in the middle of a huge quandary. The summer heat had melted their entire stock, forcing them to dispose of thousands of dollars' worth of merchandise.

Goodyear Takes Up the Challenge

In the midst of this dilemma, a down-on-his-luck merchant from Philadelphia named Charles Goodyear walked into the store. Goodyear had created a new valve for rubber life preservers, and, already a broke man who was swimming in debt, he hoped the store manager would be interested in purchasing his invention. While the manager was impressed with Goodyear's design and appreciated his initiative, he explained that the store was near bankruptcy itself due to so much lost profit. Angry customers were returning melted rubber products every day, demanding refunds and swearing off the once-prized material. And the heat wasn't the only enemy of rubber: in the winter, the material became brittle, hard, and easily cracked. It seemed the era of this wonder material was over almost as quickly as it had started.

But instead of leaving the store dejected, Goodyear was inspired; he was certain there must be a way to make rubber a stronger material, and he was determined to study the substance. Unfortunately, when he returned to Philadelphia his debts caught up to him and he was thrown in jail. Making the best of a bad situation, Goodyear asked his wife to bring him a batch of raw rubber and a rolling pin, and while in his jail cell he spent hours working the rubber and thinking about how to improve it.

Once out of jail, Goodyear began to experiment with some of his ideas. He first considered mixing the sticky substance with a dry powder, like magnesia. This resulted in a much less sticky material, and Goodyear was encouraged with the results. He and his wife and children spent weeks creating hundreds of rubber overshoes in their kitchen; but when summer arrived, the shoes began to sag and melt, just as before.

With no steady income to speak of, Goodyear was forced to sell all his furniture and move his wife and children to a boarding house. But thanks to a generous friend, Goodyear himself was able to move to an attic apartment in New York, where he continued his pursuit of a weatherproof rubber mixture. His next experiment was to add a second drying agent to the rubber, then boil the entire mixture before forming it into shape. Goodyear was extremely encouraged with the samples that resulted from this process, and others were beginning to take note, as well. This promising new form of rubber gained international acclaim, and won an award at a trade show.

A Bit of Nitric Acid Does the Trick

The determined researcher would often walk three miles a day to a mill in Greenwich Village where he had plenty of space to conduct experiments. It was at the mill one day that Goodyear used a bit of nitric acid to remove some dirt off of one of his rubber samples, and noticed that the acid had made the rubber dry and soft, almost like cloth. The rubber was an improvement over any of his past efforts, and Goodyear was certain he'd found the key to creating perfect rubber. President Andrew Jackson even sent the inventor a letter of commendation.

So with the help of a business partner, Goodyear opened a rubber factory in Staten Island. It manufactured rubber shoes, life preservers, and clothing. He was even able to move his wife and children from the boarding house to New York. But soon after things began looking up, the financial panic of 1837 hit, wiping out Goodyear's fortune. Once again a destitute man, Goodyear and his family were forced to camp out in the abandoned rubber factory, surrounded by reminders of their great loss.

With the help of some friends and financial backers, Goodyear was eventually able to pick himself up and move to Boston, where his partners got him a contract to create 150 mailbags for the post office. Confident in his nitric acid rubber mixture, Goodyear finished the bags, piled them in a warm storeroom, and then took his family on vacation, certain the bags would easily withstand the heat. But when he returned several weeks later, Goodyear was greeted by a now-familiar sight: piles of melted rubber. It seemed that the inventor would never find a way to make the sticky substance heat- and weatherproof.

Of course, one should never say never—Goodyear's luck was about to change. The details of what happened in the winter of 1839 are sketchy, but one story goes like this: A broke and desperate Goodyear had moved his family to Woburn, Massachusetts, to be close to several rubber factories that were still in operation. Still relentless in his quest to perfect rubber, he had recently created a mixture that included sulfur, and he took a sample to show the proprietor of a general store.

Sizzling Mad

Goodyear's reputation was well-known in the area, and when he walked in, store patrons began to whisper and snicker at the man who was obsessed with rubber. No doubt tired and frustrated, Goodyear rebuked the critics, gesticulating wildly. His rubber sample flew out of his hands, and landed on a hot stove, where it sizzled for several seconds before Goodyear was able to retrieve it. But instead of the gooey mess he expected to find, Goodyear was surprised to see that the substance had charred like leather, forming a dry, springy rim. The resulting material, was, at last, weatherproof rubber.

But it took several months of painstaking experimentation before Goodyear truly understood what it was he had created and was able to replicate it. He knew that sulfur and heat were the key, but it took much trial and error to discover just how much heat was needed, and for how long the mixture needed to be heated. Finally, he discovered that using steam under pressure for four to six hours at a temperature of 270 degrees Fahrenheit gave him the weatherproof material he'd been in search of for so long.

The process Goodyear stumbled upon, now known as vulcanization, makes natural rubber more elastic. When heated, the molecules in sulfur crosslink with the molecules in rubber, in effect creating one giant new molecule. This new molecule stays elastic, no matter the temperature, creating a material that stays strong yet pliable under any conditions. Today, nearly everything that is made of rubber is vulcanized, including tires, hoses, toys, rubber shoe soles, and even bowling balls.

It may be assumed that Goodyear founded the Goodyear Tire & Rubber Company and enjoyed great success thanks to his discovery, but the truth is that the inventor ended up dying thousands of dollars in debt. The company that now bears his name was created nearly 40 years after his death, founded by Frank Seiberling in 1898. But this is not to say that Goodyear died a bitter man who regretted his life; quite the contrary. The determined inventor was happy to have contributed something worthwhile to society, once saying, "Life should not be estimated exclusively by the standard of dollars and cents. I am not disposed to complain that I have planted and others have gathered the fruits. A man has cause for regret only when he sows and no one reaps."

Elevating Invention To New Heights

When Elisha Graves Otis and his sons began their elevator business in the 1850s, the solid brick buildings of America's cities had four-story height limits. By the 1920s, with the widespread adoption of safe, power-driven lifts, skyscrapers had replaced church steeples as the hallmarks of urban design.

* * * *

ELEVATORS TO LIFT cargo have been around since the pyramidal ziggurats of ancient Iraq. In 236 BC, the Greek scientist Archimedes used his knowledge of levers to deploy beast- and slave-drawn hoists. In 1743, technicians of French King Louis XV devised a special "flying chair," with pulleys and weights running down the royal chimney, to carry his mistress, Madame de Pompadour, in and out of the palace's upper floor.

An Uplifting Background

A descendant of American Revolutionary James Otis, Elisha Otis won a hard-earned path to success. Born in Vermont in 1811, Otis was a stereotype of Yankee ingenuity. In the 1840s, as a senior mechanic in a bedstead factory in Albany, New York, he patented a railroad safety brake, critical to quickly and safely hauling freight in and out of the factories of the Industrial Revolution. By 1852, Otis was a master mechanic at another bedstead firm in Yonkers, New York. He began tinkering with a safety lift for its warehouse, but the company went belly-up. Otis was mulling a move to California's Gold Rush country when a furniture maker asked him to build two safety elevators. A pair of workers at the manufacturer had died when a cable to their lift broke. Fighting off chronically poor health, Otis established his own company and set to work.

All Safe

In 1854, Otis—looking quite distinguished in a full beard and top hat—took to a platform at the Crystal Palace exposition in New York. A rope had pulled his newfangled "hoisting apparatus" high up a shaft, its side open to public view. With a flourish, he waved an ax toward the nervous onlookers crowding the hall. Then, with a quick motion, Otis cleaved the rope with the ax. The onlookers gasped as the elevator began its downward plunge—only to suddenly stop after a three-inch fall.

Elisha Otis tipped his hat and proclaimed: "All safe, gentlemen, all safe."

Otis's means of making his freight elevators "all safe" was straightforward. He attached a wagon wheel's taut springs to the elevator ropes. "If the rope snapped," explained a Smithsonian publication, "the ends of the steel spring would flare out, forcing two large latches to lock into ratchets on either side of the platform."

Otis soon patented an elevator driven by a tiny steam engine, permitting small enterprises like retail stores to purchase their own lifts. Modern department stores with multiple floors, such as Macy's, began to appear.

Despite the technical wizardry, Elisha Otis's commercial success and business sense were limited. Two years after his successful demonstration—despite a follow-up exhibit at P. T. Barnum's Traveling World's Fair—sales of Otis elevators totaled less than $14,000 a year. Even if proceeds picked up, wrote Otis's son Charles, "Father will manage in such a way [as] to lose it all," going "crazy over some wild fancy for the future." Five years later, in 1861, Otis died at age 49 of "nervous depression and diphtheria." He left his two sons a business that was $3,200 in the red.

Success

Charles and Norton Otis proved better businessmen and rivaled their father as technicians, making important improvements to their useful device. By 1873, Otis Brothers & Company, revenues soaring, had installed 2,000 elevators into buildings. Replacing steam-powered lifts, their hydraulic elevators sat on steel tubes sunk into shafts deep below the buildings. An influx of water pushed the platforms up. Reducing the water pressure lowered the elevators.

Where hotel guests previously had preferred the accessible first floor, they now opted to "make the transit with ease," boasted an Otis catalog, to the top floors, which offered "an exemption from noise, dust and exhalations of every kind."

Though taken for granted today, elevators were the height of opulence then. The Otis elevator in Gramercy Park, New York, which dates from 1883 and is still running, was made of upholstered seating and walnut paneling. Another elevator from that era in Saratoga Springs, New York, was outfitted with chandeliers and paneled in ebony and tulipwood.

Riding the skyscraper boom, the Otis firm went from one noted project to another. In 1889, the firm completed lifts for the bottom section of the Eiffel Tower. Around 1900, it bought the patents to a related invention, the escalator. In 1913, the Otis firm managed the installation of 26 electric elevators for the world's then-tallest structure, New York's 60-story Woolworth Building. In 1931, Otis installed 73 elevators and more than 120 miles of cables in another record-breaker, the 1,250-foot Empire State Building.

Setting the Ceiling

All the while, along with enhancements, such as push-button controls, came improvements in speed. Cities constantly changed their elevator "speed limits"—from a leisurely 40 feet a minute for Elisha Otis's original safety lifts, to a speedy 1,200 feet a minute in the 1930s, to today's contraptions, which, at 2,000 feet per minute can put a churning knot of G-force in the stomachs of passengers hurtling to their destinations. "That's probably as much vertical speed as most people can tolerate," says an Otis engineer.

Along the way, the elevator industry quashed early fears that speedy lifts were bad for people. In the 1890s, *Scientific American* wrote that the body parts of elevator passengers came to a halt at different rates, triggering mysterious ailments.

Like the earlier notion that fast trains would choke passengers by pushing oxygen away from their mouths, that theory has since been debunked.

The Steam Engine

Powering the Industrial Revolution, the steam engine was the first truly successful, mass-produced engine ever invented. The engine propelled factories, trains, ships, and even cars in their early days, in some cases taking a heavy burden off manpower and horse (or other animal) power. And to think that the key to it all was—and remains—good old-fashioned water!

* * * *

WATER, IN LIQUID form, can certainly be used as an energy source. Anyone who has seen the majesty of Niagara Falls has witnessed that. But heat water to its boiling point and turn it into vapor, and its volume increases by many hundreds of times. With that expansion comes raw energy—energy that scientists like Thomas Newcomen in the early 1700s and James Watt (of light bulb fame) some 50 years thereafter put to great use powering their unique engine designs.

It was British engineer Thomas Savery who patented the first steam "engine" in 1698. It was really more of a pump that solved a mine drainage problem and allowed water to be more easily dispersed to the people in his community. In 1712, Newcomen—another Brit with the same goal of delivering water from mines—invented the "atmospheric" steam engine, which became the precursor to the version of the steam engine that wound up becoming a driving force in transportation. Newcomen had been hired by Savery for his iron-forging skills and wound up improving on his mentor's invention.

Similarly, Scottish engineer Watt studied the Newcomen engine and invented the model that revolutionized transportation and factory production. Whereas prior engines operated in a stroking motion, Watt's featured a rotary motion that allowed for the generation of great energy. By the end of the 18th century, scientists and engineers were racing to apply the technology to the operation of trains and boats, along with a wide range of machinery used in manufacturing. Watt's 10-horsepower steam engine could be operated anywhere water and coal, or wood fuel, could be sourced.

Steam-powered Industry

Over the next century, steam engines evolved from 10 horsepower to 10,000 horsepower. It was rapid progress that allowed several milestone firsts along the way. Englishman George Stephenson used the technology to invent the first steam locomotive, beginning his work as a 20-year-old and completing it in 1814. His "Blucher" made its first trip on July 25 of that year, hauling eight coal wagons weighing 30 tons on a 450-foot, uphill journey at about four miles per hour. The world took notice. The widely celebrated achievement prompted others to follow suit, and spurred Stephenson to build a total of 16 different engines before he was through.

The burgeoning steam engine helped bring on the Industrial Revolution, the period in history that is usually pegged between the years 1760 and the first few decade of the 1800s—perhaps up to 1840. During this key time, steam power, mechanical tools, and factory manufacturing changed the way people lived, moved, and produced goods around the world. The steam engine was at the heart of it all, as goods became mass-produced and cross-continental travel reached new levels of accessibility.

Steam engines come in all shapes and sizes, but all of them require two main components—a boiler or steam generator, and the motor or engine unit. The boiler contains the water, heats it until it expands to steam and sometimes continues to raise the temperature of the steam for maximum energy conversion. The motor or engine, of course, becomes the object of that energy and then gets put to work powering the ship, train, car, or machine benefitting from the technology. The types of engines driven by steam power vary widely, and of course have evolved over time.

Surpassed by Gas

In the 1900s, gasoline and diesel combustion engines began to replace steam-powered engines. These engines could be built much more powerfully and efficiently, at least during times when gas and diesel fuel were widely available and not too expensive. Like their steam predecessor, these engines led to massive improvements in speed and horsepower. Cars, trains, and ships ran at much faster speeds. Factory equipment and even home machines like gas-powered lawn mowers and tools became available.

In recent years, steam engines have made something of a comeback. Many companies are putting manpower and research dollars into building on steam technologies in an effort to relinquish their reliance on other fuels. Huge strides have been made in Europe toward developing efficient steam engines using the latest in modern materials.

Some might be shocked to learn that the vast majority of nuclear plants today rely on steam engines for their power. Radioactive fuel rods replace coal in boiling water quickly and creating huge amounts of energy. Steam turbines are also employed in many places to drive solar thermal power plants. The steam engine, as it turns out, is far from extinct.

Solar Cells

We often think of clean, renewable, solar power as a modern invention, with solar cells now regularly popping up on tech gadgets, trains and buses, businesses, and even homes. But surprisingly, the very first solar cell was created more than a hundred years ago, in 1883.

* * * *

IT ALL STARTED with the "photovoltaic effect," which is the process by which we harness solar power. French physicist Edmond Becquerel first discovered the effect in 1839 when he noticed that light striking an electrode produced voltage. But it would take many more decades before this discovery—sometimes also called the Becquerel Effect—was put to good use.

First Selenium

In 1873, English electrical engineer Willoughby Smith, who was supervising the development of an underwater telegraph system from Dover, England, to Calais, France, was searching for a method of testing the cable as it was being laid. While using selenium rods as a test circuit, he noticed that the element was much more conductive when it was exposed to sunlight. Building on this discovery, a professor at King's College, William Grylls Adams, and his student, Richard Evans Day, soon proved that not only was selenium photoconductive—meaning it generated electricity when exposed to light—but it also could do so without needing heat or moving parts, exhibiting the photovoltaic effect first observed by Becquerel.

This led to the creation of the first working solar cell, constructed by American inventor Charles Edgar Fritts, who coated a plate of copper with selenium and gold leaf. The resulting solar cell created energy that was, according to Fritts, "continuous, constant, and of considerable force."

The inventor shared his solar panel with German industrialist Werner von Siemens—the founder of telecommunications company, Siemens—and the industrialist, excited by the possibilities of such a cell and considering it a huge step in harnessing clean energy, shared it with the Royal Academy of Prussia.

By the beginning of the 20th century, scientists understood more about light's energy. A trio of scientists—Heinrich Hertz, Max Planck, and Albert Einstein—began to discover in detail the properties of the photovoltaic effect, along with uncovering the properties of the "photoelectric effect." This is when a material emits electrons when it is hit with light, which, in turn, produces power. Einstein theorized that light energy is carried in "discrete wave packets" (which we now call photons) and that the power produced by these packets was dependent upon the wavelength of the light, not the intensity of the light.

Then Silicon

Although these discoveries gave researchers a better understanding of the concepts behind solar power, it would take almost another half century before scientists at Bell Laboratories in Murray Hill, New Jersey, began experimenting with silicon and discovered it was much more efficient than selenium. In 1954, they created a silicon solar cell with a 6 percent output—this was comparable to the gasoline engines of the time, and meant that this solar cell could efficiently power electrical equipment. By 1956, solar cells were available commercially, although their high cost meant that the general public rarely used them. The space program, however, jumped at the chance to use this efficient new technology. Solar panels powered almost all satellites by the 1960s. In fact, the first satellite to use solar panels—the Vanguard 1—transmitted signals for almost seven years using nothing but its solar power.

In the 1970s, the Exxon Corporation created a cheaper solar cell by using silicon from rejected semiconductors. While the lower price was still out-of-reach for most regular consumers, many businesses were now able to afford solar panels, and they began to be used to power buildings, lights, calculators, watches, and radios. President Jimmy Carter even installed solar cells on the White House in 1979. The first solar power plant was constructed in 1982, in Hesperia, California, with another soon following in Carrizo Plains. This plant consisted of 100,000 photovoltaic arrays—the largest collection in the world at the time.

In the 1980s and 90s, solar power became more accessible to everyone, with solar panels being used in desert areas, in telecommunications industries, to power remote homes, and on retractable awnings on recreational vehicles. By 2005, do-it-yourself solar panel kits became available, bringing the possibility of solar energy to just about anyone. Perhaps one of the most amazing possibilities for the use of solar cells is in aircraft: The *Solar Impulse 1* and *Solar Impulse 2* are Swiss experimental planes that have flown thousands of miles on nothing but solar power. And aerospace company Airbus has also recently constructed an unmanned, solar-powered aircraft that flew for nearly 26 uninterrupted days.

Solar power has become cheaper and more efficient than ever, with more and more homes making use of the technology and more than 1 million solar installations appearing in America alone. Solar power makes up more than 1 percent of global power use and continues to climb, with some countries—such as Germany and Greece—obtaining more than 7 percent of their power from solar cells.

With new discoveries every day and a world in need of cleaner energy, it's safe to say that we've still only scratched the surface of what solar cells can provide us.

Frequency Modulation (FM Radio)

You've heard this one before: a fearless innovator's marvelous invention is tarnished by betrayal.

* * * *

FAME AND RICHES are supposed to go to those visionaries that build the better mousetraps. But with the invention of frequency modulation (FM radio), things definitely didn't work out that way. Edwin H. Armstrong (1890–1954) invented a new transmission medium that left the former giant, amplitude modulation (AM radio), blowing static in its wake. For most people, such a lofty achievement would bring a degree of lifelong satisfaction—not to mention a nice stack of cash. For Armstrong, it would bring mostly heartache.

Armstrong the Inventor

Before this underappreciated genius found his way to FM, he made other worthy contributions. Two of Armstrong's inventions, the regenerative circuit of 1912 and the superheterodyne circuit of 1917, would set the broadcasting world on its ear. When combined, they would produce an affordable tube radio that would become an American staple. It seemed that Armstrong was on his way.

Soon afterward, the inventor turned his attentions to the removal of radio static, an inherent problem in the AM circuit. After witnessing a demonstration of Armstrong's superheterodyne receiver, David Sarnoff, the head of the Radio Corporation of America (RCA) and founder of the National Broadcasting Company (NBC), challenged the inventor to develop "a little black box" that would remove the static.

Armstrong spent the late 1920s through the early 1930s tackling the problem. Sarnoff backed the genius by allowing him use of a laboratory at the top of the Empire State Building. This was no small offering—in the broadcasting game, height equals might, and none came taller than this 1,250-foot giant, which has since been named one of the seven wonders of the modern world.

A Well-modulated Approach

In 1933, Armstrong made a bold announcement. He had cracked the noise problem using frequency modulation. With a wider frequency response than AM and an absence of background noise, the new technology represented a revolutionary step in broadcasting.

Armstrong's upgraded system also had the ability to relay programming from city to city by direct off-air pickup by 1936. Without knowing it, the inventor had effectively boxed himself in. NBC, and by extension Sarnoff, was the dominant force in conventional radio during this time. With America still mired in an economic depression, NBC wasn't interested in tooling up for a new system. Even worse for Armstrong, television loomed on the horizon, and NBC was pouring most of its resources into the new technology. Instead of receiving the recognition and financial rewards that he so rightly deserved, Armstrong was fired unceremoniously by his "friend" Sarnoff. It seemed like the end of the line—but Armstrong's battle was only just beginning.

In 1937, a determined Armstrong erected a 400-foot tower and transmitter in Alpine, New Jersey. Here, he would go about the business of perfecting his inventions. Unfortunately, without Sarnoff's backing, his operation found itself severely underfunded. To make matters infinitely worse, Armstrong became embroiled in a patent battle with RCA, which was claiming the invention of FM radio. The broadcasting giant would ultimately win the patent fight and shut Armstrong down. The ruling was so lopsided that it robbed Armstrong of his ability to claim royalties on FM radios sold in the United States. It would be hard to find a deal rawer than this.

To fully appreciate Armstrong's contribution, compare AM and FM radio stations: The difference in transmitted sound will be pronounced, with FM sounding wonderfully alive and AM noticeably flat in comparison. Even "dead air" sounds better on FM because the band lacks the dreaded static that plagues the AM medium. Without a doubt, FM technology is a tremendous breakthrough. But it came at a terrible cost. On January 31, 1954, Armstrong—distraught over his lack of recognition and dwindling finances—flung himself from the 13th-floor window of his New York City apartment.

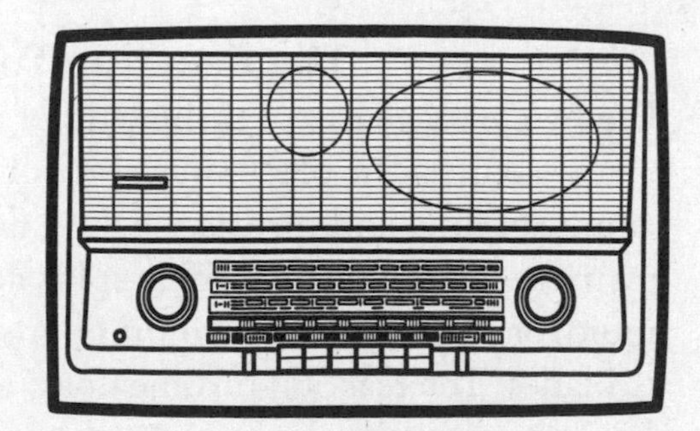

The Turbine

Almost all electrical power produced on the planet is generated by a turbine of some kind or another. Turbines are rotary mechanical devices that extract energy from a moving liquid or gas and turn it into useful work. When combined with a generator, this work can be turned into electrical power.

* * * *

A WINDMILL IS THE simplest form of turbine. It gathers kinetic energy from the moving air—or wind, as it were—and converts it to mechanical energy that can be put to use in any number of ways. The harder the wind blows, the faster the windmill turns and the more mechanical energy is produced.

More sophisticated gas, steam, and water turbines contain casing around the blades that house the working fluid. In the second part of the 19th century, Anglo-Irish engineer Sir Charles Parsons invented the reaction turbine while Swedish engineer Gustaf de Laval almost simultaneously pioneered the impulse turbine. Interestingly, their contributions to the world of energy, mechanics, and power came years after the word "turbine" had first been coined by French engineer Claude Burdin in 1822. The term came from the Latin word "turbo," for vortex or whirlpool. A former student of Burdin's, Benoit Fourneyron, is credited with building the first working water turbine. He went on to make significant contributions to his creation.

The Reaction Turbine

The two main genres of turbine differ in how they react to the energy source. The blades in a reaction turbine sit in a larger volume of fluid and turn as the fluid flows past. A wind turbine is a great example. The turbine spins as the energy source—liquid, or in the case of a windmill, rushing air—pushes past its blades. The reaction turbine does not alter the direction of the source, but simply reacts to it (hence the name). Another way to think of it is like swimming, but with an opposite flow.

Swimmers push water with their arms as long as possible so the energy gets transferred into forward propulsion of the body through water. If the water came at the swimmer with such force as to propel his or her arms and legs by itself, supplying energy to the swimmer, the dynamic would be similar to that of a reaction turbine. The longer the energy source is impacting the "machinery," the better. It's why one wants moving air hitting windmill blades for as long as possible.

The Impulse Turbine

An impulse turbine, on the other hand, has a fast-moving fluid or energy source connecting with the turbine blades and forcing them to spin. Imagine, for example, throwing tennis balls at the blades of a turbine and watching it spin faster each time a ball strikes a blade. Energy is transferred with each strike, and the source of the strike changes direction. That is, the ball loses the same amount of energy it transfers to the turbine (Newton's Law), and thus the balls bounce back at a slower rate than they moved toward their target. The more forceful the strike, the more energy that's transferred and the faster the turbine will spin. Water turbines are frequently impulse turbines. And because they do not need to be housed inside a pipe or cylinder, they are cheaper and easier to produce.

Water, wind, steam, and gas are primarily the different types of turbines powering the planet. Although they all work by spinning as fluid moves against them, there are differences between each that require them to be built and operated in varying ways. Wind turbines, for instance, are typically large because they have to turn more slowly for a variety of reasons, not the least of which is safety. Gas turbines operate at such high temperatures that special alloys are required in their manufacture. Some jet engines are forms of gas turbines.

While turbines have been around for almost two centuries, powering much of the world, some types continue to make big news for modern advancements. Two recent graduates of

Lancaster University in England concocted a small, omnidirectional wind turbine that won them the 2018 James Dyson Award, an international design prize that celebrates the designers of new problem-solving ideas.

The O-Wind, created by Lancaster pals Nicolas Orellana and Yaseen Noorani, was designed to accommodate erratic wind patterns often seen in urban environments. While most wind turbines only work when winds impact them head-on, these small globes feature wide vents facing different directions that can "grab" winds from all angles, siphon it to smaller vents and cause the devices to spin around an axis. Their spin can drive an electric generator, powering homes with green energy. "If you put a wind turbine on your balcony, you're not only going to get horizontal wind," Orellana explained. "In that position, you can also get vertical winds or diagonal winds, and it changes in every situation. With this one . . . it can work more continuously and provide much more energy to your house."

The Laser

Bet you didn't know that the word "laser" originated as an acronym to shorten the cumbersome description "light amplification by stimulated emission of radiation."

✻ ✻ ✻ ✻

TODAY, LASERS HAVE a myriad of uses, including in optical disc drives, fiber optic communication, barcode scanners, laser surgery, and devices used by military and law enforcement. But the scientist who created the first working laser wasn't even sure what the technology could be used for, calling it "a solution seeking a problem." Lucky for lasers, the world is full of problems just waiting to be solved.

The principle behind lasers dates back to 1917, when Albert Einstein described the theory of stimulated emission of electromagnetic radiation, which is the process that makes lasers possible. With stimulated emission, a photon of a specific frequency can interact with an excited molecule and cause it to drop to a normal energy level. This triggers a chain reaction in which other excited atoms return to normal, during which time, a light of a single wavelength is emitted.

This single wavelength of light is what distinguishes a laser from other forms of light. Laser light is coherent, which allows it to be tightly focused, even over great distances. But Einstein's theories about radiation didn't come into use until the 1950s, when physicists like Joseph Weber and Charles Townes researched the possibility of using stimulated emissions to create a microwave amplifier.

Townes, along with graduate students James P. Gordon and Herbert J. Zeiger, built the first microwave amplifier in 1953, creating a device that amplified microwave radiation as opposed to visible or infrared radiation. But because it operated on the same principles, it became known as the "maser," or "microwave amplification by stimulated emission of radiation." Masers are used as the timekeeping device in atomic clocks and low-noise microwave amplifiers in radio telescopes and space communication ground stations.

Townes theorized that the principles used in the maser could be used for optical light as well, and he and physicist Arthur Leonard Schawlow, working together at Bell Laboratories, began to study the concept with both infrared radiation and visible light. In 1958, Townes published a summary of their research, entitled "Infrared and Optical Masers," in the December issue of *Physical Review*. Although Townes and Schawlow had yet to build a working prototype of their idea, the article fascinated the scientific community, and other researchers scrambled to be the first to bring the concept to fruition.

With scientists at Westinghouse, RCA, Siemens, Lincoln Lab, and IBM all vying to make history, the competition was fierce.

But it was Theodore Maiman, a physicist and engineer at Hughes Research Laboratories in Malibu, California, who beat the others to the punch. Maiman, who had been hired to work on government aerospace contracts, was so interested in the idea of an "optical maser," that he convinced his reluctant supervisors to give him a nine-month budget of $50,000 to work on the project.

Looks Like a School Science Project, Dear

On May 16, 1960, Maiman debuted what was now known as a laser. The name was coined by Columbia University graduate student Gordon Gould, who published his own ideas for the creation of such a device in the late 1950s. Despite the heated competition, Maiman was the first to demonstrate a workable device, which used a ruby crystal to produce red laser light. Maiman's device was reportedly so simple that his wife remarked that "it looked like a high school [science] project." But after Maiman published an article about his invention in *Nature*, he received higher praise from Townes, who called his article, "the most important per word of any of the wonderful papers" ever published by the journal.

Maiman himself may have agreed with his wife, as he famously remarked that "a laser is a solution seeking a problem," and wasn't quite sure what to use it for. What's more, his employer, even after his success, had no desire to develop the laser further. So in 1961, Maiman left Hughes and started his own company, Korad, as researchers began to discover practical applications for his invention, such as laser sighting systems. Early lasers were also used for scientific research, to study atomic physics, and chemistry.

Korad developed and manufactured Maiman's high-power ruby lasers, and in 1968, the company was acquired by Union Carbide. Maiman went on to found the Laser Video Corporation and remained involved in developing lasers and their applications for the rest of his life. Lasers are now used in supermarket check-out scanners, in laser printers, to play CDs and DVDs, in telecommunications systems, for surgery, and in industrial assembly lines, just to name a few of their uses. Not bad for an invention that didn't have a "problem" to solve when it first debuted.

Townes was awarded the 1964 Nobel Prize in Physics for his original ideas concerning lasers. Although Maiman was never given that honor, he was continually recognized throughout his life—and even after his death in 2007—for his achievement. A laser may be a small beam of light, but its invention has had a huge impact on all of us.

Road Surface Technology

Most of us get into our cars and travel to work, school, or the grocery store at least a few times a week. But rarely do we think about the roads on which we drive, unless, of course, a pothole or speed bump serves to make the ride more bumpy. Automobiles have only been around since the late 1800s, so it's easy to assume that roads haven't existed much longer that that. But the truth is that engineers and inventors have been tackling the issue of highway transportation for thousands of years.

* * * *

THE EARLIEST PAVED roads go back all the way to 4000 BC, in the Mesopotamian cities of Ur and Babylon. And interestingly, in the southern region of Sumer, brick makers would set their bricks in place with a substance called bitumen, also known as asphalt. It would take thousands of years for Europe and America to realize the potential of this viscous liquid.

The Carthaginians, located in what is now Tunisia, developed a road system around 600 BC, but Rome saw these neighbors to the south as a threat, and destroyed Carthage in 146 BC. They took note of the city's impressive roads, however, and began building their own. The Romans were the first to create a true highway system, constructing 50,000 miles of stone-paved roads during the time of their empire. This system stretched from Syria and Jordan in the east to Portugal, Spain, and Britain in the west, and was used by the Romans to transport armies, trade goods, and link major cities.

The roads that made up this impressively large system were surprisingly complex, often consisting of three distinct layers of stones, rubble, and gravel, topped with smooth, polygonal imbedded blocks of stone. They were also cambered, or built in a slightly convex shape, to provide drainage for rainwater, which prevented erosion. In fact, these Roman roads were so well built that many of them survive even today. The fact that these roads were constructed so sturdily may explain why, after the fall of the Roman empire around 476, advancements in road construction stalled for hundreds of years. It wasn't until the early 19th century that two Scottish engineers, Thomas Telford and John McAdam, began to rethink the methods used to create roads.

The Scottish Colossus

Telford, born in rural Scotland in 1757, was a civil engineer and architect who designed numerous roads, canals, and bridges throughout Britain, including the Pontcysyllte Aqueduct in Wales which was declared a UNESCO World Heritage Site in 2009. As a teenager, Telford had also apprenticed as a stone mason, and his skill with stonecutting would eventually influence the construction of highways he designed. In the early 1800s he was tasked with improving sections of English roads. But instead of beginning with a layer of randomly chosen stones as previous road builders had, Telford carefully analyzed and cut stones so they would interlock together, then topped them with layers of gravel and pebbles. His ideas were widely adopted in the area, earning Telford the punning nickname, "The Colossus of Roads."

The Macadam Method

But Telford wasn't the only one with new ideas. Around the same time that Telford was constructing his roads, McAdam, who was born in 1756, noticed that most roads were constructed with layers of rounded gravel. McAdam felt that crushed, angular, aggregate would work better, and he began constructing roads with layers of this material, pressed together. The small stones compacted together even further over time and with wagon traffic, creating lasting roads that didn't require a base of large stones. McAdam's process became so popular that his method was given the name "macadam." It soon made its way to the United States, and the first macadam road was constructed between Hagerstown and Boonsboro, Maryland in 1823.

Tarmac

While the macadam method was economical and durable when most traffic was either by foot or horse-drawn wagon, the advent of the automobile in the late 1800s created a problem. The new motor vehicles had a tendency to kick up choking dust clouds, and their tires dislodged the gravel, causing a gradual breakdown of the roads. In 1902, a Swiss doctor living in Monaco, Ernest Guglielminetti, was approached by Prince Albert I about the dust problem caused by automobiles. Guglielminetti had spent time at a hospital in Indonesia, where tar was used on the wooden floors to prevent dirt from floating in the air, and he wondered if the same principle could be applied to dusty roads.

He developed a mixture of tar, gravel, and sand, and applied it to a road, creating the first tar-bound macadam. A combination of tar and ironworks slag was later patented by Welsh inventor Edgar Purnell Hooley, who dubbed the substance "tarmac," after "tar" and "macadam." Today, tarmac is often generically used to refer to airport runways, although the materials used are different.

Portland Cement

By the 1920s, asphalt, derived from petroleum crude oil, was becoming popular for road surfaces, although surprisingly, the first asphalt pavement was built decades earlier, in Paris in 1858. These roads were often laid on a base of gravel, but another material was gaining in popularity, as well: Portland cement, named for a quarry on the Isle of Portland in Dorset, England. This limestone power hardened after reacting with water, but once it was hardened, it was also water-resistant, making it ideal for building roads. In locations where a road needed to be built on soft or easily shifted soil, a base of Portland cement would first be poured, with a layer of asphalt on top.

But Portland cement was also found to be a durable road surface in its own right, being especially strong in high- and heavy-traffic areas. In 1913, the Lincoln Highway Association was established to oversee the construction of a transcontinental highway from New York City to San Francisco, and they were so convinced of cement's ability to hold up in heavily traveled areas that they began constructing "seedling miles" of concrete road for the future highway. Today, about 60 percent of our interstate system is paved with concrete. But asphalt roads still dominate our national highways and byways: 93 percent of all paved roads in the United States are paved with asphalt.

The Future

Like any technology, road surface technology is bound to keep changing as new methods and materials are discovered. So what does the future hold for our daily commutes?

Engineers and scientists are already testing new materials like recycled plastics and organic resins to see how they hold up under traffic, in an effort to lessen the reliance on petroleum-derived asphalt. Colder states are employing materials with built in de-icing properties, creating safer winter driving. And other ideas include road surfaces with built-in solar panels, and highways made up of piezoelectric crystals, which generate an electrical charge when under stress. Driving our cars could literally help to light up cities.

Perhaps one day, we'll finally have those flying cars that futuristic movies like to envision. But until then, just like the ancient Romans, we'll continue to make use of pavement!

Light-emitting Diodes (LEDs)

If you've bought lightbulbs lately, you've no doubt noticed many bulbs labeled "LED." An LED, or light-emitting diode, emits light in the form of photons when connected to a direct current, and operates on the principle of electroluminescence. LED lightbulbs have become popular in recent years thanks to their energy efficiency, which may make it seem as if these lights are a new technology. But in reality, the LEDs that we know today have been around for half a century.

⁕ ⁕ ⁕ ⁕

THE IDEA BEHIND LEDs is even older, beginning back in 1907 when British engineer Henry Joseph Round observed electroluminescence while experimenting with silicon carbide, a semiconductor that contains silicon and carbon. While it was an interesting phenomenon, the light emitted was too faint to be used for any practical purpose at the time, and Round wasn't even sure what to call it. In fact, it would be almost 30 years before the word "electroluminescence" was coined, by French physicist Georges Destriau. Destriau published a report in 1936 which detailed how light was produced when an electric current was run through zinc sulfide powder. But again, this was more of a curiosity than a useful, practical occurrence.

Then, in 1961, tech companies like General Electric and Bell Laboratories began searching for the best materials to use for semiconductors, experimenting with elements like germanium and silicon. At Texas Instruments, two engineers named James

R. Biard and Gary Pittman were researching another material called gallium arsenide, a compound of the elements gallium and arsenic, and noticed that when they paired it with a tunnel diode, it produced infrared light. Although the infrared light was not visible to humans, Texas Instruments received a patent, the first for a light-emitting diode, and began selling these LEDs to IBM to be used in punch-card readers.

Out of the Darkness

While the infrared LED developed by Biard and Pittman was the first modern LED, it was limited in its capabilities since its light was invisible to the human eye. But only a year later, Nick Holonyak Jr., an engineer at General Electric, continued researching the semiconductor gallium arsenide and developed an LED that emitted visible red light, and soon after, another that lit up orange. These early LEDs were often used as indicators on equipment; for instance, lighting up when a machine was "on" and going dark when it was "off." This certainly provided a helpful feature, but it was only the beginning of what LEDs would eventually accomplish.

And into Full Color

Ten years after Holonyak's invention, one of his former graduate students, electrical engineer M. George Craford, discovered a way to make yellow LEDs and to increase the brightness of red and orange LEDs. Craford was working for the Monsanto Company, which was the first company to begin mass-producing the devices in 1968. More and more applications were being developed for the lights, including indicator lights in televisions and radios, and light-up displays in calculators. By the mid-1970s, scientists had also discovered how to create violet, blue, and green lights, and LEDs went from costing a pricy $200 each, to a mere five cents, making them more popular than ever.

In the 1980s and 90s, scientists began experimenting with materials like gallium nitride and aluminum gallium arsenide, resulting in LEDs that were bright enough to be used for lighting applications. By 1987, these diodes were being used for car brake lights, turn signals, and for traffic lights, where they were especially appreciated for their reliability over previous incandescent lights.

Despite the growing usefulness of LEDs, there was still one area they were lacking: home and office lighting. Until the early 90s, LEDs were available in many colors, but none that were conducive to simply lighting up a dark room. But in 1994, Shuji Nakamura, an inventor working for the Nichia Corporation in Anan, Japan, invented an ultra-bright blue LED. Researchers soon discovered that if this bright blue light was covered with a special coating of fluorescent phosphors, some of the blue light was absorbed and the light glowed white. White LEDs were game changers in the business world because this allowed companies to light up commercial spaces with lights that used 80 percent less energy and lasted 25 times longer than incandescent bulbs.

Today, efficient, low-maintenance LEDs are used for a host of different applications. They are still used as indicator lights in airplane cockpits and on ship's bridges; they are used for automotive lighting on cars, trucks, and buses; they're found in street lights and in parking garages; and they are used for flashlights, security cameras, theater lighting, and even airport runway lighting. And of course, LEDs have made their way into our homes, providing long-lasting, energy-efficient light for everything from outdoor porches to reading lamps. As LED technology continues to improve, these lights will only get more reliable, more efficient, and more illuminating.

The Internal Combustion Engine

The principle driving the internal combustion engine is a very simple one. Put a small amount of high-energy fuel such as gasoline in a small, enclosed space and ignite it, and the expanding gas produced by the ignition will generate an enormous amount of energy. Virtually every car with a gasoline engine uses an engine with a four-stroke combustion cycle formally called the Otto Cycle, named for Nikolaus Otto. Otto was the German engineer who created the first modern internal combustion engine in 1876.

* * * *

LET'S START A little earlier than that, however, in looking at the evolution and operation of the internal combustion engine. Both the gas turbine and the first gas engine—in that order—were developed in the last decade of the 18th century. In that same decade, American Robert Street patented a version of the internal combustion engine, which was also the first engine to employ liquid gas. In the early 1800s, engines powered by dust explosions and electric sparks were introduced.

The biggest advancements in internal combustion technology, though, were still a few decades away. Belgian engineer Jean J. Lenoir is widely credited with introducing the earliest commercially successful internal combustion engine in 1858. It was gas-fired, which paved the way for Otto to build what would become the modern standard for this type of engine. Otto's four-stroke combustion cycle, just as the name would indicate, requires four strokes to complete a full cycle. They are the intake stroke, compression stroke, combustion stroke, and exhaust stroke.

Here's how it works. Intake stroke: the piston begins in an upward position, the intake valve opens, and the piston moves down to allow air and gasoline—just a small amount of gas is required—to pour into the cylinder. Compression stroke: the piston moves back up to compress this mixture of fuel and air. Combustion stroke: When the piston reaches the top, the gasoline is ignited by a spark from the spark plug, exploding the gas charge and driving the piston back down. Exhaust stroke: when the piston is back down, the exhaust valve opens and allows exhaust to leave the cylinder. In a vehicle, that exhaust typically goes out through the tailpipe.

Gas and Diesel

Two different types of internal combustion engines are currently in production—the spark ignition gasoline engine and the compression ignition diesel engine. You can probably guess which type of gas one might put into each of these at the local filling station. In the spark ignition engine, as described above, the fuel-air mixture enters the cylinder, gets compressed by the piston and gets ignited by a spark, causing combustion. Conversely, with a compression ignition diesel engine, only air is brought into the engine and compressed. Fuel then gets sprayed into the compressed hot air, combusting it.

In addition to the four-stroke engine pioneered by Otto, there are also two-stroke engines. These feature pistons that complete a full cycle with every revolution of the crankshaft. Only two strokes are required—hence the name—requiring the cycle to finish with a stroke that includes combined intake and exhaust. The first reliable two-stroke engine to gain a patent was introduced in 1879 by German engineer Karl Friedrich Benz. Several years later, Benz kicked off the first commercial production of motor vehicles with the internal combustion engine. His Benz Patent Motorcar of 1885 is widely considered the first practical automobile in the world.

Internal combustion engines of both stroke types convert the linear motion of pistons into rotational motion through the use of a crankshaft. This works well for vehicles, whose tires will mimic this same rotational motion. The core of the engine is called the cylinder. Cars have more than one—generally four, six, or eight—while machines like home lawnmowers typically have just one. Other important components of the internal combustion engine are the spark plug, valves, pistons, piston rings, the connecting rod, and a sump, which surrounds the crankshaft. One big concern with so many engines spewing exhaust far and wide has been pollution, of course. More than 250 million vehicles are regularly on the roads in the United States alone. The Environmental Protection Agency has kept a close eye on emissions and clamped down on both automobile manufacturers and owners over the last few decades.

Thankfully, the automobile industry has also recognized the issue and has made great strides toward keeping the air cleaner. Research and development have helped manufacturers reduce overall internal combustion engine emissions of criteria pollutants, such as nitrogen oxides and particulate matter, by more than 99 percent in the last 30 years. Research has also helped engine performance and efficiency, which leads to greater fuel economy. This, combined with the onset of hybrid and electric vehicles, has drivers and the EPA celebrating.

The Jet Engine

Just three decades after Orville and Wilbur Wright flew an airplane about 10 feet off the ground near Kitty Hawk, North Carolina in 1903, engineers were working on jet engines—primarily in lab settings. Germany's Hans von Ohain designed the first operational jet engine in 1937 and two years later built a functioning jet aircraft, though "official" credit for the first engine went to Englishman Frank Whittle. Whittle simply beat Ohain to the patent office.

❋ ❋ ❋ ❋

WHEN IT COMES to who gets credit for moving jet engine technology from lab settings to practical application, World War II is the unquestioned answer. With the conflict looming in the late 1930s, there was a rush to bring the tremendous power of the jet engine to the forefront. That's what prompted Ohain to design the experimental Heinkel He 178, which first took to the air on August 27, 1939. Fellow German engine designer Anselm Franz raised the technology to a new level in introducing the first jet engine to power a fighter plane. Messerschmitt put Franz's engine in the Me 262, the only jet-powered fighter to fly in World War II combat. However, its high fuel consumption kept it on the ground for much of the war.

Back in England, Whittle developed a successful jet engine for an Allied fighter, the Gloster Meteor, but that jet was used for homeland defense rather than combat. Great Britain shared the technology with the U.S., which put General Electric on the task of building the engines for its Bell XP-59.

Germany's surrender in 1945 opened a wealth of jet engine technology to the rest of the world. Since then, jet engines continue to be built more powerfully to push jet planes faster and faster through the skies. Jet technology shares some components with internal combustion engines, but there are key differentiators that make jet engines the gold standard.

Just as internal combustion engines do, jet engines burn fuel with air in a process called combustion. However, while combustion engines employ cylinders that go through four steps, jet engines carry out those same steps in a straight-line sequence down a long metal tube. In the turbojet, the simplest jet engine, air enters through an intake, gets compressed by a fan, mixed with fuel and combusted. The result is hot, fast-moving exhaust that generates incredible power.

The more fuel burned, the more power produced. The vast amounts of fuel burned by a jet engine distinguish it from a car engine and make it much more powerful. While cylinder engines can vary in the amount of power produced, jet engines are generally designed to produce maximum power at all times. Another differentiator can be seen in the way a jet engine releases its exhaust. While a vehicle engine sends exhaust through a pipe, a jet engine passes its exhaust through multiple turbine stages. This makes the engine more efficient, producing more power from the same fuel mass.

Types of Jet Engines

Other types of jet engines, aside from the turbojet, are the turboshaft (which powers helicopters), turboprop, turbofan, and the tapered ramjet and scramjet. The jet engines in production and use today make those of the World War II era look, well, slow. Instead of shooting gas out the back in one swoop, today's jet engines send it through a second turbine that powers a fan at the front of the engine. The newer models take in much more air and don't have to accelerate it as much as older jet engines did, again increasing power and efficiency. Some of the largest jet engines have fans exceeding 10 feet in diameter, but it's also worth noting that some are smaller than a human hand.

The resulting speeds generated by the world's best jet engines are truly remarkable—numbers that could never have been imagined by the likes of von Ohain and Whittle. In 1976, the Lockheed SR-71 Blackbird set the world speed record for a manned, air-breathing jet when it clocked a ridiculous 2,193 miles per hour near California's Beale Air Force Base. Manning the milestone flight were Eldon W. Joersz and George T. Morgan Jr.

It's not just aircraft that can benefit from jet engine technology. These engines are also used in cruise missiles, unmanned aerial craft and even cars (typically for drag racing). Powered by a turbofan engine, ThrustSSC—a car—set the world land speed record in 1997 when it achieved a top speed of 763 miles per hour. It also became the first land vehicle to officially break the sound barrier.

The Expanding Universe

We've all heard of the Hubble Space Telescope, the high-resolution telescope that was launched into orbit around Earth in 1990 and has sent us so many amazing visuals of space. But who, exactly, was Hubble? And why does this telescope bear his name?

❋ ❋ ❋ ❋

EDWIN HUBBLE, WHO was born in Marshfield, Missouri, in 1889, didn't set out to be a scientist, even though astronomy was a hobby when he was a boy. Instead, he studied law at the University of Chicago and The Queen's College, Oxford, fulfilling a promise to his father, who preferred he study the subject. But after his father died in 1913, Hubble went back to school, studying astronomy at the University of Chicago's Yerkes Observatory, where he earned a PhD in 1917. In a foretelling of things to come, his dissertation was entitled, "Photographic Investigations of Faint Nebulae," for which he made use of one of the most powerful telescopes in the world at time, housed at the observatory.

After serving in the United States Army during World War I, Hubble began working at the Carnegie Institution for Science's Mount Wilson Observatory, where he would work for the rest of his life. Starting out as a junior astronomer, Hubble listened to the theories his colleagues espoused about the cloudy patches of nebulae they observed through the institution's 100-inch Hooker Telescope, then the world's largest. The prevailing theory at the time was that the Milky Way Galaxy comprised the entire universe, so these nebulae, Hubble's fellow astronomers believed, were all located within the galaxy.

Just a few years earlier, Albert Einstein had begun forming his theory of general relativity, which attempted to explain the laws of gravity and their relation to other forces of nature. Just as astronomers like Isaac Newton had believed for hundreds of years, Einstein accepted the belief that our cosmos was a static, "steady-state" universe, which would appear the same from any direction an observer happened to be located. Stars would simply be suspended motionless in the void, the same today as a hundred years from now.

However, when Einstein tried applying his theory of relativity to the universe as a whole, he found the "steady-state" concept didn't seem to fit. In fact, in order for his theory to work, space-time would need to be warped and curved back on itself, which would cause matter to move as it was affected by its own gravity. But even Einstein wasn't ready to give up on the idea of a static galaxy.

The Cosmological Constant

He proposed a force that he called the "cosmological constant," which worked against the gravity in the universe to keep everything in a constant state. In this way, the Milky Way Galaxy, which was presumably the entire universe, would simply go on maintaining itself forever and always.

But when Hubble began studying the nebulae his colleagues had been observing, he wasn't so sure that they were all confined to the Milky Way. One of these nebulae, which had been named the Andromeda nebula, contained a very bright star called a "Cepheid variable." These stars pulsate in a predictable way that can help to determine their distance from an observer.

Over several months at the end of 1923, Hubble took pictures of the Andromeda nebula and measured the brightness of the Cepheid variable star, determining that its brightness varied over a period of 31.45 days. He also calculated that it was 7,000 times brighter than our own sun, and was able to determine that it was 900,000 light-years away.

Scientist Harlow Shapley, who was the head of the Harvard College Observatory from 1921 to 1952, had just a few years prior calculated the distance across the Milky Way to be about 100,000 light-years. Although Shapley would at first criticize

Hubble's findings, these measurements clearly showed that the Andromeda nebula was located far outside of the Milky Way Galaxy, proving that the universe was much more vast than previously assumed.

In fact, today, scientists know that there are actually two types of Cepheid variable stars, and Hubble's calculations were rather a bit off: the Andromeda galaxy, as it's now called, is approximately two million light-years away.

The Universe Gets Bigger

As telescopes advanced and became more powerful, Hubble began to observe more galaxies, measuring their brightness and noticing a shift in the light toward the red end of the spectrum. This "redshift" occurs due to the Doppler effect, which is a change in the frequency or wavelength of a wave relative to movement. Hubble measured as many galaxies as he could, and in 1929 published a paper that revealed his findings. The universe, Hubble realized, was not a static cosmos as had been assumed for centuries, but rather it was moving outward. All of the galaxies we can observe are flying apart from each other at great speeds, causing our universe to expand.

An Explosive Concept

What's more, Hubble calculated that each galaxy is moving at a speed in direct proportion to its distance, a concept now known as Hubble's Law. Galaxies that are farther away are moving more quickly than galaxies that are closer to us. And by measuring the rate of expansion, scientists were able to determine the approximate age of the universe, which is about 13.8 billion years old. Scientists also realized that if the universe is continually expanding, that must also mean that at one time, it was much smaller and more compact. This realization eventually gave rise to the Big Bang theory.

When Einstein heard of Hubble's discovery, he was not only happy to see that it helped to bolster his theory of relativity, but he also called his previous idea of a "cosmological constant" his "biggest blunder." But Einstein should not have been so quick to discard his "cosmological constant" theory, as many modern-day scientists believe an invisible, gravity-repellant form of "dark energy" makes up most of the universe. Could this be Einstein's "cosmological constant"? Perhaps one day, we'll know for sure; but until then, we'll continue to marvel at the images of the expanding universe made possible by technology like the Hubble Space Telescope.

Microchip Technology

It's amazing to think that computers used to be so big that they filled up entire rooms. In fact, the ENIAC, which was completed in 1946 and considered by many to be the first fully functional digital computer, was so big that it needed 1,800 square feet of space—the size of many houses! Even computers from 20 or 30 years ago look huge compared to the sleek laptops and tablets we're used to today. But in order for smaller (not to mention faster and more complex) machines to become reality, engineers have always needed to find ways to fit more components into smaller spaces. And one innovation that helped greatly was the microchip, or integrated circuit.

* * * *

SOON AFTER COMPUTERS like the ENIAC debuted, scientists noticed a problem: with tens of thousands of components working to keep these machines running (ENIAC alone had more than 17,000 vacuum tubes!), many of them reached a point where the time and energy it took to troubleshoot issues and replace broken parts exceeded the benefit derived from the computers. This problem, which scientists called "the tyranny of numbers," meant that the machines, due to their constant maintenance, were basically negating their own value. It seemed pointless to create more computers; unless, of course, a better way to construct them could be found.

After the invention of the transistor in the late 1940s, engineers began to experiment with the idea of placing several transistors on a single semiconductor. Scientists like German Werner Jacobi and American Sidney Darlington filed patents for devices that consolidated transistors in this way, but many of these early devices were never used for any practical applications. Meanwhile, electronics giants like Bell Laboratories, IBM, and General Electric searched for ways to reduce the number of components in their machines, looking to end "the tyranny of numbers."

The Integrated Circuit

Then, in 1958, Jack Kilby, a radio engineer who was working for Texas Instruments, had an idea to assemble components, like transistors, diodes, and capacitors, directly into a germanium chip. His creation became the first working integrated circuit, which he demonstrated on September 12, 1958. Kilby's invention caught the attention of the U.S. Air Force. They became the first military branch to adopt the technology for their molecular electronics program.

While it was a revolutionary invention, Kilby's integrated circuit had some shortcomings. One of these was his use of germanium, which, engineers discovered, could be damaged when exposed to high temperatures. So Robert Noyce, a physicist who co-founded Fairchild Semiconductor in San Jose, California, proposed that silicon be used to make integrated circuits as the material was more robust. In 1960, Fairchild manufactured the first silicon integrated circuit, which consisted of four transistors embedded in a single wafer of silicon.

Another engineer working at Fairchild, Jean Hoerni, developed a manufacturing process, called the planar process, which made mass-producing the microchips a simpler endeavor by eliminating the need to manually wire components together.

In 1961, Fairchild made integrated circuits commercially available, and the devices began to replace the individual transistors and parts that had made up previous versions of computers. At first the size of a pinky finger, the first integrated circuits contained only a handful of transistors, resistors, and capacitors. Called "Small-Scale Integration," or SSI, these first chips in the 1960s could contain several dozen components, which was already an improvement for the burgeoning computer industry.

From a Handful to Billions

But integrated circuit technology improved rapidly, and soon, "Medium-Scale Integration" (MSI) was possible, with hundreds of components on one chip. By the 1970s, LSI, or Large-Scale Integration, made thousands of components on one chip a possibility. VLSI and ULSI, Very-Large- and Ultra-Large-Scale Integration, increased the numbers to tens of thousands and, even more impressive, millions. Today, a chip the size of a dime can contain more than 100 million tiny components, and the most advanced computer chip has an astounding 2.6 billion transistors! These incredibly complex chips have resulted in increasingly smaller devices that can do more than ever.

While Kilby and Noyce both made important contributions to the creation of integrated circuits, only Kilby won the 2000 Nobel Prize in physics "for his part in the invention of the integrated circuit." But Noyce certainly went on to make a name for himself as well. He co-founded the Intel Corporation with Gordon Moore in 1968. And Noyce inadvertently made another huge contribution to the computer industry when he suggested silicon as an integrated circuit semiconductor. Thanks to that decision, the area of California where Fairchild and Intel are located is popularly referred to as "Silicon Valley," and Noyce earned the nickname, "The Mayor of Silicon Valley."

Intel's co-founder Moore noted in 1965 that the number of components packed into an integrated circuit was doubling every year or two. This phenomenon became known as

"Moore's Law," and amazingly, it has continued to hold true. With chips already containing billions of components, it would seem that Moore's Law is destined to eventually slow. But it's hard not to wonder what technological breakthrough might one day replace these tiny wonders.

The Particle Accelerator

There are a few scientific discoveries and creations that are so well known, even those of us who aren't scientists are aware of them.

✻ ✻ ✻ ✻

SUCH IS THE case with the Large Hadron Collider (LHC) in Geneva, Switzerland. Built by the European Organization for Nuclear Research (known as CERN, after the French *Conseil européen pour la recherche nucléaire*), between 1998 and 2008, this huge machine—the largest machine in the world—aids in the study of particle physics by accelerating particles and smashing them together in collisions. Its almost-17-mile-long circular tunnel contains two "beamlines," which are the trajectories on which the particles travel, and a total of about 10,000 magnets to propel the particles along.

The LHC gained publicity even before it was operational when it was featured in author Dan Brown's 2000 book *Angels & Demons*, and it caused a bit of uneasiness when fearmongers expressed worry that the machine might be able to cause spontaneous, microscopic black holes, which could be capable of destroying the entire Earth. (We're still here, so it's safe to say that hasn't happened.)

The LHC seems decidedly futuristic in its massive scope and abilities; so it's surprising to learn that the history of particle accelerators goes all the way back to 1929—before the invention of the ballpoint pen, the microwave oven, the transistor radio, or even seat belts. But that's not to say the early 1900s were devoid of innovation; quite the contrary. By the 1920s, access to electricity was bringing Americans a whole new world of convenience, with inventions like refrigerators, washing machines, radios, and vacuum cleaners making life easier.

Curious minds, including nuclear scientist Ernest Orlando Lawrence, found all this new technology enthralling. Lawrence had been interested in science his whole life, tinkering with radios as a teenager and then studying chemistry in college. He earned a PhD in physics from Yale University in 1925, and then became a professor of physics at the University of California, Berkeley, in 1928.

How to Build a Particle Accelerator

One evening in 1929, Lawrence was in the library perusing an article by a Norwegian engineer named Rolf Wideroe, who proposed a device that could be used to create high-energy particles by pushing and pulling ions with increasingly longer electrodes. Lawrence was intrigued, but he thought that Wideroe's design, which was laid out in a straight line, would be impractical for studying atomic particles, because it would be much too long and take up too much space for a modest university laboratory such as his.

So Lawrence began sketching out his own idea on a paper napkin. Wideroe's idea was to use alternating positive and negative electric potential to accelerate particles, and Lawrence wondered if there might be a way to use the same method more than once. His idea was to use a magnetic field to bend the trajectories of particles into a circle, enabling them to pass through the same accelerating area over and over. His design consisted of a circular, high-vacuum chamber set between the poles of an electromagnet, and two semicircular electrodes.

To test his ideas, Lawrence constructed his first device, which was only about four inches in diameter, using low-tech materials like bronze and wax. In addition, a kitchen chair and a wire hanger were used to aid in operation. It may have been rudimentary, but it worked, proving that his technique had merit.

Scaling Up the Concept

With the help of some graduate students, including M. Stanley Livingston, Lawrence set out to build a more sophisticated version of his machine. In 1931, Livingston debuted an 11-inch accelerator, and then set to work writing his PhD thesis documenting its construction. But Lawrence and Livingston knew they could go bigger; soon they set out to create a 27-inch version. In anticipation of things to come, Lawrence had already procured an empty building next to the physics department, which he dubbed the "Rad Lab," for radiation laboratory. It was a smart move, especially considering that the new 27-inch accelerator required a hulking 80-ton magnet to operate!

In 1934, Lawrence was given a patent for his device, which he called the cyclotron. The "Rad Lab" was made an official department of the university in 1936, with a $20,000 annual allotment for research, with Lawrence at the helm as director. By 1937, the 27-inch cyclotron was replaced by a 37-inch version, and two years later, a 60-inch cyclotron took over. The machine had created radioactive isotopes and the first artificial element, technetium. Lawrence also discovered the cyclotron could be used for medical applications, creating phosphorus-32 for the treatment of blood diseases and aiding in neutron therapy for cancer patients. On February 29, 1940, Lawrence was awarded the Nobel Prize in Physics for his invention.

Of course, Lawrence wasn't finished: His next cyclotron was 184 inches in diameter, and featured a magnet that weighed more than 4,000 tons. It was completed in 1946, and installed in the Rad Lab's newly constructed location on nearby Charter Hill. But the ever-larger magnets required for Lawrence's cyclotrons limited their size. Today, particle accelerators are built not with one huge magnet, but rather with many small magnets. These accelerators are called synchrotrons, and include the LHC in Switzerland and the Tevatron, located at the Fermi National Accelerator Laboratory (also known as Fermilab) in Batavia, Illinois.

Lawrence's cyclotron opened up a whole new world of physics, ultimately giving us a better understanding of the unseen world around us. More than 30,000 particle accelerators are located throughout the world, doing everything from destroying cancer tumors to helping us understand the origins of the universe. Since many of them are built underground, you could be driving or walking on top of a particle accelerator and not know it. It's amazing to think that the first accelerator could literally fit into the palm of a hand; but to cross the diameter of the LHC, it would take more than an hour of walking! There's no doubt that the inventor of the cyclotron, who always searched for ways to make his creation bigger, would be impressed.

Sighting the Higgs Boson

There's a lot more to the universe than what we see with our eyes.

* * * *

OUR UNIVERSE IS an incredibly complex place. The Big Bang, dark matter, stars, black holes, photons, galaxies, gamma rays—it's all driven by four "fundamental forces," according to the physicists who spend their lives attempting to unravel the mysteries of our vast cosmos. These forces, which include gravitational force, electromagnetic force, and strong and weak interactions, are said to encompass nearly everything in the known universe, interacting with and influencing the tiniest atoms to the largest planets.

It's All Relative

The theory of General Relativity, first proposed by Einstein in 1915, describes the properties of gravity. But the other three forces are included in the Standard Model of particle physics, which has been gradually developed since the latter half of the 20th century by many different scientists around the world. The Standard Model also classifies all elementary particles, which are subatomic particles—particles that are smaller than atoms—with no substructure, meaning they are not made up of other particles. These minute particles are classified as either fundamental fermions, particles which were first observed by physicists Enrico Fermi and Paul Dirac in the mid-1900s, or fundamental bosons. Dirac coined the name "boson" in honor of physicist Satyendra Nath Bose, who, along with Einstein, developed a theory that characterizes elementary particles.

This Thing Called Mass

In 1964, a British physicist named Peter Higgs, who was fascinated with these elementary particles, was working on one of the many mysteries of the universe: mass. Or, more precisely, where mass comes from. Mass is the resistance an object exhibits when a force is applied, and most of the time, it seems pretty obvious where that mass comes from. A feather has very little mass, whereas a car has a lot. It makes perfect sense, right? But as we break objects down into their fundamental parts, the source of mass becomes less clear. It may be obvious that an object made up of molecules and atoms has mass; it may even make sense that molecules and atoms themselves have mass, since they are made up of even smaller particles. But the smallest of these—the subatomic particles like bosons—also have mass. And where does the mass of an elementary particle, which is not made up of any other particle, come from?

Strangely, when physicists ran equations to model the behavior of elementary particles, their equations worked perfectly if they assumed the particles had no mass. But when they modified the equations to account for the mass of the particles, the perfection of their calculations fell apart. Higgs suggested that perhaps the scientists were looking at things the wrong way: Instead of assuming that the particles themselves were accounting for mass, he proposed that they were surrounded by a field—which we now call the Higgs field—that was influencing their behavior by exerting a drag force on them. This is not unlike moving a ball underwater—whether moving in air or underwater the ball is unchanged, but it becomes more difficult to move it underwater due to the additional drag.

At first, Higgs' ideas were scorned by the physics community, but over the next couple decades, scientists began to realize that the hypothesis could have merit. By the mid-1980s, most physicists embraced the idea, even confidently teaching college students about the existence of the "Higgs field," even though said existence had never been proven. But it would take another two and a half decades before the technology to test for such an invisible field would come to fruition. In 2008, the Large Hadron Collider (LHC) was completed in Geneva, Switzerland. This 17-mile-long circular tunnel lies hundreds of yards beneath the earth, and is so long that most of its underground "racetrack" is in the neighboring country of France. The LHC is the largest particle collider in the world, built to literally smash particles of matter together. Inside the LHC, protons circle the tunnel in both directions, accelerated by magnets, until they're traveling just short of the speed of light.

Finding a Needle on a Planet of Haystacks

The LHC was exactly what Higgs field proponents had been waiting for. The theory was that if the Higgs field existed, then particles colliding together should be able to disrupt it—much like two objects colliding underwater can make the water jiggle. And if these colliding particles did "jiggle" the invisible field around them, a tiny fleck of the field—called the Higgs boson particle—should be visible. But calculations predicted that it would also be unstable, almost immediately disintegrating into other particles. So scientists would need to search through the debris of the colliding particles to find a specific area of decay that pointed to the Higgs boson. What's more, the creation of a Higgs boson during particle collisions was deemed to be likely an extremely rare event, occurring only one time in 10 billion collisions or so. So trillions of collisions would need to be analyzed, to see if a pattern matching up with the Higgs boson particle emerged.

At the end of June 2012, the European Organization for Nuclear Research (also known as CERN), which operates the

LHC, announced that they would soon be holding a seminar to discuss research findings. Within a few days, the scientific community was abuzz with rumors; but when news broke that Peter Higgs himself had been asked to travel to Geneva to attend the seminar, physicists knew the news had to be what they'd been all been hoping: the Higgs boson had been found.

On July 4, 2012, CERN announced that two independent experiments, conducted separately and without sharing findings, had both reached the same conclusion: a particle consistent with the Higgs boson had been found. Although it would take several more years of tests and measurements before CERN would officially confirm the finding, scientists around the world were ecstatic, finally having proof that there is so much more to the cosmos than we can see with our naked eyes.

Higgs was awarded the 2013 Nobel Prize in Physics, and since the original discovery, experiments have only strengthened the confirmation of the existence of the Higgs boson. The Higgs field is thought to permeate the entire universe, uniformly existing on Earth, throughout the planets in our solar system, and everywhere else, and many scientists believe it played a pivotal role in the Big Bang, leading some to nickname the Higgs boson "the God particle." No matter what you call it, the discovery of the Higgs boson has drawn us one step closer to understanding our vast and complex universe.

Personal Computers

In our present age, it's nearly impossible to think of our lives without computers. While younger generations won't ever be able to recall a time without Google, older generations may reminisce about the days when libraries were the only place to find research materials. But whether young or old, it's safe to say that if computers disappeared suddenly, our lives would be much different. Personal computers have shaped the way we communicate, do business, go shopping, and even how we can order pizza. There are few inventions in our world that have so completely changed our lives.

* * * *

BUT IN THE early days of computers, the machines were anything but "personal." Those huge, cumbersome monstrosities that took up entire rooms required teams of engineers and scientists. Hardly the kind of equipment that could be propped on a desk or tossed into a backpack! The most famous of these early computers, ENIAC, was built in 1943 at the Moore School of Electrical Engineering at the University of Pennsylvania by John Mauchly and J. Presper Eckert. ENIAC cost $500,000 to build, took up 1,800 square feet of space, weighed almost 30 tons, and contained 20,000 vacuum tubes. While the computer had its own generator to provide it with the 175 kilowatts of electricity it needed to run, rumors swirled that every time ENIAC powered up, the lights dimmed in Philadelphia. Clearly, this behemoth was the antithesis of "personal computer."

But for all its shortcomings, ENIAC was still able to solve complex equations within seconds, reducing the time and manpower previously needed for such problems. This proved to universities, businesses, and researchers that creating these giant, expensive machines just might be worth the high cost. And by the late 1940s, technology was already advancing to the point that computers were smaller and less expensive. Transistors eliminated the need for bulky vacuum tubes. A decade later, the integrated circuit appeared. The stage for personal computers was quickly being set.

By the 1960s, huge, room-sized computers had given way to "minicomputers," which were cheaper and much smaller than their predecessors. They had relatively high processing power, and were often used for manufacturing businesses, telephone companies, and to control laboratory equipment. Being about the size of a refrigerator, the term "minicomputer" sounds laughable by today's standards. But minicomputers enjoyed a very short time in the spotlight: because what's smaller than a mini? A micro, of course!

The Microprocessor

In 1971, a team of engineers at Intel Corporation designed the first microprocessor, a chip that incorporates the functions of a computer's central processing unit onto a single integrated circuit. This microprocessor, the Intel 4004, held all the computing power of its grandfather, the huge, cumbersome ENIAC, in a chip that was less than an inch across. The invention was truly a game-changer, leading to increasingly smaller computers. The era of the microcomputer—soon to be known as the personal computer—had begun.

When they first arrived on the scene, personal computers were mostly a novelty purchased by hobbyists. For instance, in 1975, the January issue of *Popular Electronics* ran a story about the Altair 8800, a computer kit that was sold by Micro Instrumentation and Telemetry Systems, of MITS. The Altair used a new microprocessor, the Intel 8080, and sold for about $400. Although it had no keyboard, no screen, and was very limited in its capabilities, the Altair, considered by many to be the first true "personal computer," sold thousands of kits within a few months. In fact, the Altair was such a success that in 1975 MITS hired a couple of Harvard students, Paul G. Allen and Bill Gates, to improve the computer's software. Allen and Gates took the money they earned from the project and went on to found Microsoft, now one of the most well-known names in the computer business.

Meanwhile, another pair of engineers with now-familiar names, Steve Wozniak and Steve Jobs, was busy with a project of their own. Wozniak had designed a new computer that he shared with a computer hobbyist group in Silicon Valley called the Homebrew Computer Club. The computer was such a hit that Jobs offered to market the machine, which the pair named the Apple I, and with the money they earned from selling 200 of the computers, Wozniak and Jobs founded Apple Computer. Within a year, the Apple I was replaced by the Apple II, a complete computer with a keyboard, a plastic case, and game paddles to play the game *Breakout*. Additionally, a monitor with color graphics could be purchased separately. One of the most popular computers ever made, the Apple II sold more than four million units before production ended in 1993.

By the late 70s, companies like IBM, Tandy, and Xerox were throwing their hats into the personal computer ring, as well. The machines were becoming more and more popular for business; but the advent of faster microprocessors and cheaper parts soon made personal computers more commonplace in homes, as well. And in 1982, electronics manufacturer Commodore International released what was destined to be best-selling computer of all time, the Commodore 64. Named after its 64 kilobytes of RAM, the Commodore 64 sold for about $600 and was known for its impressive graphics. Today, it's known for another impressive feat: the Commodore 64 was listed in the 2006 Guinness Book of World Records for selling around 12 million units.

But the machine that finally succeeded in adding "personal computer" to the mainstream lexicon was the aptly named 1981 IBM Personal Computer. Also known as the IBM PC, this computer became particularly popular within the business world, and was often copied by other computer manufacturers. The IBM PC, Apple II, and Commodore 64 made such an impact on the way people lived their lives that *Time* magazine named the personal computer their "Person of the Year" (or, in this case, "Machine of the Year") for 1982.

Today, even our tiny phones hold thousands of times—sometimes even hundreds of thousands of times—the memory of the first personal computers. It's doubtful that the creators of the ENIAC ever envisioned that the computing power of their giant machine would one day easily fit onto a miniscule chip, and that chip could power a computer that fits into a student's backpack! Each new breakthrough in the computer industry has served to move us forward exponentially, changing all of our lives forever.

The "Fax" Are In

A quick scan of history reveals that the facsimile machine of the 1980s actually got its start in the 1840s.

* * * *

VISIT ALMOST ANY office supply store, and you'll find that fax machines are still plentiful. Considered de rigueur in the workplace during the business boom of the 1980s, the "facsimile" machine continues to hold its own in this day of lightning-quick e-mailing and dexterity-challenging text messaging. And why shouldn't it? The ability to "fax" a reproduced sheet of paper from here to virtually anywhere within a telephone's reach has completely changed the business landscape.

But this breakthrough invention doesn't actually hail from modern times. In fact, this gem came along when Abraham Lincoln was still an Illinois lawyer and the telephone—the yin to the fax machine's yang—had not yet been invented. How was this possible?

Scottish mechanic and amateur clockmaker Alexander Bain used existing telegraph technology, paired with a stylus functioning as a pendulum, to work such magic. The stylus picked up images from a rotating metal surface (the crude precursor of modern-day scanning), then converted them to electrical impulses for transmission. This, in turn, was sent to chemically treated paper and *voilà*—the fax-transmitted image was born. Bain was granted a British patent for his groundbreaking invention on May 27, 1843, some 33 years before the telephone was patented.

The process underwent various upgrades and improvements over the years to reach its present form. By the 1920s, a new scanning system allowed the paper original to remain in a fixed position, precisely like many modern-day machines. During the same era, the American Telephone & Telegraph Company (AT&T) came on board, thus ushering the fax machine into modern times. On March 4, 1955, the very first fax transmission was sent via radio waves across the continent. Not bad for a pre–Civil War machine produced by an amateur.

Smile: You're on Camera Obscura

Photography didn't begin with the daguerreotype (1839), but Louis Daguerre's gadget was the breakthrough step.

* * * *

Pinhole Cameras

IRONICALLY, NO PHOTOS exist of the earliest known camera pioneer, a Chinese man named Mo Ti. It would be nice if his likeness had been captured on film for the world to remember, but alas, technology in China circa 400 BC was not quite there yet. Ancient history credits Mo with uncovering a simple key principle: Light traveling through a small hole will cast an image onto a surface. Unfortunately, there was no efficient way to capture the image—which explains the lack of paparazzi photographs of Mo.

Camera Obscuras

In about AD 1015, it came to light (so to speak) that a closed box placed behind a pinhole camera makes a *camera obscura* (Latin: "dark chamber"). At that time, Arab scholar Abu Ali al-Hasan ibn Al-Haytham published a landmark optics text describing how the device could project an image onto paper, enabling one to sit and trace/draw the image. Limited, but still useful.

The situation improved in the 1500s, when clever pioneers added a magnifying lens to the pinhole. It inverted the image, but one could later flip the sketch upright and paint in the colors. This was impossible to do while tracing the image because the room's darkness made it difficult for the artist to tell the paint colors apart.

Fixing the Image

During the 1700s, researchers discovered that light could change certain substances, particularly silver compounds. In 1826, a Frenchman named Joseph Niépce used a camera obscura and about eight hours of exposure to fix an image onto bitumen (natural asphalt). Where the light hit the bitumen, it hardened, and Niépce dissolved away the unfixed remainder.

The image wasn't very clear, but it was a beginning. The oldest surviving example of what's called a *heliograph* evidently dates to about 1826, when the image was much improved. But a clever chap named Louis Daguerre grabbed onto Niépce's coattails a few years before Niépce's death in 1833. Since Daguerre also grabbed the credit, few people outside of France have ever heard of Niépce.

Daguerreotyping

Daguerre kept experimenting with chemicals until 1837 when he found a suitable light-sensitive combination involving iodine and mercury vapor. The subjects now had to stand still for "only" half an hour, which seems like forever—until you consider that the alternative was a much longer portrait sitting.

Nearly all the earliest surviving photos of American icons such as Dolley Madison, young Horatio Alger at Harvard—even early pornography—are daguerreotypes.

From Mainstream to Mass Market

George Eastman, whom you may accurately connect with Eastman Kodak, patented photo emulsion–coated paper and rollers in 1880. Farewell to cumbersome plates! Photography was about to go mainstream.

Kodak's Brownie, a $1 mass-market camera, hit the shelves in 1900. The 1900s would become photography's democratic age, where any fool could preserve whatever he or she wished on film. Before long, camera nuts would keep whole families waiting in the Model T so this or that fascinating item could be recorded for all posterity in the family album.

For whatever reason, the paparazzi wouldn't get saddled up until the 1930s. Sure took them long enough.

A Shocking Invention: The Electric Chair

Electrocution was meant to be a more humane form of execution, but things didn't exactly work out that way.

* * * *

Alfred Southwick's Lightbulb Moment

DR. ALFRED SOUTHWICK was a dentist in Buffalo, New York, but he was no simple tooth-driller. Like many of his contemporaries in the Gilded Age of the 1870s and 1880s, he was a broad-minded man who kept abreast of the remarkable scientific developments of the day—like electricity. Though the phenomenon of electric current had been known of for some time, the technology of electricity was fresh—lightbulbs and other electric inventions had begun to be mass produced, and the infrastructure that brought electricity into the businesses and homes of the well-to-do was appearing in the largest cities. So Southwick's ears perked up when he heard about a terrible accident involving this strange new technology. A man had walked up to one of Buffalo's recently installed generators and decided to see what all the fuss was about. In spite of the protests of the men who were working on the machinery, he touched something he shouldn't have and, to the shock of the onlookers, died instantly. Southwick pondered the situation with a cold, scientific intelligence and wondered if the instant and apparently painless death that high voltage had delivered could be put to good use.

Southwick's interest in electrocution wasn't entirely morbid. Death—or more specifically, execution—was much on people's minds in those days. Popular movements advocated doing away with executions entirely, while more moderate reformers simply wanted a new, more humane method of putting criminals to death. Hangings had fallen out of favor due to the potential for gruesome accidents, often caused by the incompetence of hangmen. While the hangman's goal was to break the criminal's neck instantly, a loose knot could result in an agonizingly slow suffocation; a knot that was too tight had the potential to rip a criminal's head clean off.

To prove the worth of his idea, Southwick began experimenting on dogs (you don't want to know) and discussing the results with other scientists and inventors. He eventually published his work and attracted enough attention to earn himself an appointment on the Gerry Commission, which was created by the New York State Legislature in 1886 and tasked with finding the most humane method of execution.

Although the three-person commission investigated several alternatives, eventually it settled on electrocution—in part because Southwick had won the support of the most influential inventor of the day, Thomas Alva Edison, who had developed the incandescent lightbulb and was trying to build an empire of generators and wires to supply (and profit from) the juice that made his lightbulbs glow. Edison provided influential confirmation that an electric current could produce instant death; the legislature was convinced and a law that made electrocution the state's official method of execution was passed.

William Kemmler Gets Zapped

On August 6, 1890—after much technical debate (AC or DC? How many volts?) and a few experiments on animals (again, you don't want to know)—William Kemmler, a genuine axe murderer, became the first convicted criminal to be electrocuted. Southwick declared it a success, but the reporters who witnessed it felt otherwise. Kemmler had remained alive after the first jolt, foam was oozing from the mask that had been placed over his face as he struggled to breathe. A reporter fainted. A second jolt of several minutes was applied, and Kemmler's clothes and body caught fire. The stench of burned flesh was terrible.

Despite a public outcry, the state of New York remained committed to the electric method of execution. The technology and technique were improved, and eventually other states began to use electrocution as well. Today, nine states still allow use of the electric chair, though lethal injection is the preferred option.

Whatever Happened to Pocket Protectors?

Nerdlingers across the world might like to think that the pocket protector was slowly phased out by a government conspiracy that involved underground landing strips for aliens. Alas, no.

⁕ ⁕ ⁕ ⁕

IN REALITY, IT was the portrayal of nerd culture in the 1980s—specifically in the *Revenge of the Nerds* movie series—that did in the pocket protector. What had been a mostly overlooked accessory on your chemistry teacher's shirt became a badge of public dishonor. Defiant nerds have banded together to resist this public shaming. The Institute of Electrical and Electronics Engineers, for example, has published an article that proudly chronicles the history of the geek shield while also expounding on its usefulness. Perhaps these folks take such pride in the pocket protector because it was invented by one of their own: electrical engineer Hurley Smith, who developed a prototype in 1943.

Apparently, Smith's wife had grown tired of mending and replacing the white button-down shirts that were as much a part of the engineering nerd's uniform as were horn-rimmed glasses. Technological advances that were made during World War II presented new opportunities in plastics, and Smith seized the moment.

His first model was basically a folded liner that protected the inside of the pocket and covered the lip to prevent the wear and tear that was caused by pen clips.

In March 1947, Smith obtained a patent for his handy new device, which was registered under the name "pocket shield or protector." Over the next twenty years or so, a few modifications were introduced, including a clip for an ID badge and a clear plastic design. Yes, these were the glory days for the good old pocket protector.

Fast forward to 1984 and the release of *Revenge of the Nerds*, the hit movie that brought the pocket protector to the forefront of geek fashion but ultimately led to its demise. Although the cinematic nerds ultimately won the day, the real-world ending for their treasured fashion accessory wasn't as happy. A pocket protector became akin to a scarlet letter; wearing one invited scorn and ridicule. Smith's utilitarian invention wound up in the trash bin of history.

Pangaea: Putting the Pieces Together

Pangaea was a giant supercontinent that existed on Earth some 270 million years ago. Unlike the smaller, broken pieces of contemporary continental plates, Pangaea had it all—literally.

* * * *

A History Lesson

THE ALL-ENCOMPASSING LANDMASS known as *Pangaea* straddled the equator in roughly the shape of a "C," grandly surrounded by one of the largest (if not *the* largest) expanses of water ever to exist on planet Earth: the Panthalassa Ocean. Only a few chunks of land lay to the east, including bits of what we now call northern and eastern China, Indochina, and part of central Asia. A smaller "sea," called the Tethys Sea, was located within the "C" and is thought to have been the precursor to today's Mediterranean Sea.

The climate was warm at this time, with no true polar ice caps. Because of the high temperatures, life flourished during the Pangaea years. Early amphibians and reptiles roamed the giant continent; dinosaurs and archosaurs ran amok. When the supercontinent finally broke apart, other plant and animal species arose, each developing as a direct result of being cut off from their own species. In other words, thanks to the breakup of Pangaea, our planet's organisms became extremely diverse.

The Pangaea Puzzle

German meteorologist Alfred Wegener was the first to coin the term *Pangaea* (which means "all earth" in Greek) in the early 20th century. He was also the first to publically propose and publish—much to the dismay of the scientific community—the idea that Earth's continents once lay together in the huge landmass of Pangaea. Even more shocking was Wegener's theory that the supercontinent broke apart over millions of years, with the pieces ultimately reaching their current spots on Earth millions of years later. His beliefs were based on many scientific discoveries of his time, including identical fossils found on Africa and North America—and especially on the obvious giant jigsaw puzzle "fit" of the continents.

Now known as the "father of the continental drift theory," Wegener was a pariah in his own time. He was the victim of his own ideas, as he could not come up with a logical mechanism to explain the movement of the continents. Few of his contemporaries believed his idea, many of them citing the fact that Wegener was merely a meteorologist, not a geologist. But finally—almost three decades after his death in 1930—Wegener was vindicated, thanks to additional rock and fossil evidence and, later, satellites that track the minute movements of the continents.

Getting the Drift of Continental Drift

Scientists now believe the shifting of the continental plates is caused by the moving mantle, the thick layer of viscous, liquid rock below Earth's crust. From the evidence provided by fossils, they believe two huge continents, Laurasia (to the north of the equator and made up of today's North America and Eurasia) and Gondwanaland (or Gondwana, to the south of the equator and made up of today's Africa, Antarctica, Australia, India, and South America) collided 270 million years ago during the Permian Period, forming Pangaea. Not long afterward—at least in terms of geologic time—around the Triassic Period 225 million years ago, the attraction waned.

This was the beginning of the end for the supercontinent, as a volcanic seam called a seafloor spreading rift (similar to today's Mid-Atlantic Ridge) ripped the continent apart. A second pair of continents formed, also called Laurasia and Gondwanaland; over the next tens of millions of years, they eventually plowed across the Earth's hemispheres, taking the positions we're so familiar with today. So far, there is no fossil or rock evidence that the supercontinent ever formed again. And although we will never know because our lifespans are so short, it's possible that as the continents move over the coming millions of years, another supercontinent may form.

Zero: To Be or Not to Be

You might think that zero is an insignificant little nonentity worth nothing. But you'd be wrong.

* * * *

THE ORIGIN OF the number "zero" was not as neat as the subtraction of one minus one. In fact, there was no symbol for zero for centuries. Before a symbol was used for zero, most archeologists believe various cultures used an empty space to stand in zero's stead.

It is thought that the Babylonians were the first to use a placeholder for zero in their numbering system in 350 BC. But it was not a zero; instead, they used other symbols, such as a double hash-mark (also called a *wedge*), as a placeholder. Why the need for a placeholder—or a number that holds a place? Merchants and tax collectors in particular needed such a tool. After all, their numbers became larger as trade developed and more taxes were levied. The placeholder became extremely necessary; for example, the number 5,000 implies that the three places to the right of the 5 are "empty" and only the thousands column contains any value.

The first crude symbol for zero as a placeholder may have been invented around 32 BC by Mesoamericans in Central America—most notably in a calendar called the Long Count Calendar. No one really agrees if it was the Olmecs or the Mayans who first used a shell-like symbol for zero. Either way, the idea of using a symbol for zero remained a secret for centuries, as isolation hindered the Mesoamericans, never allowing their concept of zero to be spread around the world.

More than 100 years later, another symbol for zero was invented either in Indochina or India. In this case, the culture was not averse to spreading their number system—including the concept of zero—around the world in trade and commerce. And by around AD 130, Greek mathematician and astronomer Ptolemy, influenced mainly by the Babylonians, used a symbol representing zero along with the alphabetic Greek numerals—not just to hold a place, but as a number.

But because zero technically means nothing, few people accepted the concept of "nothing" between numbers. Not every culture ignored the idea, though; Hindu mathematicians often wrote their math in verse, using words similar to "nothing," such as *sunya* ("void") and *akasa* ("space"). Finally, around AD 650, zero became an important number in Indian mathematics, though the symbol varied greatly from what we think of as our modern zero. The familiar Hindu-Arabic symbol for zero—essentially an open oval standing on one end—would take several more centuries to become accepted.

Why all the excitement after zero was "invented"? It turns out that all the wonderful, exotic properties of zero made it a necessity to the development of science and technology over time—in everything from architecture and engineering to our checkbooks. The properties are many: You cannot divide by zero (or in other words, divide something by nothing); zero only in the numerator (top number) of a fraction will always be equal to zero; it's considered an even number.

Probably its most endearing quality is that when zero is added or subtracted from any number, the result is that number. Add nothing, subtract nothing, and you get the original number—a handy device to have when it comes to counting on anything.

The Floppy Disk Story

Two engineers walk into a bar...

* * * *

THE PERSONAL COMPUTER was made possible by two crucial developments: the microprocessor and the floppy disk. The first provided the PC with its brains. The second gave it a long-term memory—one that lasted after the computer was turned off. But who invented the floppy disk? Dr. Yoshiro Nakamatsu claims he developed the basic technology—a piece of plastic coated with magnetic iron oxide—back in 1950 and later licensed the technology to IBM. IBM has never acknowledged these claims, but it does own up to an "ongoing relationship" with Dr. Nakamatsu, who holds more than 3,000 patents and also claims to have invented the CD, DVD, and digital watch.

At any rate, IBM made the first floppy disks, which were eight inches in diameter, encased in a paper jacket, and held a whopping 80 kilobytes of memory. But these disks were too big for personal computers, so in 1976 two engineers sat down in a Boston bar with An Wang of Wang Labs to discuss a new format. When they asked Dr. Wang how big the new floppies should be, he pointed to a cocktail napkin and said, "About that size." The engineers took the napkin back to California and made the floppies exactly the same size as the napkin: 5 1/4 inches. This became standard in early personal computers, but they were easy to damage or get dirty, and as PCs got smaller, the disks were still too large. Various companies released smaller formats, but it wasn't until Apple Computer incorporated a 3 1/2-inch floppy drive in its Macintosh computer that things began to change. The new disks had a stiff plastic case and a sliding metal cover to protect the data surface, and they stored 360K, 720K, and later 1.44 megabytes. The writing was on the disk, and these days, the original floppies have gone the way of carbon paper and mimeograph machines.

✳ Chapter 11

Medical Science

The First Plastic Surgeons

You might think that plastic surgery is a relatively new phenomenon, but the truth is that it's thousands of years old. No, cavewomen weren't getting tummy tucks, but the desire to improve one's looks seems to be about as old as the human race.

✳ ✳ ✳ ✳

Those Vain Egyptians

PHYSICAL APPEARANCE WAS obviously important to the ancient Egyptians—theirs was one of the first civilizations to use makeup. And if an Egyptian suffered an injury that no amount of makeup could conceal, reconstructive surgery was an option, provided that the person had a high enough social ranking. Papyrus records dating to 1600 BC detail procedures for treating a broken nose by packing the nasal cavity with foreign material and allowing it to heal—these were, in essence, primitive nose jobs. About 1,000 years later in India, a surgeon named Sushruta developed a relatively sophisticated form of rhinoplasty that eventually spread across the Arab world and into Europe.

Medical techniques continued to be improved upon in ensuing years. During the 15th century, Sicilian doctors pioneered a method of suturing and closing wounds that left minimal scarring and disfigurement, and by the 16th century, early methods of skin grafting were being created. It wasn't until the

19th century, however, that this burgeoning medical field got its name: "plastic surgery." For that, we can thank German surgeon Karl von Gräfe.

A JewelEye, Anyone?

The types of procedures that we've described thus far were typically reserved for people who had suffered horrific damage to their bodies or faces. Who, then, were the brave pioneers who gave plastic surgery a purely cosmetic bent? Well, the first silicone breast implants were developed in the 1960s by plastic surgeons Frank Gerow and Thomas Cronin; the first person to receive breast implants (not for medical reasons, such as after undergoing a mastectomy, but strictly to improve her appearance) was Timmie Jean Lindsey.

The industry seems to get a facelift every few years as new surgeries are developed. And some of them get pretty crazy. For example, the JewelEye—a procedure in which tiny platinum jewels are implanted into the eyes to create a glint—might soon be coming to an operating room near you. Makes the old Egyptian practice of stuffing junk into someone's nasal cavity seem pretty quaint, huh?

Blood Circulation

It's a simple enough concept, but medical science didn't fully understand blood circulation until the 1600s.

* * * *

MOST OF US know the feeling of stepping on a scale and lamenting the number staring back at us. But we should take some comfort in knowing that what we weigh isn't entirely dependent on how many burgers or potato chips we eat. In fact, about 7 percent of an adult's total body weight is made up of the blood circulating through arteries, veins, and capillaries, an amount that adds up to approximately five to six quarts. Our circulatory system is an amazing thing: Oxygenated blood is carried through arteries to capillaries, where oxygen is traded for carbon dioxide. The oxygen-depleted blood is then carried back to the lungs and heart through veins, and the whole process starts over again. The entire five or six quarts of blood in an average adult's body makes a complete round trip from the heart and back every minute, through a system that, if laid out end to end, would extend more than 60,000 miles—that's enough to circle the Earth twice!

Even ancient peoples knew that something remarkable must happen in the body in order for us to survive. The Ebers Papyrus, an ancient Egyptian medical text that was written in the 16th century BC, describes how air enters the mouth and travels to the heart and lungs, where the air is then carried throughout the body by arteries. Although a bit of a rough idea, the ancient Egyptians were certainly on the right track.

By the 6th century BC, Ayurvedic physicians in ancient India were aware of the circulation of vital fluids throughout the body, and in the 4th century BC, students of Hippocrates discovered heart valves. Ancient Greeks were aware of veins and arteries, but many ancient physicians and scientists still believed that many of these pathways were used to transport air through the body, not blood. But in the 2nd century, the influential Greek anatomist and physician Galen identified two different types of blood: venous, which was dark red, and arterial, which was lighter and thinner. Galen believed that the darker blood originated in the liver, while the lighter blood came from the heart, and that both types were carried throughout the body.

Galen also believed that blood in the right side of the heart passed to the left through invisible pores in the thick intraventricular septum that divides the heart. Blood was not, he thought, continually circulated, but rather converted into "sooty vapors" in the lungs and exhaled, after which, more blood would be created in the liver and heart.

Thank You, Ibn al-Nafis

Galen's theories were widely accepted for centuries. It would be a thousand years before Arabian physician Ibn al-Nafis published an accurate description of pulmonary circulation, including debunking Galen's assessment of invisible intraventricular pores. He also theorized that some kind of system must connect arteries and veins, predicting the existence of capillaries 400 years before they were actually discovered. For his impressive ideas that were ahead of his time, Ibn al-Nafis is often considered the "father of circulatory physiology."

Then, in the early 1600s, an English physician named William Harvey, who would become the personal physician of both King James I and King Charles I, began to doubt some of the long-accepted theories in his profession. Having studied under Hieronymus Fabricius, a preeminent surgeon and anatomist who identified valves in veins, Harvey was considered one of the best doctors in England, if not in all of Europe.

Thank You, Chickens

Through painstaking study of countless animals, including fish, snails, chickens, and pigeons, Harvey was able to form an accurate theory of the circulation of the blood throughout the body. In 1628 he published his findings in a 72-page book entitled *Exercitatio Anatomica de Motu Cordis et Sanguinis in Animalibus,* Latin for "An Anatomical Exercise on the Motion of the Heart and Blood in Living Beings."

Harvey was able to accurately describe the movement of the heart, including its chambers and valves, and, with the knowledge he'd gained from Fabricius on the valves in veins, correctly concluded that these were to help blood return to the heart through pulmonary veins. Although he was not yet sure of why blood traveled to the lungs and back, he was sure that the blood circulated, and was not expelled in "sooty vapors" as Galen had originally suggested.

Harvey performed his research before the microscope was available to help him see small structures in the body, but nevertheless, he predicted, just as Ibn al-Nafis had, that tiny pathways must be present to carry blood between arteries and veins, and that this blood must provide nourishment of some sort.

Combating centuries of misinformation would take time, and Harvey's theories were not immediately well received. Galen's ideas had been ingrained for centuries, so when Harvey first proposed his own concepts of blood circulation, he was met with skepticism. Some even felt that this doctor, who had literally served kings, was a quack. His private practice suffered as a result, but Harvey stuck to his theories, and by the mid-1600s, his ideas were finally accepted as fact by many physicians.

Vindicated by the Microscope

But the controversy of blood circulation continued even after Harvey's death in 1657. In the late 1600s, two scientists, Marcello Malpighi and Anton van Leeuwenhoek, would use the newly popularized microscope to see capillaries, finally confirming a centuries-old theory and strengthening the credibility of Harvey's ideas.

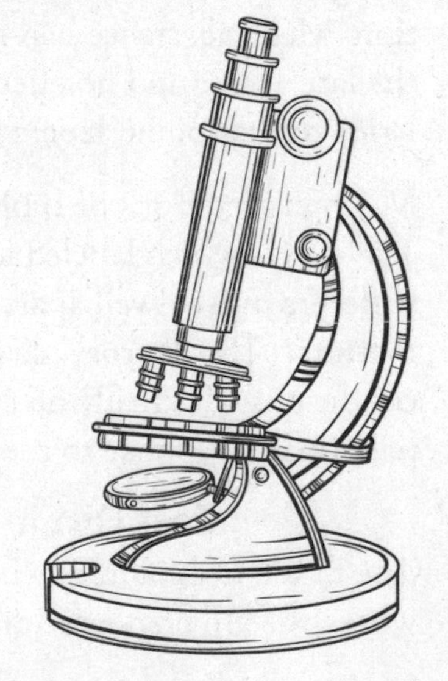

The circulation of blood is an intricate process that seems quite simple on the surface; its discovery paved the way for countless medical advancements. We could not administer medications, perform as many successful surgeries, or diagnose many diseases without the knowledge we now have of the amazing circulatory system.

History's Grim Places of Quarantine

Life has never been easy for lepers. Throughout history, they've been stigmatized, feared, and cast out by society. Such reactions—though undeniably heartless—were perhaps understandable because the disease was thought to be rampantly contagious. Anyone suspected of leprosy was forced into quarantine and left to die.

* * * *

LEPROSY HAS AFFECTED humanity since at least 600 BC. This affliction, now known as Hansen's disease, attacks the nervous system primarily in the hands, feet, and face and causes disfiguring skin sores, nerve damage, and progressive debilitation. Medical science had no understanding of leprosy until the late 1800s and no effective treatment for it until the 1940s. Prior to that point, lepers faced a slow, painful, demise.

Misinterpretations of Biblical references to leprosy in Leviticus 13:45–46, which labeled lepers as "unclean" and dictated that sufferers must "dwell apart . . . outside the camp," didn't help matters. (The "leprosy" cited in Leviticus referred to other skin conditions.) It's really no surprise that society's less-than-compassionate response to the disease was the leper colony.

Cast Out in Misery and Despair

The first leper colonies were isolated spots in the wilderness where the afflicted were driven, forgotten, and left to die.

The practice of exiling lepers continued well into the 20th century. In Crete, for instance, lepers were banished to mountainside caves, where they survived by eating scraps left by wolves. More humane measures were adopted in 1903 when lepers were corralled into the Spinalonga Island leper colony and given food and shelter and cared for by priests and nuns. However, once you entered, you never left, and it remained that way until the colony's last resident died in 1957. Still, joining a leper colony sometimes beat living among the healthy. It wasn't much fun wandering from town to town while wearing signs or ringing bells to warn of one's affliction. And you were always susceptible to violence from townsfolk gripped by irrational fear—as when lepers were blamed for epidemic outbreaks and thrown into bonfires as punishment.

Life in the American Colony

American attitudes toward lepers weren't any more enlightened. One of modern time's most notorious leper colonies was on the Hawaiian island of Molokai, which was established in 1866. Hawaiian kings and American officials banished sufferers to this remote peninsula ringed by jagged lava rock and towering sea cliffs. Molokai became one of the world's largest leper colonies—its population peaked in 1890 at 1,174—and more than 8,000 people were forcibly confined there before the practice was finally ended in 1969.

The early days of Molokai were horrible. The banished were abandoned in a lawless place where they received minimal care and had to fight with others for food, water, blankets, and shelter. Public condemnation led to improved conditions on Molokai, but residents later became freaks on display as Hollywood celebrities flocked to the colony on macabre sightseeing tours.

A Leper Haven in Louisiana

While sufferers of leprosy were being humiliated in Hawaii, they were being helped in Louisiana. In 1894, the Louisiana Leper House, which billed itself as "a place of treatment and research, not detention," opened in Carville. In 1920, it was transferred to federal authority and renamed the National Leprosarium of the United States. Known today as the National Hansen's Disease (leprosy) Program (NHDP), the facility became a leading research and rehabilitation center, pioneering treatments that form the basis of therapies currently prescribed by the World Health Organization (WHO) for the treatment of Hansen's disease. It was here that researchers enlisted a common Louisiana critter—the armadillo—in the fight against the disease. It had always been difficult to study Hansen's disease. Human nerves are seldom biopsied, so direct data on nerve damage from Hansen's was minimal. But in the 1960s, NHDP researchers theorized that armadillos might be susceptible to the germ because of their low body temperature. They began inoculating armadillos with it and discovered that the animals could develop the disease systemically. Now the armadillo is used to develop infected nerves for research worldwide.

A Thing of the Past?

In 1985, leprosy was still considered a public health problem in 122 countries. In fact, the last remaining leper colony, located in Croatia, didn't close until 2002. However, WHO has made great strides toward eradicating the disease and indicated in 2000 that the rate of infection had dropped by 90 percent. Multidrug therapies prescribed for the treatment of leprosy are available to all patients for free via WHO. Approximately four million patients have been cured since 2000.

Anesthesia

It's nearly unimaginable to consider having any sort of surgery without the aid of anesthesia. Yet for most of human history, people had no surefire way to guarantee pain-free surgery. For those of us who can barely have blood drawn without wincing, it's a terrible thing to think about; and when anesthesia was, at last, a reality, it was a relief not only for patients, but for doctors as well.

* * * *

THROUGHOUT HISTORY, HUMANS tried, of course, to find ways to dull pain. Alcohol was probably the earliest method, used in ancient Mesopotamia, where the Sumerians also cultivated opium poppy. The opium latex obtained from the poppy is about 12 percent morphine. It was described by the Sumerians as *hul gil*, or "plant of joy." The Babylonians soon became aware of opium's effects, and the plant, along with their empire, next spread to Persia and Egypt. The ancient Egyptians also made crude sedatives out of the mandrake fruit, which contains psychoactive agents that can cause hallucinations.

In India and China, cannabis incense was used to promote relaxation, sometimes along with a cup of wine. There are also legends surrounding Bian Que, an early Chinese physician, who practiced in the mid-300s BC. According to eyewitness reports, the ancient doctor was said to have used a "toxic drink" to induce a comatose-like state in two patients for several days. Of course, this story must be taken with a grain of salt, as Bian Que was also credited with successfully transplanting the heart of each man into the other; a feat which would be difficult enough for a modern doctor, much less an ancient physician.

Five centuries later, Chinese physician, Hua Tuo, is said to have used a combination of sedating herbs he called *mafeisan*, which would be dissolved in wine for a patient to drink before surgery. According to surviving records, Hua Tuo performed many surgeries with the help of this mixture, but its exact recipe was lost when the physician burned his own notes shortly before his death. At the time, Confucian teachings considered surgery to be a mutilation of the body, and such practices were discouraged. For this reason, Hua Tuo's secrets were lost, and surgery became a taboo practice in the country.

Take a Deep Breath

During the Middle Ages, Arabic and Persian physicians discovered that anesthetics could be inhaled as well as ingested. Persian physician Ibn Sina, also known by the Latinized name Avicenna, described in the 1025 encyclopedia *The Canon of Medicine* a method of holding a sponge infused with narcotics beneath a patient's nose. By the 13th century, the English were using a mixture of bile, opium, bryony, henbane, hemlock, lettuce, and vinegar called *dwale*, as a sedative. This potion was not, however, administered by physicians, who warned against its use, possibly because some of the ingredients can be poisonous; but rather, recipes to make the mixture were found in remedy books that ordinary housewives could tuck right next to their cookbooks. By 1525, Swiss physician Paracelsus had discovered that the compound diethyl ether had analgesic properties, but it would take more than two centuries before physicians would start to consider using it as an anesthetic.

Gassy Ideas

These practices all had various levels of success and plenty of risks, but nothing was foolproof when it came to surgical anesthetic. In fact, by the turn of the 19th century, surgery was performed only as an absolute last resort, usually with very little, if any, anesthetic. Not surprisingly, this caused immense fear and pain for patients, and extreme anxiety for the doctors who had to operate on them. Thankfully, the discovery of gases like oxygen, ammonia, and nitrous oxide in the late 1700s was about to give scientists new ideas, and physicians and patients some new options.

Laugh Away the Pain

English physician Thomas Beddoes was one of the first to take notice of the new discovery of gases, founding a medical research facility called the Pneumatic Institution in 1798. His aim was to find therapeutic ways to use different gases, and he hoped to treat breathing issues like asthma and tuberculosis. One of his employees was chemist Humphry Davy, who would later go on to invent the field of electrochemistry. But while working at the Pneumatic Institution, Davy discovered that one particular gas, nitrous oxide, had analgesic effects. He also noted that it produced a feeling of euphoria, which prompted him to call it "laughing gas."

While Davy suggested that the gas should be used during surgical procedures, this idea was not immediately acted upon, perhaps because Davy himself was not a physician. In 1813, Davy was joined by a new assistant, Michael Faraday (now best known for his work with electromagnetism), who began studying the inhalation of ether and found it to also have analgesic effects, as well as having the ability to cause sedation. But even after publishing his findings in 1818, ether was mostly ignored by surgeons, too.

But strangely, in the United States, these gases found another use, when people realized they could be inhaled recreationally at "laughing gas parties" and "ether frolics." By the mid-1800s, traveling lecturers and showmen would hold these unusual gatherings, where members of the audience were encouraged to inhale nitrous oxide or ether while other audience members laughed at the mind-altering results.

One of the participants of these "ether frolics" was a young dentist named William Morton, who was intrigued by the analgesic potential of the gas. After testing it on animals and then successfully using it for several patients, Morton was so confident in ether's ability to anesthetize that he offered to demonstrate his method to Dr. John Warren, the surgeon at Massachusetts General Hospital. So on October 16, 1846,-before a large audience in what is now known as the "Ether Dome," Morton administered diethyl ether to a young patient named Edward Gilbert Abbott, and then Warren removed a tumor from Abbott's neck.

To the surprise of everyone in attendance, even the surgeon, Abbott appeared to remain perfectly comfortable during the entire procedure. Afterwards, he reported that he'd felt a scratching sensation, but no pain; finally, the medical world was paying attention.

News of Abbott's surgery quickly spread around the world, and by December of that year, physicians in Great Britain were making use of inhalation anesthesia, with Scottish obstetrician James Young Simpson first using chloroform in 1847. The use of ether and chloroform made surgery a less distressing venture for patient and surgeon alike, and the success of the compounds led to more anesthesia research. The first intravenous anesthesia, sodium thiopental, was debuted for human use in March of 1934, and in the second half of the 20th century, Belgian doctor Paul Janssen synthesized more than 80 pharmaceutical compounds, including drugs used for anesthesia.

Today, anesthesia is safer than ever. This is fortunate, as ether was highly flammable, so could not be used once electricity was used to monitor patients or when wounds were being cauterized; and chloroform became associated with a high number of cardiac arrests. But modern anesthesia is among the safest of all medical procedures. Surprisingly, however, scientists are still uncertain exactly how anesthesia is able to affect our conscious minds. Anyone who is undergoing surgery, however, isn't likely to care how anesthesia works; the important thing is, it works.

Antibiotics

The term "mold juice" may be one of the most repulsive-sounding phrases ever conceived. It certainly doesn't sound like something you'd ever want to find served with your morning cereal and eggs. But believe it or not, "mold juice" is responsible for saving countless lives. And its serendipitous discovery is a reminder to scientists to never discount a finding, no matter how strange it may seem at first.

* * * *

THROUGHOUT HUMAN HISTORY, we have struggled with illness and disease, searching for the most effective ways to cure them and to alleviate symptoms. Many ancient peoples looked to the plants, herbs, and fruits that grew around them for remedies. Ancient Egyptians used honey as a medicine, as well as incense and even cannabis. In an ancient foreshadowing of things to come, they were also said to apply poultices made of moldy bread to wounds.

Ancient Greek medicine revolved around the belief that an imbalance of "humors," including blood, phlegm, yellow bile, and black bile, caused illness, so physicians attempted to bring balance to their patients by focusing on the environment. Being too hot, too cold, too wet, or too dry, could, it was believed, throw off the balance of these "humors" and result in sickness. And ancient Romans used plants like fennel, rhubarb, and aloe to treat numerous afflictions; but if these remedies failed, sick Romans could visit a "healing temple," where Asclepius, the god of healing, was said to reside.

None of these treatments was particularly effective at combating illnesses like pneumonia, rheumatic fever, syphilis, or the serious infections that could result from minor scratches and cuts. And it took centuries for theories such as the ancient Greeks' idea of "humors" or the long-held "miasma theory"—which posited that illness was caused by noxious "bad air"—to finally be debunked by the discovery of microorganisms. Yet even with this knowledge, until the 20th century, illnesses caused by bacteria were the number one cause of death in the developed world.

Multicolored Microbes

In the late 19th century, German scientist Paul Ehrlich suggested that it might be possible to target bacteria without harming other cells. He had noticed that some chemical dyes would color certain cells but not others, and he wondered if this property could be used to create chemicals that would selectively bind to bacteria. He spent years trying out different dyes and observing their effect on cells, with chemist Alfred Bertheim helping him to synthesize different compounds.

Along with Bertheim, Ehrlich was assisted by Japanese bacteriologist Sahachiro Hata, and in 1909, the trio finally caught a lucky break. While testing out a substance called arsphenamine, which Bertheim had synthesized in 1907, Hata discovered that the compound was effective at killing the bacterium *Treponema pallidum*, the microorganism that causes syphilis. In 1910, the scientists announced the discovery of their drug, which they called "compound 606" because it was the sixth substance in the sixth group that they tested. The Hoechst pharmaceutical company began to market the drug under the name "Salvarsan," and it became the treatment of choice for syphilis during the first half of the 20th century. Ehrlich referred to the discovery as "chemotherapy," since it made use of chemicals to treat disease. But today we would call arsphenamine the first true antibiotic.

Moldy Serendipity

But synthetic antibiotics are only half the story. The most famous, and naturally derived, antibiotic discovery resulted from the creation of that unappealing "mold juice." In 1928, Alexander Fleming, a Scottish bacteriologist working at St. Mary's Hospital in London, was getting ready for a summer vacation with his family. Having been researching the properties of *staphylococci* bacteria, the notoriously untidy Fleming hastily stacked several cultures of the bacteria on a bench in his laboratory and left for an August vacation.

When he returned in September, Fleming found something very surprising on one of his cultures. A bit of fungus had contaminated the sample, and the area directly surrounding it was free of bacteria. But further away, on the other side of the culture, the *staphylococci* grew unencumbered. Curious, Fleming grew more of the mold in another culture and discovered that it produced some sort of substance, which he simply dubbed "mold juice," that effectively killed a number of other disease-causing bacteria, including *meningococcus* as well as the bacteria that causes diphtheria.

Penicillin

After more research, Fleming identified the mold as *Penicillium notatum*. And since no one would ever want to be treated with a substance called "mold juice," he named the *Penicillium* secretion *penicillin*. But discovering penicillin and finding a way to practically use it were two different things. Although Fleming published his findings in the *Journal of Experimental Pathology* in 1929, the newly found antibiotic properties of penicillin didn't cause an immediate stir. And Fleming himself wasn't sure that penicillin could ever be used in mass quantities, as it was difficult to cultivate enough of the mold to extract the beneficial agent.

For a decade, penicillin was mostly just used for research, as Fleming and other bacteriologists tried unsuccessfully to find a way to easily isolate and purify the antibiotic. Then, in 1938, a professor of pathology at Oxford University named Howard Florey came across Fleming's article when he was thumbing through old issues of the *Journal of Experimental Pathology*. Florey, who had more funding and a much more well-stocked laboratory than Fleming had ever had, assembled a team of scientists to investigate the properties of this interesting antibacterial mold.

Florey and his team were impressed with the findings of their research, discovering that the fluid taken from the *Penicillium* fungus cured mice infected with *streptococcus* bacteria. The results were so promising that Florey wanted to start testing penicillin on humans, but it took an astounding amount of "mold juice"—about 2000 liters—to create enough pure penicillin to treat one case of sepsis in a human.

By this time, World War II was sweeping across Europe, and even Florey's well-stocked lab was suffering from equipment shortages. The team of scientists began using every available container they could find, from food tins and milk churns to bottles and bedpans, to grow enough of the mold and its antibacterial liquid to be useful. They even hired several "penicillin girls" to look after the fermentation process. Florey's laboratory had basically been converted to a penicillin factory.

But even so, Florey realized that the amount of work he and the team were putting into cultivating penicillin could not last long, especially with the war using up so many resources. If they ever wanted to mass-produce penicillin, they would need to find another strain of the *Penicillium* fungus that produced more antibacterial fluid. And then one day, in a case of remarkable coincidence, a lab assistant named Mary Hunt arrived to work from the market, where she'd picked up a cantaloupe that was covered in yellow mold. Florey tested it, and, by chance, it turned out to be *Penicillium chrysogeum*.

With the help of X-rays and filtration, the *Penicillium chrysogeum* strain of mold produced 1,000 times the amount of penicillin as the *Penicillium notatum* strain. Florey flew to the United States in 1941 to recruit American scientists to help mass-produce the drug, and it immediately began to prove its worth as it cured infections in injured soldiers. Five months after the United States joined the war, 400 million units of penicillin had been manufactured, and by the end of the war, American pharmaceutical companies had fully embraced the treatment, pumping out 650 billion units a month.

Antibiosis

French bacteriologist Jean Paul Vuillemin first described the property exhibited by penicillin and similar drugs as "antibiosis," meaning "against life." And American microbiologist Selman Waksman used the same description when he named the drugs themselves "antibiotics" in 1942. Penicillin wasn't the only antibiotic discovery, of course; but the work of Fleming, Florey, and their teams of researchers helped to pave the way for future scientists, who continue searching for cures for disease, sometimes in the most unexpected places.

Shaping the Practice of Medicine

Peter Mere Latham was a physician who revolutionized medical education. In the 19th century, he championed the idea that medical practitioners should also be teachers of clinical medicine.

⁂ ⁂ ⁂ ⁂

Physician to the Queen

LATHAM WAS BORN in 1761 to a medical family; his grandfather, John Latham, was the President of the Royal College of Physicians and the founder of the Medical Benevolent Society. He followed in his grandfather's footsteps, becoming a medical resident at St Bartholomew's Hospital and attending Oxford University, where he received a Doctor of Medicine and joined the College of Physicians. Latham began working at St. Bartholomew's as a resident physician in 1815, but he periodically returned to Oxford to deliver annual lectures. His reputation as a physician grew, and he would eventually go on to become the personal physician to Queen Victoria.

On Medical Education

In 1836, Latham published *Lectures on Subjects Connected with Clinical Medicine*, which primarily discussed diseases of the heart and pulmonary system. Latham's intent in publishing the book was to provide a medical manual for students, which was a fairly new idea at the time; until then, medical students primarily attended lectures given by physicians on medical topics. Latham's *Lectures* could therefore be considered the forerunner of medical college textbooks of today.

At that time the concept of a full-time professor and medical researcher also did not yet exist. Medical students would attend college and listen to lectures, but did not go through any kind of residency program before beginning their practice. There was also no uniform means of collecting or analyzing clinical data, and so the bulk of medical knowledge that was taught was based in existing and largely untested doctrine.

During his tenure at St Bartholomew's, Latham taught undergraduate medical courses, and he pioneered a number of educational techniques that are taken for granted today. Latham stressed the importance of what he called "self-learning" in clinical instruction, or observational learning. He believed that as a teacher he would be most effective if he presented his students with medical cases that they could then see with their own eyes, thereby building a foundation of experience that would serve them when they entered practice. He emphasized the value of lifelong learning, telling his students that "pathology is a study for your whole life." Latham believed that the essential purpose of learning was to gain a base of wisdom that could equip the clinician with keen problem-solving ability.

Latham was also a pioneer of the use of teaching aids. He encouraged his students to interact with patients, and said there were "a multitude of things" that could be learned at the patient's bedside. He had students attend postmortem autopsies and view pathological specimens, and thought that books and lectures should be used in medical education only sparingly and to supplement hands-on learning.

"The All-Pervading Poppy of Oblivion"

In spite of his extraordinary professional success and his contributions to the field of medical education, John Latham is not widely known, even to students of medicine or clinicians. William Bennett Bean, a 20th century internist and medical historian, described Latham's legacy as being "a classic victim of the all-pervading poppy of oblivion" due to his obscurity.

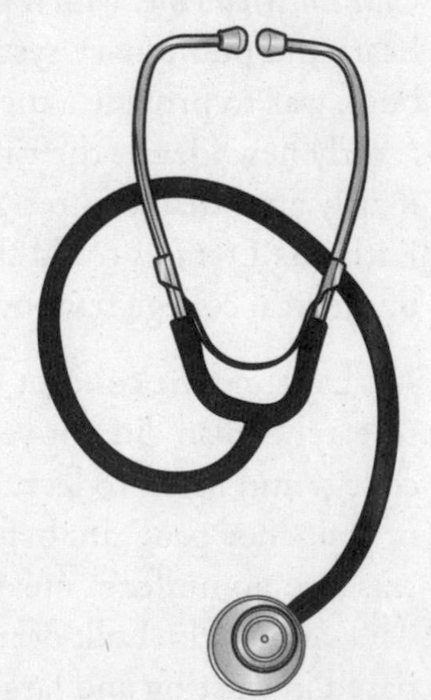

The Birth Control Pill

In late 1916, Margaret Sanger was arrested and thrown in jail. Her crime? Distributing contraception at a Brooklyn family planning and birth control clinic. During the trial that followed, the judge declared that women did not have "the right to copulate with a feeling of security that there will be no resulting conception." Needless to say, this sexist opinion did not sit well with Sanger.

* * * *

But Sanger's experiences in the early 1900s would eventually lead to a huge breakthrough in medicine, one that would change women's lives forever. Sanger herself coined the term "birth control" in 1914, but she was far from the first to attempt pregnancy prevention. Women in ancient Egypt used honey and acacia leaves as a sort of diaphragm, while ancient Greek and Roman women used many different plants and herbs, including fennel, willow, date palm, pomegranate and myrrh, to create oral contraceptives. In China and India, Queen Anne's lace was often used as an oral contraceptive, a method still used in parts of India. But none of these methods was as effective as "coitus interruptus," also known as the "withdrawal method" (which, unfortunately, was still notoriously ineffective). But despite its reputation for being unreliable, the withdrawal method was, for centuries, the best that couples could do if they wished to avoid ending up with a gaggle of children.

By the late 1800s, in part due to English economist Thomas Malthus' views on the dangers of unchecked population growth, the idea of contraception became a popular topic in Great Britain. In 1877, the Malthusian League was founded to educate the public about family planning and methods of birth control, and while they were met with some resistance at first, by the 1880s, birth rates began to drop in Britain. This was also helped by the advent of vulcanized rubber, which made condoms and diaphragms inexpensive and relatively effective.

The Comstock Act

Meanwhile, in the United States, the Comstock Act of 1873 prohibited the distribution of information deemed "obscene," which included information about contraception. Comstock laws also made contraception illegal in order to prevent "illicit" sex. This, of course, had no such effect, and by 1911, Sanger, a New York City nurse who often visited slums on the East Side of the city, was struck by the number of women she tended to who had undergone frequent childbirth, miscarriages, and even self-induced abortions because they had no information about how to prevent pregnancies.

Sanger took matters into her own hands, defying the Comstock Act and publishing two newsletters, *The Woman Rebel* and *Family Limitation,* which candidly detailed various contraception methods. Her efforts landed her in trouble with the law, and she fled to England for two years. On her return, she smuggled diaphragms back into the country, and when she opened her birth control clinic in 1916, she handed them out to many of the women who came to her seeking information about contraceptives. But once again, her activities were considered illegal, and she was arrested, where she faced a judge who said she had no right to be doing what she was doing.

Grassroots Change Begins

But the trial sparked a huge wave of activism throughout the country, and slowly but surely, the restrictions on birth control in the United States were lessened. By 1918, women could use contraceptives for "therapeutic purposes," and by 1936, doctors could distribute contraceptives across state lines, making it legal to send birth control throughout the country.

However, Sanger's fight was still not over. Although she'd succeeded in making birth control and information about it available to everyone, Sanger thought there was still room for improvement. The contraceptive methods available were good, and certainly better than nothing, but they still required some planning and forethought and were susceptible to far-too-frequent failure. What if there was simply a pill that women could take that would prevent pregnancy?

In 1951, Sanger was attending a dinner hosted by Abraham Stone, who was the medical director and vice president of Sanger's clinic, now known as Planned Parenthood Federation of America (PPFA). At the dinner, Sanger met a biologist by the name of Gregory Pincus, who had been studying hormones and their effect on the reproductive system since the early 1930s. Hearing Pincus speak of his experiments, Sanger was fascinated by what he'd already discovered and the potential of what could possibly be accomplished with more research. She convinced Stone to provide Pincus with a modest grant from PPFA, and asked the biologist to look into the possibility of a hormonal contraceptive.

Hormones Prove the Key

Along with reproductive biologist Min Chueh Chang, Pincus was able to theorize that the hormone progesterone, which had already been shown to inhibit ovulation, could be the key to creating a contraceptive pill. But funding from PPFA was meager, and he was unable to conduct experiments on animals to confirm his hunch. In 1952, Sanger was speaking to her friend Katherine McCormick, an heiress and philanthropist who had also helped Sanger to smuggle diaphragms into the country in the 1920s. When McCormick heard about the struggles Pincus was having with funding, she agreed to donate to his project, greatly increasing his ability to conduct research. Thanks to McCormick's funding, by 1953 Pincus and Chang had demonstrated that ovulation could be stopped in mammals through repeated injections of progesterone.

With continued help from McCormick—who, in total, would invest about $2 million ($23 million in today's dollars) of her own money in the project—and pharmaceutical companies Syntex and Searle, Pincus was finally ready to conduct trials of a progesterone pill on humans. In 1956, large-scale trials of the birth control pill were conducted in Puerto Rico, which had no anti-birth-control laws on the books to hinder the study.

The pill was deemed a huge success, and in 1957, the FDA approved it for "severe menstrual disorders" (not surprisingly, a huge number of women were suddenly afflicted with such disorders), and by 1960, it was approved for contraceptive use.

Although controversial and not without side effects, within five years, 6.5 million women were taking oral contraceptives. Over the decades, pill formulations have improved, and there are now around 35 different types of oral contraceptives available, all with varying amounts of hormones to best fit each individual. Although Sanger is often vilified by those who oppose abortion, Sanger herself was never an advocate of the procedure. Rather, she saw prevention as the answer to unwanted pregnancies. "Do not kill, do not take life, but prevent," she would tell the women who came to her for counsel. And thanks to the work of Sanger, McCormick, Pincus, and others, prevention is now a realistic option for women everywhere.

The Evolution of the Cesarean Section

Today the cesarean section is considered a fairly routine surgery. But there was a time when it was a tragic procedure of last resort.

* * * *

What's in a Name?

THE HISTORY OF the cesarean section is shrouded in mystery. Legend has it that Julius Caesar was born by cesarean section in an era when the procedure was only done as a last-ditch effort to save the infant if its mother was dead or dying. However, historians debate whether he was born via cesarean. There are references to his mother, Aurelia, being alive during his life, which would probably have been impossible had she undergone a cesarean. The procedure's name might alternatively derive from a Roman law called "Lex Caesarea," which dictated that if a woman in labor was dead or dying, the baby must be saved so the population of the state would continue to grow.

A Look Back

The cesarean's history is difficult to trace because two lives are at stake during the procedure, so the exact circumstances were either intentionally kept hush-hush or were mixed up with religious birthing and funereal ceremonies. In ancient times, cesareans were usually postmortem procedures: Deceased babies were removed from deceased mothers so that both could have a proper funeral. It was probably pretty rare for an emergency cesarean operation to be done on a dying woman during labor, since few would want to be responsible for determining when and if the mother's chance of survival was hopeless.

References to cesarean sections can be found in ancient pictures and texts the world over. There is some evidence that the surgery was performed on living women; scattered accounts indicate that some women may even have survived the procedure. Cesareans have also been reported in hunter-gatherer and tribal cultures. In fact, one Western report dated to 1879 describes a Ugandan tribal healer who used banana wine to sanitize his hands and the woman's abdomen as well as to intoxicate the woman before practicing a strategy to massage the uterus and make it contract. She survived, and the Western observer concluded the procedure was old and well established.

The cesarean's Western rise to popularity began with the Renaissance, when precise anatomical studies improved all surgeries. It was the development of anesthesia in the 19th century that made the cesarean possible as a procedure intended to preserve the mother's life during a difficult labor: If the mother was not conscious during the procedure, the risk of her dying of shock was removed. Additionally, the development of antiseptics lowered the risk of infection. The cesarean slowly began to replace a tragic procedure called a *craniotomy.* During long and painful childbirths—or in cases when it was clear that the infant had died in the womb—the baby had to be removed without killing the mother. A blunt object would be inserted through the woman's vagina in an effort to crush the baby's skull. The infant was then removed, piecemeal. This was a dangerous procedure that the woman often did not survive.

At Last, Success

Although the cesarean ultimately replaced the craniotomy, maternal mortality remained high. A groundbreaking change came with the realization that the uterine suture—stitching up the uterus—which had once been considered dangerous, was actually the key to a successful cesarean. This, combined with other advances, made the cesarean a feasible option during problematic labor. In fact, the cesarean has become so common that women sometimes elect a cesarean over a vaginal birth, even if there is no difficulty with their labor. This modern practice has led to a peculiar problem: Assuming there is no medical reason for the cesarean, it is safer for a woman to stick with a vaginal birth. The cesarean has thus completed its amazing transformation from an almost impossible option to one that, if anything, is used too frequently.

DNA

Swiss physician and biologist Friedrich Miescher's amazing discovery began in 1869.

* * * *

ALTHOUGH HE'D EARNED his medical degree the year before, Miescher was hearing impaired due to a bout with typhoid fever and felt his deafness would be a hindrance when dealing with patients. So instead, he focused on research, studying white blood cells called neutrophils.

The scientist obtained used bandages from a nearby hospital, carefully washing them with sodium sulfate to remove cells without damaging them. His plan was to then isolate and identify the various proteins within the blood cells.

A Substance with Unique Properties

Miescher experimented with different salt solutions to study the cells, and then tried adding an acid. When he did so, he noticed a substance that separated from the solution, but its properties were unlike those of the proteins he'd been studying. This substance, which he dubbed "nuclein," contained phosphorus and nitrogen, and dissolved when he added an alkali solution. Miescher knew he'd stumbled upon something important, but the tools and methods available to him at the time limited much further research. It would take decades before his discovery, which was later called "nucleic acid" and finally "deoxyribonucleic acid" and "ribonucleic acid," or DNA and RNA, was appreciated as a major scientific breakthrough.

In 1881, German biochemist Albrecht Kossel began to further Miescher's research, isolating the organic compounds found in the nearly identical molecules of DNA and RNA: adenine, cytosine, guanine, thymine, and uracil. Kossel was awarded a Nobel Prize for his work in 1910, just a year after another scientist, Russian biochemist Phoebus Levene, identified some fundamental differences between DNA and RNA when he discovered the phosphate-sugar-base components of the molecules. DNA has a thymine base and deoxyribose sugar; RNA has a uracil base and ribose sugar.

Although other scientists had proposed that these molecules might have something to do with heredity, it was still widely believed that proteins served to carry genetic information. But British bacteriologist Frederick Griffith was the first to demonstrate that DNA could be responsible for this property. In 1928, Griffith mixed two forms of the *Pneumococcus* bacteria, a "smooth" form and a "rough" form, transferring the properties of the "smooth" bacteria to the "rough" bacteria. Fifteen years later, Canadian-American physician Oswald Avery and his coworkers, Colin MacLeod and Maclyn McCarty, identified DNA as the substance that causes genetic transformation, verifying Griffith's findings.

Crick and Watson

Almost ten years later, in May 1952, a PhD student at King's College London, Raymond Gosling, working under the supervision of his instructor, Rosalind Franklin, was studying samples of DNA by hydrating the substance and using X-ray diffraction photography to take images. One of these images, called Photo 51, drew the attention of two scientists, Francis Crick and James Watson, who were working together at the Cavendish Laboratory at the University of Cambridge. The photo displays a fuzzy, X-shaped image, with each leg of the "X" appearing like the rung of a ladder. After studying the blurry photograph and consulting many other sources of information, Crick and Watson proposed the three-dimensional, double-helix model of DNA structure. Although slight changes have been made to Crick and Watson's initial suggestions, the main features of their model turned out to be correct.

The pair's theory was that DNA was a double-stranded helix, connected by hydrogen bonds. Adenine bases were always paired with thymine, and cytosine was paired with guanine. Now that they understood more about the structure, Crick and Watson proposed another theory about DNA replication. They believed that when DNA replicated, it would be "semi-conservative," meaning that the two strands of DNA would separate during replication, with each strand acting as a template for the synthesis of a new strand.

In 1958, molecular biologists Matthew Meselson and Franklin Stahl ran an experiment to test this theory. In what has been called "the most beautiful experiment in biology," the Meselson-Stahl experiment extracted DNA from *E. coli* bacteria and allowed it to replicate. The resulting DNA strands were studied and found to be consistent with Crick and Watson's semiconservative replication hypothesis.

While Crick and Watson are often credited with "discovering" DNA, it is clear that the pair had plenty of help along the way. Thanks to the discoveries and research of everyone from Miescher to Meselson and Stahl, the study of DNA has led to a myriad of practical uses. DNA profiling can now be used to convict or exonerate an individual accused of a crime; genetic genealogy can use DNA to help us find ancestors we never knew we had; and we may even one day be able to genetically modify DNA to eradicate certain diseases or conditions. With barely 150 years of research behind us, the science behind DNA is still fairly new; no doubt more discoveries are just over the horizon.

X-ray Imaging

A discovery that seemed like magic changed medicine forever.

* * * *

IN THE LATE 1800s, scientists were experimenting more and more with the properties of electricity. Cathode rays, which are streams of electrons that can be created in a vacuum outfitted with two electrodes, were first observed in 1869, and this led to the 1875 invention of the Crookes tube. The Crookes tube was created by English physicist William Crookes, and it consisted of a partially evacuated glass bulb with two electrodes, called the cathode and anode, at either end. When voltage was applied between the electrodes, the cathode rays became visible.

One of the most interesting aspects of a Crookes tube was its ability to radiate X-rays. X-rays are electromagnetic waves that act similarly to light waves, but X-rays are 1,000 times shorter than wavelengths of light. Crookes tubes emitted X-rays when the high voltage applied to the electrodes ionized the bit of air inside the tube, creating free electrons. This voltage then accelerated the electrons to such a high velocity that they created X-rays as they struck the side of the glass tube.

Invisible Light

At first, the researchers who used Crookes tubes didn't even realize this was happening, but there had been hints of strange occurrences. The first to make note of "invisible light" emanating from a partially evacuated glass tube through which electric currents were passed was British physician and physicist William Morgan in 1785. Although the glow he observed was no doubt a fascinating phenomenon, the majority of Morgan's efforts were soon concentrated on actuarial science, and his physics experiments took a back seat.

Almost a hundred years later, in 1893, a physics professor at Stanford University named Fernando Sanford discovered something he called "electric photography." Unbeknownst to Sanford, he, too, had stumbled upon X-rays during experimentation with cathode rays. His "electric photography" drew publicity when the *San Francisco Examiner* published an article about it entitled "Without Lens or Light, Photographs Taken With Plate and Object in Darkness." And in 1894, famed inventor Nikola Tesla noted that film in his laboratory was damaged after he'd been experimenting with Crookes tubes, and concluded that the tubes must be emanating some kind of "invisible" energy.

Wilhelm Röntgen

On November 8, 1895, German physicist Wilhelm Röntgen was experimenting with a Crookes tube when he noticed that a chemically coated screen several feet away from the tube was glowing. Intrigued, he covered the Crookes tube with cardboard to try to block any light emanating from it, but the screen continued to glow. He then tried blocking the tube with books and papers, but the same glow occurred. Röntgen then realized that there must be rays of some sort of radiation escaping from the tube to cause the glow. He called them "X-rays" since it was an unknown type of radiation, and obviously, the name stuck.

X-rays and light are both parts of electromagnetic radiation, but X-rays are higher energy, having a very short wavelength. It is this high energy that allows them to pass through many objects. But Röntgen wanted to find out just how many objects these X-rays were able to penetrate, so he holed himself up in his lab and began bombarding anything and everything with X-rays. He discovered that the rays easily passed through material of low density, but higher-density materials, including lead and bone, could block the rays. He also discovered that X-ray images could be recorded on photographic plates. One of the very first images he made was of his wife's hand, resulting in a picture of her wedding ring and the bones of her fingers. Reportedly, she was so shaken by the image that she exclaimed, "I have seen my death!" and refused to ever visit her husband's lab again.

I See Through You

While Röntgen's wife may have found the picture macabre, Röntgen himself immediately recognized the potential of his new discovery. In fact, because he wanted this technology to be easy for those in the medical field to access, he decided not to patent his discovery of X-rays. He sent a description of his findings to many physicians around Europe, and by early 1896, the use of X-rays for medical reasons had already made its way to North America. This "medical miracle" allowed doctors to see broken bones, inflamed joints, or deformities, and during the Balkan War in 1897, the technology was used for the first time on a military battlefield to find bullets and other injuries.

Meanwhile, in America, Thomas Edison heard of Röntgen's discovery and began his own experiments with X-rays, eventually inventing the first mass-produced X-ray imaging device which he called the "Vitascope." Now called a fluoroscope, these machines provide live X-ray images and are not only used for medical purposes, but also for security in airports. X-rays were also found to shrink cancer tumors, giving them another important medical application.

Not Without Risks

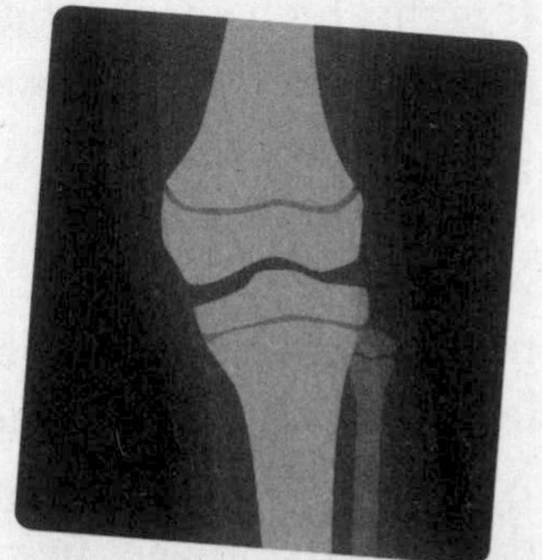

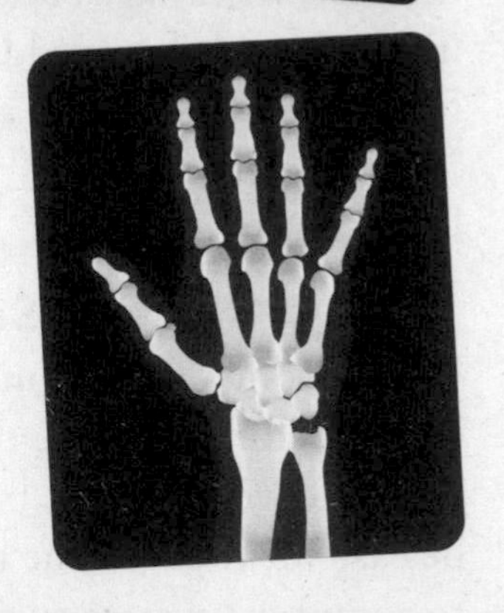

The novelty of seeing inside your own body was fascinating to the general public, and at first, X-ray machines were used for entertainment, with some photography studios offering "X-ray portraits" and "X-ray slot machines" giving customers a quick look at the bones in their hands. The parts needed to create X-rays were so simple and cheap that some people even set up homemade X-ray machines in their homes. But soon, people who exposed themselves to X-rays on a regular basis began noticing unpleasant side effects, like burns, hair loss, and blisters. When one of Edison's assistants, 39-year-old Clarence Dally, died of skin cancer, scientists realized that they needed to start taking more care with the still not-well-understood X-rays. Today, X-ray machines are no longer used for entertainment, and scientists have been careful to document how much radiation a person is exposed to when being X-rayed, leading to smaller doses and more efficient machines.

Röntgen was awarded the first Nobel Prize in Physics in 1901 for his discovery. Although "X-ray" is now the common term for this medical imaging technology, the unit of measurement for the radiation in an X-ray dosage is called a roentgen, and in Germany, X-rays are still known as Röntgen rays.

Vaccination

Throughout history, humans have been plagued with all manner of maladies and contagions. Some, like the common cold, are mostly a nuisance; but many have proven to have devastating, and deadly, consequences. Thanks to the creation of vaccines, we now have a fighting chance against them.

✻ ✻ ✻ ✻

ILLNESSES LIKE CHOLERA and typhoid have killed millions, and still occur in less developed parts of the world, but incidences of these diseases declined significantly with the advent of cleaner food, water, and hospitals. But there are some illnesses, like the pesky cold, that are transmitted even more easily from person to person. Unfortunately, in the case of the common cold, there's not much we can do if we're affected except stock up on chicken soup and tissues. But there are far worse contagions than colds. One of the most insidious afflictions in human history was smallpox. This viral disease began with a fever and vomiting, common to many illnesses. But the hallmark of smallpox was a horrible skin rash that soon developed as the disease progressed, which would cover most of the body with fluid-filled bumps. Those who recovered were left with permanent and extensive scarring, and sometimes blindness. But a third of those who contracted smallpox never recovered, and the risk of death was even higher for babies.

The earliest recorded case was found in ancient Indian medical writings dating back to 1500 BC, and archeologists have uncovered Egyptian mummies with signs of the illness. There is even a Hindu goddess, Sitala Mata, worshipped for her ability to cure smallpox. In Japan, a 7th-century outbreak was thought to have killed off a third of the country's population, and over the centuries, the disease slowly made its way from Asia into Europe. By the 16th century, smallpox was well entrenched throughout Europe, and as explorers from the region traveled to the Caribbean and the Americas, they brought the terrible illness with them. Smallpox became common nearly everywhere in the world by the mid-1700s, becoming the leading cause of death in Europe.

The Variolation Method

For hundreds of years, humans attempted to prevent the spread of this illness through a practice called "variolation." A precursor to vaccines, this less-than-pleasant technique involved scraping the scabs or blisters of a person infected with smallpox and introducing this material into a cut or scratch made on the skin of a healthy person. The hope was that this would result in a less severe version of the disease in the variolated patient, and they would then acquire immunity. First used in the 1500s in China and India, the practice then spread to Africa, Turkey, and Europe. At first, the practice was thought to be a great success; many who were variolated were spared the worst effects of smallpox, and were far less likely to die of the disease. But it soon became clear that variolation posed a major problem: although those who received this treatment were less likely to suffer from the disease, they could still pass it along to other people, ultimately resulting in even more cases of smallpox.

But in the late 1700s, an English physician named Edward Jenner noticed something unusual. Milkmaids were occasionally infected with cowpox, a viral disease found in cows that is transferable between species. The illness is caused by the *orthopoxvirus* genus of viruses, which also includes the *variola* virus that causes smallpox. But after recovering from cowpox, which is less severe than smallpox and often caused nothing more than a few blisters on the hand, the milkmaids never seemed to contract the smallpox virus that often plagued their communities. Jenner knew the two must somehow be linked, so he came up with a bit of a risky plan.

On May 14, 1796, drawing upon the idea of variolation, Jenner took fluid from a cowpox blister on a milkmaid named Sarah Nelmes and scratched it into the skin of a healthy eight-year-old boy named James Phipps, confident that the same protection afforded the milkmaids would work for others as well. Phipps soon developed a few blisters in the scratched area and became mildly ill, but after several days he fully recovered. On July 1, Jenner then took fluid from a blister of a recently infected smallpox patient and scratched it into Phipps' skin, but this time, the boy remained perfectly healthy. Jenner subsequently performed the same procedure on Phipps twenty more times, and the boy never contracted smallpox.

Something Called "Vaccination"

Jenner's discovery was not immediately well-received by the scientific community, so in 1798 he self-published a pamphlet under the wordy title, *An Inquiry into the Causes and Effects of the Variolae Vaccinae, a Disease discovered in some of the Western Counties of England, particularly Gloucestershire, and known by the name of the Cow Pox*. In it, Jenner coined the term "vaccination," and described the technique he used on Phipps and more than 20 other patients. When word of Jenner's discovery reached London, his vaccination technique quickly took off, and within three years, 100,000 people had been vaccinated for smallpox.

Jenner's new technique was soon adopted by other scientists searching for ways to prevent diseases, including Louis Pasteur, who created vaccines for rabies and anthrax; Jonas Salk, who developed the polio vaccine; and Maurice Hilleman, who developed more than 40 vaccines, including those for measles, mumps, chickenpox, meningitis, and pneumonia. Jenner's vaccine was so successful that the last naturally occurring case of smallpox in the world occurred in 1977. A disease that had killed more than 300 million people in the 20th century alone, smallpox was officially declared eliminated in 1980 by the World Health Organization.

Today, vaccination prevents up to three million deaths per year, and although some diseases, including measles, have not been completely eradicated, their incidence has been sharply reduced. Experts are hopeful that polio, which has already been eliminated in the Americas, will soon go the way of smallpox and no longer pose a threat for anyone in the world. With only 22 reported cases in 2017, it looks as if vaccination may soon succeed in annihilating another disease, making the world a healthier place for all of us.

The Lobotomy: A Sordid History

There's a reason why lobotomies have taken a place next to leeches in the Health Care Hall of Shame.

* * * *

Beyond Hollywood

FEW PEOPLE HAVE firsthand experience with lobotomized patients. For many of us, any contact with these convalescents comes via Hollywood—that searing image at the end of *One Flew Over the Cuckoo's Nest* of Jack Nicholson, as Randle Patrick McMurphy, lying comatose. Hopefully, we've all experienced enough to know that Hollywood doesn't always tell it like it is. What would be the point of a medical procedure that turns the patient into a vegetable? Then again, even if Hollywood is prone to exaggeration, the fact is that a lobotomy is a pretty terrible thing.

Dissecting the Lobotomy

What exactly is a lobotomy? Simply put, it's a surgical procedure that severs the paths of communication between the prefrontal lobe and the rest of the brain. This prefrontal lobe—the part of the brain closest to the forehead—is a structure that appears to have great influence on personality and initiative. So the obvious question is: Who the heck thought it would be a good idea to disconnect it?

It started in 1890, when German researcher Friederich Golz removed portions of his dog's brain. He noticed afterward that the dog was slightly more mellow—and the lobotomy was born. The first lobotomies performed on humans took place in Switzerland two years later.

The six patients who were chosen all suffered from schizophrenia, and while some did show post-op improvement, two others died. Apparently this was a time in medicine when an experimental procedure that killed 33 percent of its subjects was considered a success. Despite these grisly results, lobotomies became more commonplace, and one early proponent of the surgery even received a Nobel Prize.

The most notorious practitioner of the lobotomy was American physician Walter Freeman, who performed the procedure on more than three thousand patients—including Rosemary Kennedy, the sister of President John F. Kennedy—from the 1930s to the 1960s.

Freeman pioneered a surgical method in which a metal rod (known colloquially as an "ice pick") was inserted into the eye socket, driven up into the brain, and hammered home. This is known as a transorbital lobotomy.

Freeman and other doctors in the United States lobotomized an estimated 40,000 patients before an ethical outcry over the procedure prevailed in the 1950s. Although the mortality rate had improved since the early trials, it turned out that the ratio of success to failure was not much higher: A third of the patients got better, a third stayed the same, and a third became much worse. The practice had generally ceased in the United States by the early 1970s, and it is now illegal in some states.

Who Got Them?

Lobotomies were performed only on patients with extreme psychological impairments, after no other treatment proved to be successful. The frontal lobe of the brain is involved in reasoning, emotion, and personality, and disconnecting it can have a powerful effect on a person's behavior. Unfortunately, the changes that a lobotomy causes are unpredictable and often negative. Today, there are far more precise and far less destructive manners of affecting the brain through antipsychotic drugs and other pharmaceuticals.

So it's not beyond the realm of possibility that Nicholson's character in *Cuckoo's Nest* could become zombie-like. If the movie gets anything wrong, it's the idea that a person as highly functioning as McMurphy would be recommended for a lobotomy in the first place.

The vindictive Nurse Ratched is the one who makes the call, which raises a fundamental moral question: Who is qualified to decide whether someone should have a lobotomy?

Why Do Doctors Hit Your Knee with a Hammer?

If you're naturally paranoid, you may have considered the possibility that doctors hit your knee just because they can. After all, they could do all sorts of malicious things to us in the name of health, and we would be none the wiser. But thankfully, there's a valid reason for your doctor to whack you on the knee.

* * * *

THE DOCTOR IS timing a stretch reflex, a type of involuntary muscle reaction. While you're sitting on a table, the doctor taps a tendon of the quadriceps femoris, the muscle that straightens your leg at the knee. This tendon stretches the muscle suddenly, and sensory neurons send a message to motor neurons in your spinal cord. These motor neurons send a signal to the muscle in your thigh, which contracts.

The result is that your leg jerks forward. The reflex is highly efficient—the sensory neurons in your knee are wired directly to the motor neurons in your spinal cord that control the reaction, and the brain isn't even involved.

The body reacts this way to keep you balanced while standing and walking without your having to think about it. Putting weight on the leg as you move or shift your balance causes the muscle to contract to support you. Similar stretch reflexes make the rest of the muscles in your legs and feet do what they're supposed to, as well.

Doctors have been banging on knees to test for spinal cord and nerve disorders for more than a century. A diminished reflex reaction can indicate a serious nerve problem, such as tabes dorsalis—the slow degeneration of nerve cells that carry sensory information to the brain. So rest assured—your doctor isn't knocking your knee simply for the entertainment value.

The Microscope

When you were a kid, you probably had at least a few chances to look through the lens of a microscope during a science class. Or maybe you were lucky enough to get your own microscope for your birthday or Christmas, after which you no doubt started using it to look at whatever tiny thing you could get your hands on: hair follicles, pond water, salt crystals, bread crumbs, you name it! There are few tools that have fascinated us and taught us as much about the world around us as the microscope.

✻ ✻ ✻ ✻

MANMADE GLASS AND objects that resemble lenses date back to at least 3500 BC, although it was a slow and arduous process to create the material. But by the first century AD, the process had been perfected enough that the Romans began to experiment with different shapes, diameters, and widths. When they created a piece of glass that was thick in the middle and thin on the edges, they realized that the glass would magnify an object it was held over. But no practical use was found for this phenomenon until the 13th century, when these lenses began to be turned into eyeglasses to magnify written words on a page.

Flea Glasses

The earliest "microscopes" were really just magnifying glasses used for entertainment. Pieces of glass would be fitted into a small tube, and an observer could use the device to examine small objects or insects. A common subject was the flea, so these simple microscopes became known as "flea glasses." Their use was more fun than scientific, but curious minds were starting to take note of the possibilities of glass lenses, including spectacle-makers Zacharias Janssen and his father Hans Martens. They are often credited with inventing the first compound microscope around 1590, although no one knows for certain who truly deserves that recognition. But it is known that Janssen and Martens were fascinated with the way that convex lenses could magnify objects, and they experimented with several lenses in a single tube. The result was a microscope that magnified objects 20 to 30 times their actual size.

Two decades later, Galileo Galilei, who constructed a telescope in 1609, discovered that with a bit of tweaking, he could use his telescope to view small objects and use it as a microscope. Galileo improved on the design of early microscopes by adding a focusing device, but even with this contribution, microscopes remained mostly a novelty for many more decades. It wasn't until the late 1600s that naturalists, interested in studying biology and cell structure, began to use microscopes for scientific research. But as fascinating as microscopes were, they were still rather limited in their magnification power.

Antony van Leeuwenhoek

But in the late 1660s, a Dutch businessman by the name of Antony van Leeuwenhoek was working with magnifying glasses at a dry goods store. He used the magnifying glasses to count threads in woven cloth, to measure their quality. He became so interested in the properties of the curved glass that he decided to try making his own lenses, grinding and polishing glass into a much rounder shape. This greater curvature provided a much higher magnification with only a single lens. Van Leeuwenhoek fit the glass lens into holes between two metal plates, and his simple microscope was able to achieve up to 300 times magnification.

Van Leeuwenhoek's invention helped to solidify the microscope's place in the scientific world. With such a high magnification, van Leeuwenhoek was able to observe red blood cells, bacteria, and yeast, and he also discovered microorganisms in drops of water. Scientists would embrace this new microscope for studying biological structures, and even as new microscopes were created, entomologists would continue to use van Leeuwenhoek's design to study insects for centuries.

Magnification Increases

Throughout the 1700s and 1800s, scientists and researchers began to use microscopes more often, increasing demand for the devices and leading to more advancements. Purer glass was used to craft lenses, increasing image resolution and reducing color distortion, and mirrors were added to focus light on the subject. By the early 20th century, electric lamps were used for light, and after hundreds of years of progress, microscopes were able to magnify objects by thousands of times.

But microscopes rely on the reflection of light off whatever object is being observed, and at a certain point, magnification is so great that wavelengths of light can no longer create a clear image for the naked eye. Scientists were not content to stop at this limit, which is about .275 micrometers, so they began to think about wavelengths that are shorter than light. Their solution was to use beams of electrons instead of light to produce a magnified image. The first prototype of an electron microscope was developed in 1931 by German physicist Ernst Ruska and electrical engineer Max Knoll. Their invention used a beam of electrons in place of light, aimed at the subject and absorbed or scattered. They then form an image that is captured by an electron-sensitive photo plate. Electron microscopes can magnify a subject up to one million times its actual size.

You may think that such a dramatic magnification would be the end of a microscope's capabilities, but scientists are still working on devices that can see smaller structures even more clearly. Fluorescence microscopes use fluorescence and phosphorescence to give tiny substances more contrast and better resolution; scanning probe microscopes can study surfaces at an atomic level; and super resolution microscopes work to create clearer images than we ever thought possible. Just because something is small doesn't mean it's not important; and these fascinating tools have shown us that there's still much to learn about the miniscule world around us.

Magnetic Resonance Imaging

If you've never heard of "zeugmatography"—from the Greek for "join together" and "record" —you're certainly not alone.

* * * *

BUT NO DOUBT you've heard of its much more common name—magnetic resonance imaging, or MRI. An MRI scanner uses magnetic fields and radio waves to see inside the body without the need for invasive surgery or radioactive X-rays. The strong magnetic field created by the machine temporarily realigns hydrogen atoms within the body, while the radio waves cause these atoms to produce a signal—the signal is then recorded, and an image is created. When these images are stacked together, a cross-sectional picture results.

Almost a hundred years before MRI technology was created, scientist Nikola Tesla envisioned a rotating magnetic field. According to his own autobiography, the idea hit Tesla in 1882 when he was walking in a park with a friend, where he picked up a stick and drew a diagram in the sand to illustrate his concept. It would take many more discoveries, breakthroughs and, strangely, a McDonald's restaurant, before MRI technology became the mainstream diagnostic tool it is today.

Nuclear Magnetic Resonance

In 1937, a physicist at Columbia University in New York City, Isidor Rabi, developed what he called "nuclear magnetic resonance," or NMR, which measured the movements of atomic nuclei using magnetic fields and radio waves. This NMR method—for which Rabi was awarded a Nobel Prize in 1944—was originally used to study the structures of chemical substances, but by the 1960s, some scientists began to experiment with the same method on living organisms. In 1971, doctor and research scientist Raymond Damadian discovered that when using Rabi's method, cancerous tissue emitted longer response signals than healthy tissue, due to its higher water content. In this way, Damadian proved that NMR could be useful in the field of medicine.

Damadian's findings were published in March 1971 in the journal *Science*. Miles away on Long Island, at the State University of New York at Stony Brook, chemistry professor Paul Lauterbur read about Damadian's discovery and was immediately intrigued. Lauterbur had been running his own experiments with NMR, running scans on objects like jars of water, clams, and peppers. But the idea that this technology could be used in biomedical applications was too exciting to ignore. Lauterbur began to consider the idea of a gradient magnetic field that would pass linearly over a subject, creating two-dimensional "slices" that could then be stacked to create a three-dimensional image—a technique he dubbed "zeugmatography."

With this new method catching on in the scientific world, researchers began experimenting with gradient magnetic fields. This early MRI method was extremely time-consuming, with scans of small body parts taking hours to complete. But across the Atlantic, a physicist in England named Peter Mansfield was addressing the problem. He developed the "echo-planar imaging" technique, or "line scan imaging," which expedited scans to minutes instead of hours.

The Launch of the "Indominable"

Meanwhile, back in New York, Damadian was granted a U.S. patent for "Apparatus and Method for Detecting Cancer in Tissue." Issued in 1974, it was the world's first patent in the field of MRI. Damadian next set his sights on building a full-body scanner, something that Lauterbur and Mansfield were also working on. Damadian was the first to complete his scanner. The 1.5-ton machine, dubbed "Indominable," was completed in May 1977. The scientist was determined to be the first guinea pig for the machine, climbing in and allowing his assistants to run the scan. But after hours of attempts and adjustments, no readable image resulted.

One assistant suggested (hopefully in a tactful way) that Damadian's body fat percentage might have been a bit too high for a clear scan; consequently, a skinny grad student, Larry Minkoff, volunteered to have a go at it. It took five hours, but the machine finally produced a two-dimensional image of Minkoff's chest.

Although rewarded with this successful scan, Damadian's method was ultimately deemed too slow for general use, and the faster methods that Lauterbur and Mansfield had developed were embraced. The first full-body MRI scanner to use these techniques—the MRI Scanner Mark I—was built at the University of Aberdeen in Scotland in 1980. Over the next few years, MRI scanners began popping up at universities and research centers but were still considered experimental.

Do You Want Fries with That?

But in 1983, Robert Kagan, a pathologist at Holy Cross Hospital in Ft. Lauderdale, Florida, recognized the immense potential of MRI technology. While his hospital refused to purchase the $1.2 million piece of equipment, Kagan found dozens of other doctors who also believed that MRI could provide valuable diagnostic insight, and many of them donated up to $25,000 to cover the high price tag of a machine.

But there was another problem: Since the hospital had already refused the machine, Kagan needed to find a home for the huge piece of technology. So he walked into the nearest freestanding building—which happened to be a McDonald's restaurant—and offered to buy it. The restaurant was actually owned by McDonald's founder Ray Kroc. He agreed to sell the building to Kagan as long as it was not used for another restaurant.

Kagan retrofitted the building, adding $30,000 of copper radio-frequency shielding, and opened his outpatient MRI center—one of the first in the U.S.—on April 1, 1984, the same year the Food and Drug Administration approved the technology. Medicare began covering the procedure in 1985, and by the late-1990s, MRI scans became commonplace.

In 2003, Lauterbur and Mansfield were awarded the Nobel Prize for their "discoveries concerning magnetic resonance imaging."

The technique has come a long way since Lauterbur first called it "zeugmatography," and today, more than 60 million MRI scans are performed every year worldwide, saving lives and bringing peace of mind to countless patients.

The Cardiac Pacemaker

In December 2001, an 86-year-old Swedish engineer named Arne Larsson passed away in Stockholm after a battle with skin cancer. This may not seem very remarkable, but Larsson was actually a bit of a medical marvel; the very fact that he died of skin cancer is notable. Because for the last 43 years of his life, Larsson's approximately 1.5 billion heartbeats were assisted by pacemakers, preventing a death that easily could have been caused decades earlier by his malfunctioning heart.

❋ ❋ ❋ ❋

OUR HEARTS, INTERESTINGLY enough, are powered by electricity. Within the cardiac muscle lies the sinoatrial node, a group of cells that sends out electrical signals that causes a contraction of the heart muscle. Normally, these electrical impulses are spontaneous and serve to maintain the pace of the heart. But if the sinoatrial node is damaged or malfunctioning, this can result in an arrhythmia, or an abnormal speed or rhythm in the heart. If the sinoatrial node fails altogether, the atrioventricular junction, another part of the heart's electrical system, can take over, setting a pace of about 40 to 60 beats per minute. And if neither of these areas is functioning correctly, the ventricles of the heart can set a "default" rate of about 30 beats per minute.

In Larsson's case, a viral infection in the 1950s left his heart scarred and disrupted the normal electrical impulses, resulting in a heart rate that dipped as low as 28 beats per minute. He often passed out due to a lack of blood flow to the brain, a potentially fatal occurrence known as Stokes-Adams syndrome. Larsson's wife and home health aides, needed for his round-the-clock care, kept him alive by pounding on his chest dozens of times a day. Knowing that this was no way to live, Larsson and his wife were desperate for some kind of help.

Scientists had known since the late 1800s that electrical impulses could cause the heart to contract. Scottish physiologist John Alexander MacWilliam wrote about just such a thing in an 1889 issue of the *British Medical Journal,* where he described experimenting with electricity applied to a stopped human heart that caused the heart to beat. The first practical application of this phenomenon occurred in the late 1920s, when an Australian physician named Mark Lidwell devised a machine that could directly deliver electrical impulses to a heart by way of a needle. As unpleasant as it sounds, this device was able to successfully revive a stillborn baby, even after many other methods of revival had failed.

Around the same time in the United States, cardiologist Albert Hyman and his brother Charles invented a similar device, but one that had to be powered by a hand-turned motor. Although Hyman's machine never gained much popularity, he was the first to refer to such a device as an "artificial pacemaker."

Effective but Unwieldy

Twenty years later, in the early 1950s, doctors and scientists were able to create the first external pacemakers. But these bulky machines, which used vacuum tube technology, needed to be plugged in to a wall socket, limiting the patient's movement. And even though they eliminated the unsettling needle-in-the-heart, they could be painful for the patient and ran the risk of causing electrocution. An improvement, perhaps, although certainly not ideal. But the newly invented transistor, which helped to reduce computers to a manageable size, was about to change pacemakers, as well.

Several inventors, including American cardiologist Paul Zoll and Columbian electrical engineer Jorge Reynolds Pombo, had created smaller external pacemakers that were powered by batteries, but these still could weigh around 100 pounds. But in 1952 Zoll's device not only successfully resuscitated a man but kept his heart beating steadily for 50 hours. This success sparked a renewed interest in the potential of these devices. This proof that they could save lives, coupled with transistor technology, led to the creation of smaller pacemakers over the next few years.

Small and Portable

In 1958, American surgeon C. Walton Lillehei and engineer Earl Bakken created the first wearable, transistorized pacemaker. Lillehei had noticed that children who underwent cardiac surgery often suffered from heart rhythm disturbances, so to counteract this effect, he implanted pacemaker leads directly into the myocardial muscle and connected these to a large external pacemaker. The method was successful in maintaining normal heart rhythm, and Bakken improved on the design by making the pacemaker small and portable. These were a huge improvement over the bulky, restrictive pacemakers of the earlier part of the decade, but there was still a drawback: the leads for the pacemaker came through the skin, and often caused the patient burns and infections.

Success from Desperation

But the same year, in Solna, Sweden, engineer Rune Elmqvist and surgeon Ake Senning were working on a breakthrough device: a fully implantable cardiac pacemaker that weighed only six ounces. Although their design was extremely promising, it was still in its testing phase when a desperate Larsson and his wife, Else-Marie, visited Senning and begged him to implant the device in Larsson's chest. Hesitantly, Elmqvist and Senning assembled the best version of their device they could manage, and Senning performed the surgery to implant the pacemaker. After only eight hours, the device failed, and Larsson was whisked back to the operating room, where Senning implanted their only backup device. This time, the surgery was a success, and the pacemaker functioned for three years.

Throughout Larsson's life, he received 25 different pacemakers, each a bit smaller, more durable, and more efficient than the last. Surgeons were also able to improve the method of implantation, threading leads through a large vein in the arm until the electrode could be secured to the heart muscle, which eliminated the previous need for open-heart surgery.

The first implantable pacemakers used a nickel-cadmium battery, but in 1960, engineer Wilson Greatbatch, now a member of the National Inventors Hall of Fame, created a pacemaker with a mercury battery. This led to medical device company Medtronic, now one of the leading manufacturers of pacemakers in the world, to further their development of the new technology.

After lithium batteries, which are prized for their long life, were patented in 1968, Greatbatch created a lithium iodide battery for pacemakers, which was introduced in 1972. With an average lifespan of ten years, these batteries became the go-to power source for pacemakers, and are often still used today. By the 1980s, pacemakers were able to administer steroids to decrease swelling around the implants, and in the 1990s, microprocessors kept track of patients' hearts to allow the pacemaker to adapt to changes in pace or rhythm.

Wireless and Noninvasive

Today, pacemakers weigh less than an ounce and can transmit information about a patient's heart via Wi-Fi, and self-adjust depending on a person's activity. Trials are underway which are testing new devices about the size of a pill, which would make pacemakers about as noninvasive as possible!

There's no doubt that cardiac pacemakers save countless lives every year. After Larsson received his pacemaker in 1958, he was finally able to resume his normal life and career, traveling the world to supervise the repair of electrical systems aboard ships. He frequently acted as an ambassador for the pacemaker industry, grateful for the extra time he'd been given. Larsson and Else-Marie often told their story over the years, to encourage the advancement of this life-saving technology, and until his health began to decline in 2000, the pacemaker-advocate remained active, swimming, riding his bike, and dancing with his wife—enjoying life for an extra 43 years. One final note: Larsson's pacemakers worked so well that he outlived both Elmqvist, who died in 1996, and Senning, who died in 2000.

The Medical Uses of Bee Venom

Believe it or not, that pesky bee buzzing around your fruit salad could one day save your life.

* * * *

All-Natural Apitherapy

DO YOU SUFFER from apiphobia? In other words: Do you freak out when you see a bee? You might want to get over that—it turns out bees can make you better. For more than 4,000 years, the common honeybee has been used to treat everything from multiple sclerosis to rheumatoid arthritis, gout, asthma, impotence, epilepsy, depression, bursitis, shingles, tendonitis, and even some types of cancer. As part of a unique alternative-medical approach called apitherapy, practitioners use natural bee by-products such as raw honey, beeswax, pollen, royal jelly, and even bee venom to treat medical conditions that may be unresponsive to traditional medicine.

Facing the Stinger

Bee venom is currently being used to reduce inflammation and pain and to treat resistant skin diseases such as psoriasis. Although researchers have already identified more than 40 pharmacologically active substances in bee venom, very little is known about them or what they can do. One protein that is understood is *melittin*, which has been shown to stimulate the adrenal glands, which produce *cortisol*. Cortisol is a naturally occurring anti-inflammatory that promotes healing in the body. In multiple sclerosis patients, bee venom is thought to dissolve scar tissue on myelin sheaths, improving nerve transmissions.

Bee venom can be administered directly from the bee's stinger or via injection by a trained healthcare professional. After identifying the affected area, the venom is injected, often at key acupuncture points, hence the term, "bee acupuncture."

While few published studies exist supporting the use of bee venom, there are hundreds of stories that attest to its effectiveness. A man in India accidentally disturbed a hive and was stung more than 20 times. While no doubt the punctures were painful, the psoriasis on his scalp and the arthritis in his knees disappeared in less than three months.

Birkenstocks

Birkenstock sandals may be icons of the 1960s granola-munching crowd, but there's more to them than you might realize—and the company that makes them is more than 200 years old.

❋ ❋ ❋ ❋

THE BIRKENSTOCK COMPANY traces its roots to the German village of Langen-Bergheim, where in 1774 Johann Birkenstock was registered as a "Shoemaker." In 1897, his grandson, Konrad Birkenstock, introduced a major advance—the first contoured shoe lasts, which enabled cobblers to customize footwear.

At the time, there was a debate regarding whether it was healthier to train your feet to fit your shoes or to wear shoes that were made to support your foot's natural shape. The Birkenstock company was dedicated to promoting the second idea. By 1902, Birkenstock's flexible arch supports were being sold throughout Europe. During World War I, Birkenstock employees worked in clinics to design shoes especially for injured veterans.

It took the aching feet of a tourist to bring the Birkenstock sandal to America. In 1966, Margot Fraser came across Birkenstocks during a visit to a German spa. The shoes soothed her foot ailments, and she was hooked. Fraser secured the distribution rights and set out to sell these strange-looking German sandals back at her California home.

At first, the only places that would carry her Birkenstocks were health food stores. In the 1970s, as health food became more popular, people discovered Birkenstocks at the same time that they discovered tofu and alfalfa sprouts. Birkenstocks' association with "granola" and "hippies" came directly out of this.

Although the company has attempted to bring its sandals into the realm of high-end fashion (model Heidi Klum has designed her own line of Birkenstocks), this is one brand of footwear that can't shake its "crunchy" connotations. Interestingly, researchers have discovered shoes more than 8,000 years old in a cave in Missouri, and the modern shoe that these most resembled was the Birkenstock! Crunchy or not, comfort never goes out of style.

Grandpa, Stick Around a While: The Origins of Mummification

Turns out that the Egyptians—history's most famous embalmers—weren't the first. By the time Egyptians were beginning to fumble with the art, Saharans and Andeans were already veterans at mortuary science.

* * * *

Andes

IN NORTHERN CHILE and southern Peru, modern researchers have found hundreds of pre-Inca mummies (roughly 5000–2000 BC) from the Chinchorro culture. Evidently, the Chinchorros mummified all walks of life: rich, poor, elderly, didn't matter. We still don't know exactly why, but a simple, plausible explanation is that they wanted to honor and respect their dead.

The work shows the evolution of increasingly sophisticated, artistic techniques that weren't very different from later African methods: Take out the wet stuff before it gets too gross, pack the body carefully, dry it out. The process occurred near the open-air baking oven we call the Atacama Desert, which may hold a clue in itself.

Uan Muhuggiag

The oldest known instance of deliberate mummification in Africa comes from ancient Saharan cattle ranchers. In southern Libya, at a rock shelter now called Uan Muhuggiag, archaeologists found evidence of basic semi-nomadic civilization, including animal domestication, pottery, and ceremonial burial.

We don't know why the people of Uan Muhuggiag mummified a young boy, but they did a good job. Dispute exists about dating here: Some date the remains back to the 7400s BC, others to only 3400s BC. Even at the latest reasonable dating, this predates large-scale Egyptian practices. The remains demonstrate refinement and specialized knowledge that likely took centuries to develop. Quite possibly some of this knowledge filtered into Egyptian understanding given that some of the other cultural finds at Uan Muhuggiag look pre-Egyptian as well.

Egypt

Some 7,000 to 12,000 years ago, Egyptians buried their dead in hot sand without wrapping. Given Egypt's naturally arid climate, the corpse sometimes dehydrated so quickly that decay was minimal. Sands shift, of course, which would sometimes lead to passersby finding an exposed body in surprisingly good shape. Perhaps these surprises inspired early Egyptian mummification efforts.

As Egyptian civilization advanced, mummification mixed with views of the afterlife. Professionals refined the process. An industry arose, offering funerary options from deluxe (special spices, carved wood case) to budget (dry 'em out and hand 'em back). *Natron*, a mixture of sodium salts abundant along the

Nile, made a big difference. If you extracted the guts and brains from a corpse, then dried it out it in natron for a couple of months, the remains would keep for a long time. The earliest known Egyptian mummy dates to around 3300 BC.

Desert Origins

It's hard to ignore a common factor among these cultures: proximity to deserts. It seems likely that ancient civilizations got the idea from seeing natural mummies. Ice and bogs can also preserve a body by accident, of course, but they don't necessarily mummify it. Once exposed, the preservation of the remains depends on swift discovery and professional handling. If ancient Africans and South Americans developed mummification based on desert-dried bodies, it would explain why bogs and glaciers didn't lead to similar mortuary science. The ancients had no convenient way to deliberately keep a body frozen year-round without losing track of it, nor could they create a controlled mini-bog environment. But people could and did replicate the desert's action on human remains.

Today

We make mummies today, believe it or not. An embalmed corpse is a mummy—it's just a question of how far the embalmers go in their preservation efforts. If you've attended an open-casket funeral, you've seen a mummy.

Head Like a Hole: The Weird History of Trepanation

There aren't many medical procedures more than 7,000 years old that are still practiced today. Trepanation, or the practice of drilling a hole in the skull, is one of the few.

* * * *

An Ancient Practice

HAS ANYONE EVER angrily accused you of having a hole in your head? Well, it's not necessarily an exaggeration.

Trepanation (also known as "trephination") is the practice of boring into the skull and removing a piece of bone, thereby leaving a hole. It is derived from the Greek word *trypanon,* meaning "to bore." This practice was performed by the ancient Greeks, Romans, and Egyptians, among others.

Hippocrates, considered the father of medicine, indicated that the Greeks might have used trepanation to treat head injuries. However, evidence of trepanning without accompanying head trauma has been found in less advanced civilizations; speculation abounds as to its exact purpose. Since the head was considered a barometer for a person's behavior, one theory is that trepanation was used as a way to treat headaches, depression, and other conditions that had no outward trauma signs. Think of it like a pressure release valve: The hole gave evil spirits inside the skull a way out of the body. When the spirits were gone, it was hoped, the symptoms would disappear.

How to Trepan

In trepanning, the Greeks used an instrument called a *terebra,* an extremely sharp piece of wood with another piece of wood mounted crossways on it as a handle and attached by a thong. The handle was twisted until the thong was extremely tight. When released, the thong unwound, which spun the sharp piece of wood around and drove it into the skull like a drill. Although it's possible that the terebra was used for a single hole, it is more likely that it was used to make a circular pattern of multiple small holes, thereby making it easier to remove a large piece of bone. Since formal anesthesia had not yet been invented, it is unknown whether any kind of numbing agent was used before trepanation was performed.

The Incas were also adept at trepanation. The procedure was performed using a ceremonial tumi knife made of flint or copper. The surgeon held the patient's head between his knees and rubbed the tumi blade back and forth along the surface of the skull to create four incisions in a crisscross pattern. When the incisions were sufficiently deep, the square-shaped piece of bone in the center was pulled out. Come to think of it, perhaps the procedure hurt more than the symptom.

Trepanation Today

Just when you thought it was safe to assume that the medical field has come so far, hold on—doctors still use this procedure, only now it's called a craniotomy. The underlying methodology is similar: It still involves removing a piece of skull to get to the underlying tissue. The bone is replaced when the procedure is done. If it is not replaced, the operation is called a *craniectomy*. That procedure is used in many different circumstances, such as for treating a tumor or infection.

However, good ol'-fashion trepanation still has its supporters. One in particular is Bart Hughes, who believes that trepanning can elevate one to a higher state of consciousness. According to Hughes, once man started to walk upright, the brain lost blood because the heart had to frantically pump it throughout the body in a struggle against gravity. Thus, the brain had to shut down certain areas that were not critically needed to assure proper blood flow to vital regions.

Increased blood flow to the brain can elevate a person's consciousness, Hughes reasoned, and he advocated ventilating the skull as a means of making it easier for the heart to send blood to the brain. (Standing on one's head also accomplishes this, but that's just a temporary measure.) Some of his followers have actually performed trepanation on themselves. For better or gross, a few have even filmed the process. In 2001, two men from Utah pled guilty to practicing medicine without a license after they had bored holes into a woman's skull to treat her chronic fatigue and depression. There's no word as to whether the procedure actually worked, or if she's just wearing a lot of hats nowadays.

A Cure for Bedroom Blues

Viagra began as a suprising breakthrough and ultimately became a genuine pharmaceutical phenomenon. Not since the birth control pill has a drug had such an astounding social impact.

* * * *

PFIZER'S "LITTLE BLUE pill" has brought relief to millions of men worldwide who suffer from erectile dysfunction, which is defined as an inability to maintain an erection. But that wasn't what its developers originally had in mind.

Sildenafil citrate, the active ingredient in Viagra, was originally designed as a heart drug. Because it acts as a *vasodilator* (a drug that helps blood vessels dilate), researchers thought it would be an effective treatment for high blood pressure and conditions such as angina.

However, sildenafil had an unusual side effect: The drug made it easier for men—especially those with erectile dysfunction—to get an erection. Pfizer, knowing a lucrative breakthrough when it saw one, changed direction and began studying sildenafil as a treatment for one of the most common sexual problems in the world. The rest, as they say, is history.

The response to Viagra has been phenomenal. Nearly 570,000 prescriptions were written during the drug's first month on the market in early 1998. Viagra remains one of the world's most frequently prescribed drugs and in recent years has been joined by competitors like Cialis (*tadalafil*) and Levitra (*vardenafil*).

Viagra works by boosting blood flow to the penis, blocking an enzyme known as *phosphodiesterase type* 5 (PDE5). Approximately 70 percent of men who take Viagra for erection problems with a physical cause report success, noting that their erections develop faster, are harder, and last longer.

But Viagra is not perfect: Potential side effects include headache, a blue tint to vision, facial flushing, indigestion, and dizziness. Satisfied users say it's a small price to pay for a big boost in the bedroom.

I Vant to Transfer Your Blood!

Basically, a blood transfusion is the process of transferring blood from one person to another, and it's a life-saving procedure that is performed about every two seconds somewhere in the world. More than 4.5 million patients need blood transfusions each year in the United States and Canada. That's a lot of red stuff. Read on for some interesting facts about blood and the history of transfusions.

* * * *

- One of the earliest recorded blood transfusions took place in the 1490s. Pope Innocent VIII suffered a massive stroke and his physician advised a blood transfusion. Unfortunately, the methods used were crude and unsuccessful; the Pope died within the year and so did the three young boys whose blood was used. But good effort!
- Women receive more blood than men: 53 percent of all transfusions go to women, while 47 percent to men.
- Richard Lower, an Oxford physician in the mid-1600s, performed blood transfusions between dogs and eventually between a dog and a human. The dog was kept alive via the transfusion, and the experiment was considered a success. Several years later, however, cross-species transfusions would be deemed unsafe.
- Donating blood seems to appeal to folks with a sense of civic duty: 94 percent of blood donors are registered voters.
- In 73 countries, many of which are undeveloped and have a great need for blood, donation rates are less than 1 percent.

* One unit of blood can be separated into several components: red blood cells, plasma, platelets, and cryoprecipitate, which is a substance helpful in the clotting process.
* If you're older than 17 and weigh at least 110 pounds, you may donate a pint of blood (the most common amount of donation) about once every two months in the United States and Canada.
* If only 1 additional percent of Americans would donate blood, shortages would disappear.
* In 1818, British obstetrician James Blundell performed the first successful transfusion of human blood. His patient, a woman suffering postpartum hemorrhaging, was given a syringe-full of her husband's blood. She lived and Blundell became a pioneer in the study of blood transfusions.
* About one in seven people entering a hospital need blood.
* More than 85 million units of blood donations are collected globally every year. About 35 percent of these are donated in developing and transitional countries, which makes up about 75 percent of the world's population.
* Those with type O blood are considered "universal donors," which means their blood can be cross-matched with all blood types successfully.
* Only 38 percent of the U.S. population is eligible to donate blood and less than 10 percent actually do on an annual basis. There are 43,000 pints of blood used each day in the United States and Canada.
* In 1916, scientists introduced a citrate-glucose solution that allowed blood to be stored for several days. Due to this discovery, the first "blood depot" was established during World War I in Britain.

* Thirteen tests (11 of which test for infectious diseases) are performed on every unit of donated blood.
* If you need a blood transfusion, try not to need one in the summer or around the holidays. Shortages of all blood types happen during these times.
* The first screening test to detect the probable presence of HIV was licensed and implemented by blood banks in the United States in 1985. Since then, only two people have contracted HIV from a blood transfusion.

Open Up and Say "Ugh"!

One of the last things a patient wants to see as they look across a sterile operating room are leeches, maggots, and scum-sucking fish. But all three have earned a solid place in the medical community—simply by doing what comes naturally.

* * * *

The Flies Have It

MAGGOTS ARE NOTHING more than fly larvae—one of the most basic forms of life. But to many patients with wounds that refuse to respond to conventional treatment, they are a godsend. For the majority of people recovering from life-threatening wounds, contusions, and limb reattachments, antibiotics provide much of the follow-up care they need. But for a small percentage of patients who do not respond to modern medicines, maggots slither in to fill the gap.

Applied to a dressing that is made in the form of a small "cage," maggots are applied to almost any area that does not respond well to conventional treatment. The maggot thrives on consuming dead tissue (a process called "debridement"), while ignoring the healthy areas. After several days, the maggots are removed—but only after they have consumed up to ten times their own weight in dead tissue, cleaned the wound, and left an ammonialike antimicrobial enzyme behind.

While maggot therapy may not be everyone's cup of tea, it is effective in treating diseases like diabetes where restricting circulation for any reason can often result in nerve damage and even loss of limb.

Golden Age of Leeches

Similar to the maggot, leeches are small animal organisms that have been used by physicians and barbers (who, in the olden days, were also considered surgeons) for over 2,500 years for treating everything from headaches and mental illnesses to—gulp—hemorrhoids. And while they might appear to be on the low end of the evolution scale, leeches actually have 32 brains!

Leeches are raised commercially around the world with the majority coming from France, Hungary, Ukraine, Romania, Egypt, Algeria, Turkey, and the United States. Used extensively until the 19th century, the "Golden Age of Leeches" was usurped by the adoption of modern concepts of pathology and microbiology. *Hirudotherapy,* or the medicinal use of leeches, has enjoyed a recent resurgence after their demonstrated ability to heal patients when other means have failed.

Leeches feed on the blood of humans and other animals by piercing the skin with a long proboscis. Oftentimes this is the most effective way to drain a postsurgical area of blood, and it can actually facilitate the healing process. At the same time leeches attach to their host, they inject a blood-thinning anticoagulant; they continue until they have consumed up to five times their body weight in blood. The host rarely feels the bite because the leech also injects a local anesthetic before it pierces the skin.

The Doctor (Fish) Is In

Another unlikely ally to the medical community is the doctor fish, found in bathing pools in the small Turkish town of Kangal. The therapeutic pools in Kangal are a popular destination for people suffering from fractures, joint traumas, gynecological maladies, and skin diseases. While the pools themselves have a number of beneficial qualities such as the presence of selenium (a mineral that protects against free radicals and helps with wound healing), they are most famous for the doctor fish that live there.

At only 15 to 20 centimeters in length, doctor fish are relatively small and do not physically attach themselves to their host like leeches or maggots. Instead, they surround a person's skin, striking and licking it. They are particularly fond of eating psoriatic plaque and other skin diseases that have been softened by the water, eating only the dead and hyper-keratinized tissue while leaving the healthy tissue behind. While many people might be uncomfortable at the thought of being surrounded by a school of fish feasting on their skin, many actually enjoy the pleasant and relaxing sensation of getting a "micro-massage."

Dreamland

In an average lifetime, people spend approximately 2,100 days (almost 6 years) dreaming. Everyone dreams every night, though some of us can't remember our dreams.

⁕ ⁕ ⁕ ⁕

- Blind people dream. If they became blind after having sight, visual images appear in their dreams. If they were born blind, their dreams, like their lives, are made up of feelings, smells, movements, and sounds.
- Developmental psychologists say that toddlers never dream about themselves. Children are not believed to appear in their own dreams until a stage that occurs when they are three or four years old and realize they are separate from other people.
- Rapid eye movement (REM) sleep is the stage during which we have our most vivid dreams, characterized by bizarre plots involving unlikely people or things. In contrast, non-REM dreams are more like waking thoughts. They have less

imagery and tend to repeat a thought obsessively (for example, "I've lost my keys!").

* It was once believed that dreams occurred only during REM sleep. Improved technology has allowed researchers to discover that dreams are less frequent in non-REM sleep phases but still exist. In fact, it is likely that we dream during every single moment of sleep.
* Color in dreams is a constant source of speculation. Some monochrome dreams can have a single image that's in color, such as a bright pink poodle. Other dreams seem to speak a language of colors (e.g., red or blue lights) and shapes (repeated circles or squares). Sometimes, natural colors pervade the dream, as in waking life.
* In the late 1950s, scientists proved that external stimuli can be incorporated into dreams. When researchers sprinkled water on sleeping volunteers and woke them up seconds later, 14 out of 33 subjects said they had dreamed of water.
* It is believed that we rarely feel pain in dreams. When we do, though, our bodies perceive it as a signal that something is wrong, and we react by waking up.
* Studies conducted by Harvard University reveal that dreams exhibit five strange features. They have the qualities of hallucinations (seeing things that don't exist), delusions (believing something imaginary), emotional intensity (extreme feelings about a situation), amnesia (forgetting our lives and even who we are in those lives), and cognitive abnormalities (having thoughts that differ from the waking norm).
* What is the purpose of dreams? Some experts speculate that the primitive part of the brain is overloaded during the day and cannot process all of our experiences. Dreaming gives us a way to sort through our memories and eliminate the ones that aren't useful for our growth.

To Pee or Not to Pee

Some alternative medicine proponents claim urine therapy can cure a long list of ailments.

* * * *

Urine Business!

URINE IS SOMETHING even moderately healthy people see on a daily basis. A fluid carrying waste from the kidneys out of the body, urine has an ammonialike smell due to the nitrogenous wastes that make up a small percent of it (the remaining 95 percent being water.) The chief constituent of the nitrogenous wastes in urine is *urea,* a product of protein decomposition. When your body doesn't need urea or can't use it, it's processed through your system and comes out in urine.

Pee as Medicine

In ancient Rome, there are accounts of people to the west that brushed their teeth with urine for a whitening effect. In India, a Sanskrit text promotes the benefits of "pure water," or urine, for treating afflictions of the skin. The prophet Muhammad is said to have advocated drinking camel urine when sick, and some point to a line in the Bible as proof urine is supposed to be used medicinally: "Drink waters from thy own cistern, flowing water from thy own well." (Proverbs 5:15)

Urine *is* sterile (though only fresh urine qualifies) and does contain various elements that are found in medicines, such as urea and whatever vitamins the body decided it wasn't going to absorb, but scientists and doctors have found that it doesn't contain enough of any of these to really be an effective medicinal tool. Many folks will tell you that urine can be used to treat athlete's foot, but there's no scientific evidence that this is true.

However, urine therapy *is* helpful on a battlefield, when water is unavailable to clean a wound. In those cases, using urine in place of water is better than nothing, but it's still a last resort.

Research shows that while you're unlikely to die if you drink your own pee, there have been cases where folks became ill after doing so. It's always a good idea to consult your doctor before experimenting with any home remedies, including this one.

Great Achievements in Medical Fraud

Have you fallen for pills, ointments, and gadgets that swear to make you thinner or more muscular, with thicker hair and rock-hard abs? Well, you're not alone. Read on for a look at some of history's medical shams and charlatans.

* * * *

To the Rescue

If you were diagnosed at the turn of the century with lumbago, puking fever, black vomit, consumption, decrepitude, falling sickness, milk leg, ship fever, softening of the brain, St. Vitus's dance, trench mouth, dropsy, or heaven forbid, dyscrasy, then chances are you were in big trouble. Not only did the "modern" medical community misunderstand most of these diseases, they were also clueless as to how to treat them.

Facing a life of interminable pain and suffering, many sufferers of these diseases resorted to hundreds of unfounded medical treatments, which sometimes worked and sometimes didn't. Here's a brief list of some of the more popular medical treatments and the claims by their originators:

The Battle Creek Vibratory Chair: Many people who enjoy a bowl of Corn Flakes in the morning are familiar with their inventor, Dr. John Harvey Kellogg of Battle Creek, Michigan. Dr. Kellogg also designed a number of therapeutic devices, including this contraption. After strapping the patient in, the chair would shake violently and "stimulate intestinal peristalsis" that was beneficial to digestive disorders. Prolonged treatments were also used to cure a variety of maladies from headaches to back pain.

The Toftness Radiation Detector: If people thought the Toftness Radiation Detector looked suspiciously like the PVC piping and couplings found at a hardware store, that's because it was. By passing PVC tubing outfitted with inexpensive lenses over the patient's back, chiropractors listened for a high-pitched "squeak" that meant that the device had detected areas of neurological stress, characterized by high levels of radiation. The device was widely used until 1984 when it was deemed worthless by the Food and Drug Administration.

The Foot Operated Breast Enlarger Pump: In the mid-1970s, silicone breast implant technology was still in its infancy. Instead, many women pining for larger breasts spent $9.95 for a foot-operated vacuum pump and a series of cups that promised "larger, firmer and more shapely breasts in only 8 weeks." As it turned out, more than four million women were duped into buying a device that produced nothing more than bruising.

The Crystaldyne Pain Reliever: In 1996, one of the most popular pain relievers on the market was nothing more than a gas grill igniter. When the sufferer pushed on the plunger, the device sent a short burst of sparks and electrical shocks through the skin to cure headaches, stress, arthritis, menstrual cramps, earaches, flu, and nosebleeds. After being subjected to FDA regulations, however, the company disappeared with thousands of dollars, falsely telling their consumers that "their devices were in the mail."

The Prostate Gland Warmer and The Recto Rotor: Even someone without the slightest bit of imagination would cringe at the idea of inserting a 4-inch probe connected to a 9-foot electrical cord into their rectum. However, for thousands of adventurous consumers in the 1910s, the Prostate Gland Warmer (featuring a blue lightbulb that would light up when plugged in) and the Recto Rotor promised the latest in quick relief from prostate problems, constipation, and piles.

The Radium Ore Revigator: In 1925, thousands of unknowing consumers plunked down their hard-earned cash for a clay jar with walls that were impregnated with low-grade radioactive ore. The radioactive material was nothing more than that found in the dial of an inexpensive wristwatch, but the Revigator still promised to invigorate "tired" or "wilted" water that was put into it—"the cause of illness in one hundred and nine million out of the hundred and ten million people of the United States."

Hall's Hair Renewer: For as long as there's been hair loss, there has probably been hair-loss cures. One of the better-known snake oils in the 19th century was Hall's Vegetable Sicilian Hair Renewer, which Reuben P. Hall began selling in 1894. According to the inventor, an Italian sailor passed the recipe onto him; the results promised hair growth and decreased grayness. The first version was composed of water, glycerine, lead sugar, and traces of sulfer, sage, raspberry leaves, tea, and oil of citronella.

Eventually, the formula was adjusted to include two kinds of rum and trace amounts of lead and salt. Of course, lead is poisonous, and the ingredients had to be changed once again. Still, the product sold into the 1930s. Perhaps it was its promise that "As a dressing it keeps the hair lustrous, soft and silken, and easy to arrange. Merit wins."

The Relaxacisor: For anyone who hated to exercise but still wanted a lithe, athletic body, the Relaxacisor was the answer. Produced in the early 1970s, the Relaxacisor came with four adhesive pads that were applied to the skin and connected by electrodes to a control panel. The device would deliver a series of electrical jolts to the body, "taking the place of regular exercise" while the user reclined on a sofa. All 400,000 devices were recalled for putting consumers at risk for miscarriages, hernias, ulcers, varicose veins, epilepsy, and exacerbating preexisting medical conditions.

The Timely Warning: In 1888, one of the most embarrassing and debilitating experiences a man could endure was an "amorous dream" or "night emission." Fortunately, Dr. E. B. Foote came up with the "Timely Warning," a circular, aluminum ring that was worn to prevent "the loss of the most vital fluids of the system—those secreted by the testicular glands." For better or for worse, no diagrams have been found to illustrate exactly just *how* the device was worn.

Body Talk: Anthropometrics and Human Engineering

Every day, anthropometry and its related fields, ergonomics and biomechanics, directly affect your life. In fact, you'd be hard pressed to name one modern man-made device that isn't preceded by years of research regarding its size, shape, function, color, and how marketable it is to its consumers.

* * * *

The Measurement of Body Parts

ANTHROPOMETRY IS THE science of measuring human body parts. It's typically done for the fields of architectural, industrial, and clothing design—in short, any field that can benefit from understanding how a body moves through space. Alphonse Bertillon, born in 1853, was one forerunner in the field of anthropometry.

Bertillon's system, measuring and cataloguing facial and bodily characteristics to help identify criminals, was being used by 1883. Eventually, his methods were replaced by modern fingerprinting. In the 1940s, William Sheldon took Bertillon's work one step further by identifying three *somatotypes* or basic human body types: the ectomorph, mesomorph, and endomorph. According to Sheldon, every human fits into one of these basic body types.

In the early 20th century, anthropometry was used to characterize the nuances between various human races to identify those deemed "inferior." It also played an important role in so-called human intelligence testing, in which physical measurements such as height, width of the head, foot length, and width of the cheekbones were taken using crude measuring devices.

Today, anthropometric measurements are taken using computerized 3-D scanners. General Dynamics Advanced Information Systems (GD-AIS) has used 3-D scanning to improve products such as clothing. By scanning the human body, the scan can show how it fills a garment and how a jacket or blouse effectively moves with its wearer. GD-AIS also invented the "Faro Arm" to analyze how commercial and military pilots move within their cockpits and what happens to a pilot's concentration when their eyes wander from one instrument to another. This data affects where the controls need to be placed.

Biomechanics: The Body Moving Through Space

Another area of study concerned with the physical characteristics of the human body is *biomechanics,* or how the body moves in the home, workplace, and in everyday activities. Have you ever wondered how they came up with the cupholders that hold your drink while you're driving? How about the weight-training equipment at the gym? All of these were developed after years of painstaking research using biomechanics.

A number of new inventions that make our lives more comfortable are also based on biomechanical principles. Take the Reach toothbrush: The toothbrush existed for thousands of years as a series of fibers glued to a straight spine. It wasn't until recently that someone took a look at the design of the human mouth and determined through biomechanical and anthropometrical measurements that the back teeth could be cleaned more efficiently by tilting the end of the toothbrush.

Making Life Easier

If you've ever seen one of those odd, curved ergonomic computer keyboards used to eliminate long-term stress effects that result in carpal tunnel syndrome, then you've seen an example of how science can work with medicine. *Ergonomics* is the science that determines how man-made objects "fit" human beings. Ergonomists evaluate specific tasks in the home or workplace and determine the demands they put on the worker, the equipment being used, and how they are performed.

Ergonomists look at the safety, comfort, performance, and aesthetics of commonly used products. Their ratings directly imply how well the product will sell. For instance, medicine bottles must come with labels that are easy to read by people with limited vision. Alarm clocks use "contrast principles" that make them easy to read in the dark without keeping the owner awake, and poorly designed VCRs spend years blinking "12:00" because their owners can't figure out how to set the time.

Frozen Stiff

People can learn a lot from a cadaver, especially if said corpse has a body temperature hovering around –100°F and has single-handedly inspired an annual festival in its honor. Don't believe us? Check out Nederland, Colorado's Frozen Dead Guy Days, a festival guaranteed to ruin your taste for Popsicles.

* * * *

Dead Man Thawing

JUST AFTER BREDO Morstoel passed away in 1989, his doting grandson Trygve Bauge transported his body from Norway to Colorado with plans of putting him on deep ice for eternity. But Trygve's luck (and visa) ran out, and he was deported in a fairly cold manner. So his mother, Aud, stepped in to keep her deceased dad on the cryogenic rocks.

Unfortunately for Aud, her dead dad wasn't the only one with ice in his veins. In 1993, when local authorities learned she was living in a home with no electricity or plumbing, they kicked her to the curb, citing violations of local ordinances. Aud decided to go public, and soon, her neighbors heard of the plight of the cryogenically frozen man slowly thawing in a backyard shed.

The case brought about an ordinance forbidding residents from keeping human body parts in their homes. But Bredo was exempted through a "grandfather" clause, and someone in town even built him a climate-controlled shed for good measure.

Lucky Stiff

In 2002, a festival was held in Bredo's honor, and since then the event has been held annually each March. One highlight of the event is the Grandpa Look-Alike Contest, wherein homely humans try to emulate Bredo, now rigid for nearly two decades.

Another popular event is the Polar Plunge. Participants jump into a frigid Colorado pond to simulate the cryogenic experience. There's also a coffin race, and at night the Blues Masquerade Ball offers up a chance to dance 'til you drop as proof that no matter how old—or dead—you are, you can still have fun.

Why Was Smallpox So Deadly for Indians, but Not Europeans?

The Europeans were not exactly good guests in the New World.

❋ ❋ ❋ ❋

WHETHER IT WAS conquistadores in the Caribbean, Pilgrims in New England, sailors in Fiji, or settlers in Australia, they left a calling card no one wanted: diseases that killed thousands of people. Some experts think that smallpox and other diseases, such as measles and influenza, killed up to

95 percent of the native populations of these locales—in other words, only one in twenty people survived.

Yet the Europeans remained ridiculously healthy. And when they sailed back home, they brought no new illnesses with them. Why?

The Europeans had already been exposed to epidemic diseases—or at least their ancestors had. Smallpox was known in ancient Egypt, and a smallpox epidemic killed millions of Romans in the second century AD. The disease hit Europe so frequently that the folks who had no natural immunities died off. Those who lived passed their immunities on to their children. Over the centuries, with so many nasty plagues hitting big population centers, the surviving Europeans became more resistant to the killer microbes.

Where did these diseases originate? Was there a Patient Zero? No. Most of the epidemic bugs—smallpox, measles, influenza, and even tuberculosis—came from livestock. When Asians and Europeans began herding cattle and penning up ducks and pigs thousands of years ago, they breathed in the strange germs that hung around the animals. Once humans started living in cities in large numbers, these germs were able to spread like wildfires. Europe suffered through the same plagues that killed so many Indians and islanders, but Europe's experience took place hundreds of years earlier, and its populations recovered.

The conquistadores, Pilgrims, sailors, and settlers who crossed the seas during the Age of Exploration came from families that had survived waves and waves of disease. Without realizing it, they brought smallpox, measles, and influenza germs with them to infect people who had never seen cattle, never herded animals, and never, ever been exposed to any of these diseases.

You know the result: Millions died. Exactly how many millions is unknown because experts aren't sure about the sizes of pre-encounter populations. The first wave of smallpox to hit

Mexico's Aztec Empire in 1520 killed half the kingdom. Up to ten million perished, including the emperor. More disease followed, and a century later, the area's native population numbered only 1.6 million.

Here's another infamous example: In 1837, smallpox hit the Mandan, an Indian tribe in North Dakota. The disease, brought by someone who was on a steamboat traveling up the Missouri River, almost destroyed the tribe. Within weeks, the Mandan population of one village dropped from two thousand to forty.

And since no one back then knew about germs, microbes, or how sicknesses spread, the Europeans weren't even aware of what they'd done.

What Was the Black Death?

The Black Death, also called the Great or Black Plague, first swept through Europe in the 1340s, killing nearly 60 percent of the population. It returned periodically, spreading panic and death, and then disappeared into history. Was it bubonic plague, as many people believe, or was something else to blame?

⁕ ⁕ ⁕ ⁕

"The Great Mortality"

BROUGHT TO ITALY in 1347 by Genoese trading ships, the Black Death spread through Europe like wildfire. Contemporaries described scenes of fear and decay—the sick were abandoned and the dead were piled in the streets because no one would bury them. Anyone who touched the infected, the dead, or even the infected people's belongings also caught the disease.

It could take as long as a month before symptoms showed, which was plenty of time for the infection to spread. But once the dreaded blackened spots began to appear on a victim's body, death was quick—usually within three days. At least one autopsy recorded that the person's internal organs had almost liquified and that the blood within the body had congealed.

Transmission

One major problem with the bubonic plague theory is that bubonic plague doesn't transmit from person to person—it can only be transmitted through the bites of fleas that have left an infected rat after its death. The signature symptom—the black swellings, or *buboe*—begin showing within two or three days. Accounts of the Great Plague often mention people who became infected merely by touching an infected person. Some writings describe the infection as spreading via droplets of body fluid (whether sweat, saliva, or blood), which isn't possible for bubonic plague but is a defining characteristic of hemorrhagic fever.

Incubation and Quarantine

It must be mentioned that it is possible for a bubonic plague infection to spread to the lungs and become pneumonic plague, and this kind of infection is transmittable from person to person. However, pneumonic plague is extremely rare—it occurs in only 5 percent of bubonic plague cases. It also isn't easily transmitted, and it certainly isn't virulent enough to have been responsible for the widespread person-to-person infection rates during the Black Death.

Like bubonic plague, pneumonic plague also has a very short incubation period, taking only a few days from infection to death. However, historical records show that the Black Plague took as long as a month to do the same thing. It was no coincidence that cities started to mandate a strict 40-day quarantine. Officials had observed from multiple cases that that much time was needed to determine whether someone was infected. If it had been pneumonic plague, such a long period of quarantine would not have been needed. Those infected would have been dead within a single week.

Another anomaly that casts doubt on the bubonic plague theory is how the Plague spread in Iceland. There were no rats in Iceland and there wouldn't be until 300 years later. However, the Plague still ravaged the island, killing nearly 60 percent of the population.

Rapid Spread

The Black Death spread throughout Europe faster than any disease people had ever seen. It made the trip from Italy to the Arctic Circle in less than three years and is recorded as having traveled 150 miles in England within six weeks. Rats can't travel that quickly, but people can. Frightened citizens fled from cities where the epidemic raged, not knowing that they were infected, and they spread the disease as they went. Many parish records indicate that after strangers arrived in town, the Plague emerged there within a few weeks. In contrast, studies of confirmed bubonic plague outbreaks show that it spreads very slowly. One outbreak in India in 1907 took six weeks to travel only 100 yards, and another in South Africa from 1899 to 1925 moved only eight miles per year.

Why Do Doctors Have Such Lousy Handwriting?

An estimated seven thousand Americans die each year due to incorrect medication or dosages brought about by doctors' sloppy handwriting on prescriptions.

* * * *

WHY CAN'T WELL-EDUCATED, literate, responsible, confident medical professionals write legibly? Because doctors are likely to be men—and men in executive positions have lousy handwriting. That's according to a study posted on bmj.com, an online medical journal. It turns out that doctors' handwriting is no worse than that of their peers in other important professions.

Why are doctors singled out? Because what they write—it can be a life-saving prescription—is in a sense powerful and magical. So are doctors themselves, with their special knowledge and exalted position. In this context, their chicken scratches seem oddly infantile and out-of-sync with their training and standing. It's made worse, experts agree, by the arcane symbols that doctors have long used to indicate dosages and schedules, and now by the huge array of drugs with similar-looking names that are easily confused on a sloppily written prescription.

There has been a movement for years to implement prescription-writing computer software that transmits letter-perfect scrips to pharmacies. But this process continues to be slow in coming. Critics argue that it's just as easy to choose the wrong drug on a drop-down menu as it is to write "Celebrex" so illegibly that a busy pharmacist sees "Cialis," and some poor dude's arthritis doesn't get better at all.

Here's some sound advice: Look at your prescription and ask your doctor to repeat it to you. Then make sure your pharmacist understands it the same way. If there's a discrepancy, alert your pharmacist to call your doctor. Your life may depend on it.

Maass-ively Brave

Today the name Clara Maass is not well known. But at the turn of the century, practically everyone in the United States knew her as the nurse who risked her own life to help defeat the dreaded yellow fever epidemic.

* * * *

Young Girl Grows Up

CLARA LOUISE MAASS was born in East Orange, New Jersey, on June 28, 1876. She was the daughter of German immigrants who quickly discovered upon their arrival in America that the streets were not paved with gold. The family was just barely getting by.

Clara began working while she was still in grammar school. At around age 16, she enrolled in nursing school at Newark German Hospital in Newark, New Jersey. Graduating from the rigorous course in 1895, she continued working hard; by 1898, she was the head nurse at Newark German.

Yellow Jack War

The Spanish-American War began on February 15, 1898. As wars go, the war was more period than paragraph, lasting just four short months. However, there was something more deadly to American troops than Spanish bullets: yellow fever.

Since its arrival in America in the summer of 1693, yellow fever had ravaged the country. For two centuries, "Yellow Jack," as the disease was dubbed, killed randomly and indiscriminately. The disease would devastate some households, while those next door would go untouched. One in ten people were killed by yellow fever in Philadelphia in 1793. The disease caused President George Washington and the federal government (Philadelphia was then the U.S. capitol) to flee the city. In 1802, Yellow Jack killed 23,000 French troops in Haiti, causing Napoleon to abandon the New World (and eventually agree to the Louisiana Purchase). An 1878 outbreak in Memphis killed 5,000 people there, and 20,000 total died in the Mississippi Delta. No one knew what caused yellow fever or how to stop it.

The Experiments

In April 1898, Maass applied to become a contract nurse during the Spanish-American War; in Santiago, Cuba, she saw her first cases of yellow fever. The next year she battled the disease in Manila, as it ravaged the American troops there.

No one knew why Maass kept going to these danger zones when she could have remained safely at home nurturing her burgeoning nursing career. Yet when Havana was hit by a severe yellow fever epidemic in 1900, Maass once again answered a call for nurses to tend the sick.

In Havana, a team of doctors led by Walter Reed was trying to find the cause of yellow fever. They had reason to support a controversial theory that mosquitoes were the disease carrier, but they needed concrete proof. Desperate, they asked for human volunteers, offering to pay them $100—and another $100 if they became ill. Maass volunteered for the tests, though no one is sure why she put herself up to it. Possibly she hoped that contracting a mild case of the disease would give her immunity to it and allow her to better treat patients.

On August 14, 1901, after a previous mosquito bite had produced just a mild case of sickness, a willing Maass was bitten once again by an *Aedes aegypti* mosquito loaded with infectious blood. This time the vicious disease tore through her body. She wrote a feverish last letter home to her family: "You know I am the man of the family but pray for me . . ."

On August 24, Maass died. Her death ended the controversial practice of using humans as test subjects for experiments. But it also proved, beyond a doubt, that the *Aedes aegypti* mosquito was the disease carrier—the key to unlocking the sickness that scientists had been seeking for centuries. Yellow fever could finally be conquered, in part because of Maass's brave sacrifice. The next day, a writer for the *New York Journal* wrote, "No soldier in the late war placed his life in peril for better reasons than those which prompted this faithful nurse to risk hers."

Collecting Toenails in the Name of Science

The record for the largest number of toenail clippings may seem like a ridiculous record, but the science behind it is incredibly important in order to better understand diseases like cancer.

* * * *

THE ATLANTIC PARTNERSHIP for Tomorrow's Health, or PATH, is part of the Canadian Partnership for Tomorrow

Project, the largest study of its kind ever undertaken in Canada. Led by a group of Dalhousie University researchers, Atlantic PATH is investigating the various factors that contribute to the development of cancer and chronic diseases, including environmental factors, genetics, and lifestyle. The study has recruited thousands of men and women from across Canada.

The 30,000 participants have provided the PATH team with the usual sorts of samples you might see in a study like this, such as body measurements and blood samples. But what's more unusual is that the study participants have also given the team their toenail clippings—tens of thousands of them.

"Toenails are an important part of our research," explains Atlantic PATH's Principal Investigator Dr. Louise Parker, professor in the Departments of Pediatrics and Medicine, as well as the Canadian Cancer Society Chair in Population Cancer Research. "By the time you trim the end of your toenails, they've been on your body for about six-to-nine months and during that time they're exposed to everything that you're exposed to. What we're particularly interested in, in this context, is the extent to which environmental exposure affects our risk of disease."

Dr. Parker and her team didn't set out to break a record, but nevertheless, collecting toenails from 24,999 individuals was enough to earn Atlantic PATH the Guinness World Record of owning the world's largest collection of toenail clippings.

"My colleagues David Thompson and Trevor Dummer had the idea," says Dr. Parker, when asked about how they decided to submit the collection to the Guinness organization. "It's a heck of a collection: a quarter of a million toenail clippings altogether. So they looked at the Guinness website and while there were other toenail records—for example, the longest toenails—there wasn't a record for the largest collection of clippings. We thought it was a great opportunity to have a bit of fun after everyone's hard work and commitment to the project."

Because It Does Sound Gross

Does cracking your knuckles give you arthritis?

* * * *

THIS MYTH MAY have been started by friends of knuckle-crackers who think knuckle-cracking is gross. They aren't wrong about that, but there's no evidence that cracking your knuckles gives you arthritis.

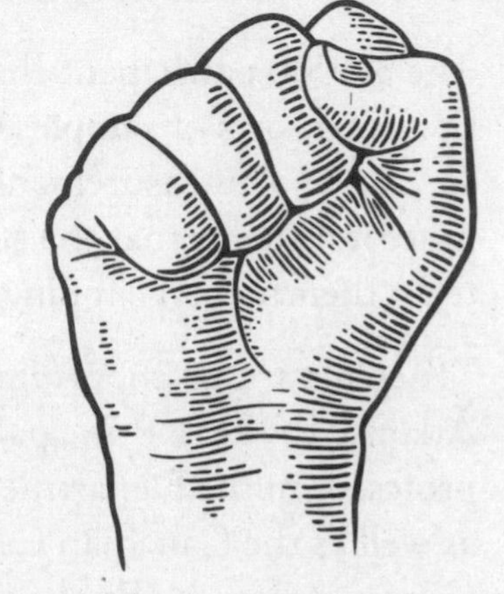

Floating around on the internet somewhere you can find a slow-motion X-ray view of a knuckle being cracked, and you can see a pocket of gas moving quickly from one place to another within the knuckle. This is a simple mechanism that doesn't seem to cause any harm to the knuckles themselves or the tissue around them.

Lifelong knuckle crackers may show off their chronically swollen knuckles and claim that this is the result of cracking them for years, but this has also not been studied. Swelling is usually reserved for fluid-filled, temporary inflation of tissue because of an injury, not simply that a body part may change over time. If the knuckles don't feel painful or stiff, there's probably nothing to worry about.

✳ Chapter 12

Food

Junk Food Verdict: Not Guilty

Your complexion may be bumpy, but if you eat a lot of junk food, you can at least have a clear conscience.

✳ ✳ ✳ ✳

PIMPLE-PRONE ADOLESCENTS ARE often told to skip the pizza, fries, and potato chips to keep their faces acne-free. But greasy foods only affect your appearance if you're a messy eater. Research shows that breakouts come from a surge in hormones, which stimulates the secretion of oils. When the oil glands in the skin are overactive, pores get clogged from the secretions and become perfect hosts for a bacterium called *Propionibacterium acnes*. So bacterial activity produces pimples.

Chocolate has also been blamed for causing acne. But a University of Pennsylvania study compared people who ate a bar of chocolate with those who ate a bar with similar amounts of fat and sugar. The study found no evidence that chocolate had an effect on producing acne. (Besides, chocolate has an abundance of antioxidants, which may help prevent wrinkles.)

You might want to think twice about eating dairy, though. A recent study found that women who drank two or more glasses of skim milk a day were 44 percent more likely to report severe acne as teenagers. Researchers believe that hormones or whey proteins found in dairy products might be the cause.

A Sweet Lie

Does sugar contribute to hyperactivity in children or adults?

⁕ ⁕ ⁕ ⁕

THIS MAY BE a myth and may not be a myth, but it was probably caused by ambiguity between saying a child is "acting hyper" and saying a child "probably has attention deficit hyperactivity disorder." Sugar can spike any person's blood sugar and give them a brief little thrill, but that's a far cry from a psychiatric diagnosis.

It also doesn't make intuitive sense in the body that any amount of sugar in a child could alter their brain chemistry to induce a lifelong condition. Most *chemical medications* can't even cross the blood-brain barrier.

Wake Up!

Why does turkey make you sleepy?

⁕ ⁕ ⁕ ⁕

IT DOESN'T. THINK about your average lunchtime turkey sandwich, which probably doesn't make you feel sleepy. On Thanksgiving, when most Americans eat their peak amount of turkey for the year, we also surround the turkey with literal piles of different delicious and buttery side dishes and bread, with bonus points for the snacking that usually happens through Thanksgiving days spent with family.

Tryptophan is found in many foods besides turkey, and you probably can't think of any other single one. Maybe it's making you sleepy instead, but probably not! Next Thanksgiving, if you have a vegetarian at your gathering, ask them nicely if they get sleepy after dinner. Then you'll know you can blame the mashed potatoes.

Putting the "C" in "Panacea"

Vitamin C is an essential vitamin. It plays a critical role in the formation and repair of collagen, the connective tissue that holds the body's cells and tissues together. It promotes the normal development of bones and teeth, and it is a potent antioxidant. But it is not a cure-all.

* * * *

VITAMIN C WAS catapulted into the limelight in the 1960s when Nobel Prize–winning scientist Linus Pauling began touting vitamin C megatherapy, claiming high doses could reduce the frequency and severity of colds. Sales of vitamin C exploded after Pauling's book *Vitamin C and the Common Cold* was published in 1970. Pauling went on to claim that megadoses of vitamin C could also prevent or slow the growth of cancer, among other diseases. Even though Pauling's methods (and findings) have been soundly debunked, the myth that vitamin C is a cure-all persists.

Found especially in uncooked fruits and vegetables, vitamin C is a water-soluble substance that humans don't produce naturally. The body's tissues can't absorb more than 100 milligrams a day, so it's a waste of money to take megadoses. High doses are not known to be toxic, but too much vitamin C (more than 2,000 mg/day) can cause gas, diarrhea, and kidney stones, as well as impede the absorption of other vitamins.

However, there is still debate over vitamin C's role in disease prevention and treatment. Some studies show a positive impact on heart disease, cancer, and other illnesses, while others do not—a few even indicate that it can be harmful. More medical research is needed for definitive answers.

Why Does My Microwave Heat Food Inside-Out?

Food does not heat "from the inside out" in a microwave—at least, not usually.

* * * *

MICROWAVES WORK BY making the water molecules inside of any food move rapidly and therefore generate heat. (Yes, your food, like anything else in this world, is just vibrating nanothingies all the way down.) Most foods have the highest water concentration and most freedom of movement in their centers: imagine spending a dance party standing next to the wall versus in the very middle of the room. This is how the infamously nuclear-hot center of a fast-food fruit pie gets that way. Watch your mouth!

Microwave ovens were invented after the discovery that microwaves, the bandwidth one step up the spectrum from radio waves, generated noticeable heat. Radio waves are able to travel through walls—that makes them sound a bit like superheroes!—but microwaves are confined to so-called "line of sight," meaning they can be harnessed and safely contained within a microwave oven. Microwaves are also used to cauterize surgical cuts. Infrared light is the next step up the spectrum from microwaves. It was discovered serendipitously when a scientist wanted to measure the temperature of different colors of light. Astronomer William Herschel placed thermometers in each color of light projected by a prism (as seen on the cover of *Dark Side of the Moon*) and included one final thermometer outside the visible spectrum as his control. His due diligence in the scientific method paid off when this control thermometer registered a very high temperature, indicating to Herschel that some unseen phenomenon was at play. The heat of infrared radiation is used in consumer laser cutters, among many other things.

Beta Keep Guessing

Do carrots help you see in the dark?

⁕ ⁕ ⁕ ⁕

CARROTS TASTE GOOD and they're good for you, but they won't give you superpowers. Beta carotene, or vitamin A, is important to your vision and your optical health. But not even vitamin A can help a human's eyes see in the dark. We're simply not able to, and it makes sense from the way our eyes are shaped and how they function. All eyes evolved beginning as patches that responded to light, and light remains the fundamental ingredient in almost every creature's form of vision. Some, like bees or butterflies, can see outside the spectrum of what humans consider to be visible light. Some animals can see in the dark, although this is often more like infrared vision.

What a Pill

Am I healthier because I take a daily vitamin?

⁕ ⁕ ⁕ ⁕

ALMOST ALL PHYSICIANS recommend that people try to get their recommended daily amounts of vitamins and trace minerals from food. Supplements are costly and often they're made of forms of vitamins or minerals that are less understood. Something about the way vitamins are carried into our bodies as they naturally occur in food makes them absorb into our systems and be more effective. Usually, you'll be told that it probably doesn't hurt to take a vitamin. That's true if you stick with a typical consumer vitamin, although research is examining the role of vitamins all the time and our understanding could change. The known danger now is for vitamins that offer very high doses, like the B vitamin supplements that boast excessive amounts of the daily recommended value percentage. Can taking 10,000% of anything really be good for you?

The (De-) Hydration Myth

Do I need to drink eight glasses of water a day? Am I dehydrated and I don't know it?

* * * *

THIS BEWILDERING MYTH is, nonetheless, entrenched in our lives. But unless you feel thirsty, you don't need to drink any more water than you naturally would. The only demonstrable way more water can potentially help your health—and everyone is different, so this isn't a blanket rule—is that drinking water instead of caloric beverages like soda means you likely consume fewer calories overall.

Many weight-loss programs strongly incentivize drinking a lot of water, because it can help you feel fuller and eat less overall. Digestion can feel easier when we're hydrated, and it's easy to picture a slice of pizza becoming a gluey mess inside our stomachs. Certainly it doesn't hurt to drink more water if that's what you want to do, as long as you stay moderate. Overhydrating can actually poison your body if taken to extremes like drinking a gallon of water in a short time.

People who don't have kids or spend much time with kids may also not know that babies *can't* drink water. This is one reason it's very important to keep babies cool, covered, and comfortable when they're outside on hot days. Options to rehydrate babies are very limited because of their tiny, growing systems.

As for dehydration, the symptoms of this condition are noticeable and alarming. Feeling thirsty does not mean you're "already a little dehydrated"—a frightening headache, dizziness, weakness, blacking out, and other nightmare symptoms mean that you're dehydrated. Feeling thirsty means you should just drink some water.

The Tall and Short on Kids and Coffee

A cup of java may keep kids up at night, but it shouldn't affect their height.

❋ ❋ ❋ ❋

IN PAST GENERATIONS, parents didn't allow their children to drink coffee, believing that it would stunt their growth. But today, kids are consuming coffee in record numbers and at younger ages. In fact, young people are now the fastest-growing coffee-drinking group in the United States.

Does this trend indicate a corresponding shrinkage in the younger generation's adult height? No, say researchers. There is no evidence that drinking coffee affects growth or a person's eventual height.

At one time there seemed to be a link between caffeine consumption and the development of osteoporosis, and that may be how coffee originally got blamed for inhibiting growth. Early studies suggested that drinking lots of caffeinated beverages contributed to reduced bone mass.

More recent studies have debunked that idea. Dr. Robert Heaney of Creighton University found that much of the preliminary research on caffeine and bone loss was done on elderly people whose diets were low in calcium. Other researchers have noted that even if caffeine does affect bone mass, its influence is minimal and can easily be counteracted with a sufficient amount of calcium-rich foods.

The myth that coffee stunts growth was laid to rest by a study that followed 81 adolescents for six years. At the end of the study, there was no difference in bone gain or bone density between those who drank the most coffee and those who drank the least.

In other words, don't worry about letting your kids have the occasional cup of joe, but unless you want to be up all night while they're bouncing off the walls, make sure they drink it in the morning.

Is Chicken Soup Really Good for a Cold?

Well, an Egyptian rabbi, physician, and philosopher named Moshe ben Maimonides seemed to think it was a good remedy.

⁂ ⁂ ⁂ ⁂

MAIMONIDES WAS THE first to prescribe chicken soup as a cold and asthma treatment, way back in the twelfth century. Since then, mothers and grandmothers worldwide have been pushing bowls of homemade broth to cure everything from colds, flus, and stomach problems to severely broken hearts. It's no surprise, then, that chicken soup is often referred to by another name: Jewish penicillin.

Until recently, there was little scientific literature to explain how or why chicken soup seems to make us feel better. Some suspected that hot steam from the soup worked to open congested airways. Others believed that it was simply a matter of receiving some much-needed attention and TLC.

In 2000, however, a team at the University of Nebraska Medical Center provided evidence that chicken soup can, in fact, cure what ails you. It began when Dr. Stephen Rennard, a researcher and specialist in pulmonary medicine, brought a batch of his wife's homemade chicken soup into the lab. It was her grandmother's recipe—a medley of chicken, onions, sweet potatoes, parsnips, turnips, carrots, celery stems, parsley, and matzo balls.

After running numerous laboratory tests on it, Rennard and his colleagues determined that chicken soup contains several ingredients with "beneficial medicinal activity." Specifically, the soup blocks the movement of inflammatory cells called neutrophils. Why is this important? Neutrophils are responsible for stimulating the production of mucus. By limiting the movement of neutrophils, chicken soup helps reduce the horrid inflammation and congestion associated with colds and upper respiratory infections. While it is not a cure, a bowl of chicken soup can make your nose less stuffy, your throat less sore, and your cough less hacking.

And the good news is that chicken soup doesn't have to be homemade to help you out. As a point of comparison, Rennard tested thirteen different commercial brands of chicken soup commonly found at the grocery store. He discovered that all except one (chicken-flavored ramen noodles) relieved the inflammation associated with colds. But, we ask: were any as good as Grandma's?

Deadly Nightshades

Do nightshade vegetables like tomatoes and eggplant contain a toxic acid? Or a toxic alkaloid?

⁎ ⁎ ⁎ ⁎

THE NIGHTSHADE FAMILY goes far beyond tomatoes and eggplants, including potatoes, bell and chile peppers, and tobacco. Also: deadly nightshade, petunias, and the real-life mandrakes featured as shrieking nuisances in the Harry Potter series. Do nightshades deserve their bad rap? That depends who you ask, although evidence suggests nightshades are fine.

Supermodel Gisele Bundchen and New England Patriots quarterback Tom Brady, a celebrity married couple who look very much like siblings, made news in 2016 with their radically healthy private-chef diet. Their self-imposed limitations are extensive, including no fruit, nightshade vegetables, gluten, caffeine, "fungus," dairy, or MSG.

This long, mystifying list might register with some readers as bordering on orthorexia, an eating disorder in which the pursuit of a so-called healthy diet leads to obsessive tracking, avoiding, and phobic reactions to allegedly unhealthy foods.

But Bundchen and Brady have a dedicated private chef who offers up tailored meals to fit their purported needs, which definitely must make it easier not to choose foods that aren't part of your plan. Strangely, despite their laundry list of no-nos, Bundchen and Brady do still eat meat.

Their chef's shoutout to nightshades drew attention to this whole family of vegetables. Are they the new gluten, another food that is not harmful at all for about 99% of people? Should we take the word of someone who swaps a gluten-containing whole grain for trendy quinoa that actually doesn't have any more protein or less carbs than that comparable whole grain? Let's address some claims.

Nightshades cause migraines. Migraines are enigmatic and individual sufferers should choose for themselves when it comes to avoiding trigger foods. But there isn't a demonstrable link between nightshades and migraine. Other common triggers like chocolate or garlic have longer paper trails to back them up.

Nightshades contain a toxic alkaline. Certain specific nightshades, like potatoes that have turned green and need to be thrown away, do contain solanine. But almost all other vegetable nightshades are fine.

Nightshades contain a toxic acid. Nightshades have oxalic acid in small quantities, while the herbs parsley and chives contain much more of the acid by volume. Oxalic acid can stop your body from absorbing calcium but only if you consume very little calcium to begin with—and a *lot* of nightshades and other vegetables with oxalic acid.

Nightshades make arthritis worse. Arthritis is a frustrating condition that can lead sufferers to grasp for possible triggers, very similar to migraine. There's no evidence that nightshades worsen arthritis inflammation or pain. But again, individual sufferers can do their due diligence to find if they experience discomfort around any kind of food or substance.

Nightshades cause general inflammation. Inflammation has become a health buzzword without a clear definition. Acetaminophen is an anti-inflammatory drug; does that mean it cancels out a nightshade? In any case, there's no evidence that nightshades cause inflammation in people without an allergy to them. Chiles may make your mouth feel inflamed, but you can douse that with milk, unless you're Gisele.

Nightshades can upset your stomach. Certainly you may find that specific nightshades, like starchy fried potatoes or hot chiles, upset your stomach. And everyone's digestive system is different. If you find that you feel sick or bloated after meals where you've eaten these foods, you can try avoiding them for a couple of weeks to see if you notice an improvement.

Nightshades aren't worth eating. The nightshade vegetables represent a huge portion of the human diet during the agricultural age. For a surprising number of Americans, the slice of tomato they get on a sandwich embodies one of their most consumed vegetable groups—after lettuce. This huge family of very nutritious vegetables should be part of the diet of everyone who doesn't have an explicit health reason to avoid it.

It's interesting to find the "unhealthy" potato among the eggplants, tomatoes, and peppers. Potatoes may be caloric, but they contain many nutrients, and they were even carried onto ships to help prevent scurvy because of their high levels of vitamin C. The alkaline substance that green potatoes produce is a natural defense against predators, which is what the spiciness of peppers and the other nightshade idiosyncrasies are probably for. It may be part of why this family of plants is so robust and durable.

The danger with drawing a circle around an entire family of foods and declaring them off limits for no reason, as many people have done with gluten as well, is that all humans must always eat every day. Placing restrictions on foods that are good for us only makes choosing foods feel even more frustrating overall, and that feeling of confusion and helplessness can create a bad feeling around eating. Food nourishes us and keeps us going, even nightshades.

In 2017, Tom Brady released a book that forms the cornerstone of his new pyramid scheme of wellness products. In between complete common-sense nonstarters like how people should eat more vegetables and get enough sleep, he throws in plugs for his Tom Brady brand of Tom Brady supplements. We can only hope one of them contains some nightshades to mix things up a little.

What Causes Food Cravings?

Got a hankering for some steamed carrots and Brussels sprouts? Didn't think so. Most cravings are of the sweet, salty-crunchy, super-high-fat varieties. But just what is it that prompts us to make a mad dash to the 7-Eleven for barbecue chips and Ding Dongs in the middle of the night?

✻ ✻ ✻ ✻

Researchers aren't sure, but one theory that's gaining acceptance speculates that food cravings are actually addictions. How so? Brain image studies conducted by Marcia Pelchat, a sensory psychologist at the Monell Chemical Senses Center in Philadelphia, show that food cravings activate parts of the brain that are typically involved with habit formation. Known as the caudate nucleus, this is the same region of the brain that's affected by cocaine, alcohol, and cigarettes. "Think of food cravings as a sensory memory," says Pelchat. "You remember how good it felt the last time you had that food."

Happy Chomping

It all has to do with the biological and emotional resonance some foods can bring us. Brian Wansink, a food psychology expert and the author of *Mindless Eating: Why We Eat More Than We Think*, agrees that people crave foods that connect them to pleasant experiences.

Men, he says, are drawn toward hearty meals—such as barbecue ribs, burgers, meat loaf, pasta, pizza—because they associate those foods with a nurturing wife or mother. Women, on the other hand, connect those same savory meals to long hours spent in the kitchen. Wansink notes that chocolate and ice cream don't involve any prep work or cleanup, which may help to explain why women are drawn to those types of sweets. That's right, ladies—just flip the lid off that pint of Häagen-Dazs and you've got one quick euphoria fix.

But what about the "wisdom of the body" theory, which states that our bodies simply crave what we nutritionally need? Pelchat says that wisdom doesn't apply, unless you're a sodium-deficient rodent: "When rats are salt deprived, they show a sodium appetite; they seem to be able to detect amino acids when they're protein deprived. But there's actually very little evidence for that in people. A lot of people in our society crave salty foods, but very few are actually salt deficient."

So that sudden urge to hit the A&W drive-thru isn't exactly motivated by nutritional necessity. You're really just addicted to the chili cheese fries.

Oysters: Nature's Viagra

One of the most enduring food myths is that oysters have aphrodisiac properties. And the thing is, it's actually true.

❋ ❋ ❋ ❋

MANY PEOPLE SWEAR that a plate of raw oysters can put even the coldest fish in the mood for love. Rumor has it that Casanova would dine on oysters before an amorous encounter. Some nut-jobs have even gone so far as to feed Viagra to oysters to increase the sexual power of the shellfish.

For years, scientists attempted to show that oysters don't have any libido-increasing abilities. A simple chemical analysis of an oyster reveals that it is made of nothing more than water, carbohydrates, protein, and trace amounts of sugars, fats, and minerals. None of these elements, whether taken separately or together, have been shown to have any effect on sexual desire or prowess.

But in 2005, at a meeting of the American Chemical Society, the oyster's sexual secrets were finally revealed. A group of American and Italian researchers discovered that the oyster belongs to a family of shellfish that has been shown to increase the release of certain sexual hormones. Oysters apparently contain amino acids that, when injected into rats, increase the levels of testosterone in males and progesterone in females. Elevated levels of these hormones in their respective genders have been linked to an increase in sexual activity.

So, surprisingly, this myth appears to be grounded in fact. If you want to keep your libido humming, it might not be a bad idea to chow down on a dozen or so oysters.

Apples and Oranges

Does an apple a day keep the doctor away?

✻ ✻ ✻ ✻

YES AND NO. Certainly apples, which are especially good for you if you eat them with the skin on, are part of a healthy diet that can help prevent disease and support your health. People who regularly eat apple are also more likely to have other good health indicators. But the old folk-saying may date from a time when apples were the only fresh fruit many people had access to.

Apples were actually one of the major staple foods in the English colonies of the 1600s, when frightened settlers were too afraid to try the local foods they saw the scary native Americans eating, so they planted their own apple trees. Apple pie is almost unheard of in Britain but it was one of the only celebration foods in the early colonies. And centuries later, it's the go-to dessert for any true-blue American potluck.

How Sweet It Is

High fructose corn syrup is worse for us than sugar, right? Well, it's complicated.

✻ ✻ ✻ ✻

MYTHS AROUND SUGAR and other sweeteners are invasive and insidious because of the complicated politics involved. The United States grows more corn than any other country—more than the next three top producers combined, in fact. When given the choice between tariffs on imported sugar or subsidies on domestic corn syrup, American food manufacturers made the financially logical decision to turn to corn, and their powerful lobbyists followed with the charm offensive.

Birth of an Industry

The same qualities that bakers value in traditional corn syrup—ease of use, soft-baked texture, and moisture—made it ideal for processed and packaged foods as well. But the problems began when manufacturers realized they could make the same quantity of corn syrup taste sweeter by using natural chemistry to convert some of the existing glucose to fructose. This added to the existing market incentive to use corn sweeteners, and high fructose corn syrup is nutritionally almost identical to much more expensive refined white sugar.

So what's the problem?

High Fructose, Low Veracity

There is a huge amount of misinformation about high fructose corn syrup swirling around. Isn't fructose naturally found in fruit and therefore not unhealthy? Yes and no.

Fresh fruits have varying amounts of fructose but little overall. Dried fruits have higher concentrations of fructose and are very calorie dense—you could easily eat a harmful amount of sugar if you're indiscriminately eating dried fruit or drinking fruit juice. But the amount of sugar in whole fresh fruit is not impactful for the average person.

Fruit is high in water and fiber, both of which help to slow down how quickly sugar is absorbed by the body. There's a world of difference between a juicy summer peach and the added sugar in a cherry yogurt or a granola bar. Even the starchy, dense banana is over half water.

Case Study: Weight Watchers

In fact, the false equivalence of calories in fresh fruit versus calories in processed foods led the Weight Watchers corporation to make a landmark change to its plan in 2010.

Local leaders noticed that members were choosing so-called "healthy" packaged snacks over fruit because these foods had the same value in proprietary "points," but food science was demonstrating that fruit calories are digested differently. Weight Watchers wanted to incentivize choosing whole foods like fruit, and they changed their points system in order to make this change. Members were encouraged to fill up on fruit with no impact on their daily counting.

Drawing Conclusions

So what's the bottom line with high fructose corn syrup? Nutritionally, it isn't any different from white sugar. But its cheapness has made it ubiquitous in even savory packaged foods, and studies show that a diet with too much added sugar (where "too much" is a quantity far below what the average American eats now) increases risk for obesity and life-shortening diseases.

Choosing a convenient whipping boy has allowed the packaged food industry to rally around "pure cane sugar" as its savior, when really, all added sugars are created pretty equal. Handle with care.

But, the Alcohol All Cooks Off!

No, the alcohol does not "cook off" completely when used in recipes like risotto or bombe Alaska.

❋ ❋ ❋ ❋

OVER A COOK time of several hours, such as a very long braise of a huge cut of meat, virtually all the alcohol you use in a recipe does cook off. But for bombe Alaska, bananas foster, and other flambéed recipes, the very brief cook time leaves most of the alcohol in. Osso buco and risotto retain about a quarter to a third of the alcohol.

Most cooks prefer to add wine to risotto very early in the cooking anyway in order to let the flavor steep and mellow as they continue to add broth for the rest of the cook time. In a single serving of any of these dishes, there shouldn't be enough alcohol to impair anyone—but some people may want to avoid alcohol altogether for health or other personal reasons. The safest bet (in many senses!) is to ask before you light everyone's desserts on fire.

Does Searing Lock in Moisture?

Many recipes for cooking meat or poultry start with a sear. But searing does not "seal in moisture"—it only adds tasty browning.

* * * *

IN FACT, SEARING can dry out your food if done for too long. There are different kinds of reactions and processes that cause what we think of as browning. Fruits and vegetables turn brown due to oxidation, and we won't get into that because it is an accidental and gross kind of browning. (Scientists have patented some genetically modified varieties of apple that purportedly don't brown after cutting! Packaged sliced apples use safe, natural preservatives like lemon juice or citric acid to stanch the browning effect.)

The Maillard Reaction

Searing meat triggers the Maillard reaction, where simple sugars and amino acids react and rearrange each other's molecular structures to create new brown compounds on the surface of the meat. Avid omnivore cooks know that meat has to be dry in order to sear in hot fat—water disrupts the Maillard reaction—and dredging meat in flour helps to create an even drier surface that quickly breaks down into helpful simple sugars. This is one reason why simply dredging meat once in flour yields such different results than dredging followed by an egg wash and a second dredge. And a fun fact: dulce de leche is made not by caramelization but by the Maillard reaction!

Golden Brown

Fast-food French fries use Maillard-friendly additives to get predictable, fast browning. Some franchises add a tiny amount of starch, whether by itself or in the form of some kind of flour.

Some use sugar or dairy instead. Foodie culture places a high premium on whole and simple foods with fewer additives, but major fast-food chains don't have the option to get a bad batch of potatoes when trying to deliver thousands of consistent meals each day. Certainly their harried employees can't look into the deep fryer to check for doneness on a variable French fry. But "fast casual" chains like Five Guys offer a simpler potato product if that's what you're looking for, or you can make your own at home.

Layers of Flavor

What about caramelization? Are caramelized onions different from the browning on seared meat or French fries? The results of all three are delicious, but yes, onions undergo a completely different chemical process. Remember that the Maillard reaction involves protein—caramelization is instead the controlled chemical change of just sugar. If you heat granulated white sugar in a pan, it will eventually go through all the stages of caramelization used in candymaking. Compare this with melting white sugar into water to make simple syrup, where there is no color change.

Where do onions fit in? Most onions don't taste sweet, but they contain enough sugar, trapped inside of caramelization-friendly insulated layers, to brown over time and create the delicious base for French onion soup. But when you caramelize onions, you must add liquid very slowly to avoid interrupting the caramelization process. This is also why recipes call for browning onions before you add any liquid, because that's the only window for browning.

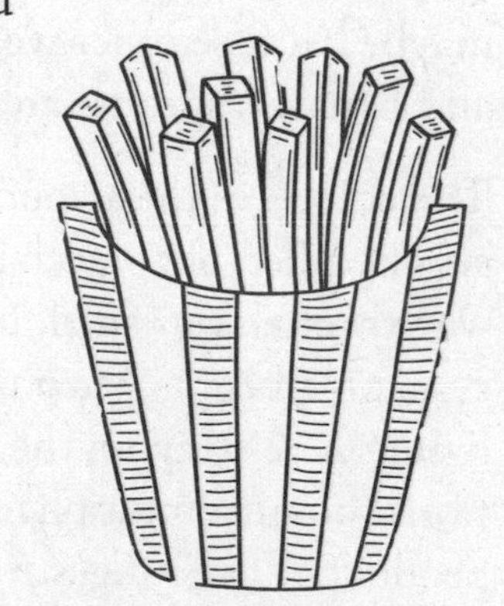

If Beans Cause Gas, Why Can't We Use Them to Power Our Cars?

There are two ways to consider this question: the high road and the low road. And you'd like to hear about both, wouldn't you?

* * * *

TAKING THE HIGH road first, we can discuss the technology that transforms biomass into ethanol, a proven fuel for cars. Beans, like corn or virtually any other organic material, contain starches and complex carbohydrates that can be refined into ethanol, a combustible alcohol blended with gasoline to become that "green" E85 fuel you've heard about. Through fermentation—and with a lot of help from science—beans and their organic cousins can also find their way into methane gas, another proven biomass automotive fuel.

But how boring is the high road?

What's the Stink All About?

By "causing gas," this question really refers to the process by which the consumption of beans produces that bloated feeling that escapes us as . . . flatulence. High-fiber foods tend to cause intestinal gas, but beans seem to bear most of the blame, maybe because other world-class gas-promoters like cabbage and Brussels sprouts aren't as big a part of our diet.

The culprit in these foods is a natural family of hard-to-digest sugars called oligosaccharides. These molecules boogie their way through our small intestine largely unmolested. The merrymaking begins when they hit the large intestine. Bacteria living there strap on the feedbag, chomping away at this nutritional bounty, multiplying even. Our intestinal gas is the by-product of their digestive action.

Most of this gas is composed of odorless hydrogen, nitrogen, and carbon dioxide. In some humans—about 30 percent of the adult population—this process also produces methane.

Ethanol isn't a part of the oligosaccharide equation. But hydrogen and methane are, and they're flammable gases. In fact, hydrogen is another player in the fuel-of-the-future derby and already powers experimental fuel-cell vehicles.

So order that chili and fill 'er up. Beans in your Beemer! Legumes for your Lexus!

No so fast, burrito buddies. Setting aside the daunting biotechnical hurdle of actually capturing bean-bred flatulence from a person's, um, backside, the challenge becomes one of volume and storage.

Human flatulence simply doesn't contain hydrogen or methane in quantities sufficient to fuel anything more than a blue flame at a fraternity party. Even if we did generate enough of these gases to power a car, they'd have to be collected and carted around in high-pressure tanks to be effective as fuels.

Human biochemistry is a wonderful thing, but it isn't yet a backbone of the renewable-energy industry. For that, breathe a sigh of relief.

All Ground Up

Who said coffee grounds make a great fertilizer?

❋ ❋ ❋ ❋

THIS OLD WIVES' tale offers a perfect storm of repeatability: it uses a waste product, makes a surprising claim, and would be difficult to disprove. But there are some immediate red flags about the idea, and they turn out to be valid as well.

Overall, this myth is a big question mark. Another myth claims that dairy is good for growing moss, when in reality, gardening with dairy is just good for attracting mold. Maybe we can combine these myths into one big cappuccino of deception.

Percolating Objections

Coffee is caustic enough that plenty of people either forgo it altogether or mellow it out using insulating dairy. The grounds are still very acidic and contain caffeine, both qualities that are generally harmful to plants. For these reasons, coffee grounds don't make sense as part of a blanket rule shared with all gardeners or plant owners.

Guardian reporter James Wong wrote about an experiment he did in his own garden in 2016, where two identical vegetable patches could be compared. He followed the folk wisdom of sprinkling coffee grounds onto one vegetable patch each day and kept the other one coffee-free as he usually would. The results, where the coffee-treated plants became visibly sick or died, caused him to think again.

Wong found reasons why this happened, including the acidity and caffeine. The best insight from his experiment is the idea that caffeine is a way for the plants that produce it to kill off their competition. Psychoactive alkaloids—nicotine and opium are also in this category—are great ways for plants to defend themselves. The ways they alter the human mind aren't the same ways they work on plants, or at least plants don't have minds to alter. But the chemical ways these substances affect our brains and body chemistry speaks to how they can equally or more severely impact plant life.

A Cup by Cup Basis

Experts say you can use coffee grounds to accomplish good things in your garden, but they should be applied carefully, sparingly, and never in direct contact with the stems or roots of your plants. Some plants love acidity enough that you might use coffee grounds, in small quantities, to make the soil more acidic. But too much and the caffeine will still slow down these plants in both growing and flowering.

It's wonderful to think you might turn many people's most regular source of garbage into a secret dynamo of plant nutrition. But proponents are coffee cherry-picking and offering a variety of competing pretend facts and anecdotes. If this all makes you think again about coffee grounds in your garden, maybe you can use those grounds in a compost pile instead.

Overkill: A Matter of Degrees

Boiling water kills bacteria, but it also wastes energy. Here's how you can have your clean water and save the environment, too.

* * * *

When in doubt about the potability of your water supply, you can ensure its cleanliness (well, at least its "antimicrobialness") by boiling it. But if you don't have an updated wilderness manual, you may not know that boiling water for extended period of time doesn't make it any cleaner, and in fact, boiling isn't even necessary.

It's a misconception that water must reach the boiling point—212 degrees Fahrenheit—to kill pathogens. The temperature needed to knock off most critters (excluding extreme varieties such as bacteria living in volcanoes) is just 185 degrees Fahrenheit. Bacteria, microbes, viruses, and parasites are killed off after just a few minutes at that temperature. Disease-causing pathogens, then, are already dead by the time the water begins to boil.

In developing countries where firewood is scarce and water is filled with bacteria, it is imperative to adhere to recommended boiling times. Since you're unlikely to have a cooking thermometer handy, the prudent route is simply to wait for the water to come to a boil.

Yet various safety guides recommend boiling water for 5, 10, even 20 minutes. At this point much of the water will evaporate, and fuel will have been wasted. In light of criticism about wasting energy to boil water, the Centers for Disease Control and the Environmental Protection Agency recently lowered their suggested boiling times. Both now recommend heating water to a rolling boil for only one minute.

A Refined Industry

Plastics are ballyhooed as one of the great bogeymen of modern life, but use of plasticky materials dates back thousands of years and continues to evolve.

⁕ ⁕ ⁕ ⁕

Roots in the Ancient World

ANCIENT HUMANS USED natural rubber, wool, and animal horns in ways we recognize in modern life. These materials are strong and elastic because of cellulose, a naturally occurring, strong, fibrous compound that gives wood (and subsequently paper) its utility and durability as well.

The Modern Age

The age of plastics we live in now was enabled by scientists who found ways to turn natural plastic-like materials into distilled, uniform versions that could be mass produced, usually relying on fossil fuels. Climate change and the impending "peak oil" have pushed some scientists to explore a return to the natural roots of plastics, whether for sustainability, recyclability, or simply our inevitable future with less petroleum.

Casein Point

One early modern plastic was made with milk, and schoolchildren still often make a version of this for science fairs and other experimental fun. Milk has a strong, consistent structure called a colloid, which means a drop of milk is always the same as any other drop of milk—nothing settles at the bottom and there are no "particles" moving around. Melty cheese is not just stretchy but elastic, meaning it bounces back into shape at least a little. This is because of a protein called casein, which does for dairy products what gluten does for wheat: adds strength, stretch, and elasticity. Casein is the reason milk can be turned into a usable plastic.

Making plastic from milk is simple, and in fact it's the same way many world cultures make fresh cheeses like ricotta, quark, paneer, mascarpone, cottage cheese, and queso blanco. Vinegar or lemon juice is added to milk, which is usually heated before or after the acid is added to speed up the curdling. This process creates curds and whey. (Sorry, Miss Muffet, you'll need to bring your own tuffet.) If you want to eat the fresh cheese, this is the time.

But to make it into plastic, you should drain the whey and press the curds dry in a clean dishtowel or other absorbent material, and then knead it until it forms a dough that you can mold like modeling clay. Google it and double check the details before you dive into this experiment—only you can prevent dairy wastage.

Glutenous Maximus

Twenty-first-century supervillain gluten, enemy number one in trendy restaurants and yuppie homes, plays the same role for wheat that casein plays for dairy. Bread bakers mix ingredients and let the dough rest and rise, which both activates the yeast that will leaven the bread and lets gluten begin to form the long structural chains that make bread strong and elastic.

Making gluten-free bread is a challenging task for a lot of reasons, but the main one is that gluten is simply impossible to mimic. Really talented bakers have found great substitutes that often mean adding different starches, like those from potatoes, corn, or tapioca, and sometimes more scientific ingredients like xanthan gum. (Despite its sinister name, xanthan gum is a harmless food additive made by fermenting sugars with bacteria. But it does have a laxative effect, so be careful.)

The upshot to all this gluten talk is that scientists are exploring ways to use gluten as the foundation for more sustainable plastics. Especially as more and more people seek to avoid gluten, which has historically been added to other processed foods as a stabilizer, there's a real *glut* of gluten on the commodities market. Plastics made with gluten may be more biodegradable or recyclable than the plastics made with petroleum.

The Future of Plastics

Plastics have symbolized the world's future, first as an optimistic and cutting-edge convenience and later as the primary culprit bulking up our landfills. But a changing relationship to plastic could be emblematic of the next step in our future, especially if scientists can make biodegradable natural plastics that are easy to manufacture. Eco-friendly restaurant and grocery chains may already use biodegradeable flatware made from corn or other strong starches, so there's an existing market for this category of product. Plastics may be our future one way or the other, but science lets us see a less polluted vision.

Interview with a Banana Gasser

Q: You gas bananas?

A: That's the short version of it. My work is much more precise and technical than you might think, because bananas are so perishable. Have you ever wondered how bananas can make it from other continents to your grocery store in edible shape, and yet they turn black and disgusting after three days on your kitchen counter?

Q: Now that you mention it, yes. But they are refrigerated during their cross-continental trip, right?

A: By itself, refrigeration wouldn't bring home the banana, so to speak. Banana pickers harvest them when they're green, and they're shipped in chilled containers. That's where I come in. New shipments are placed in hermetically sealed chambers, where I spray the fruit with gas to catalyze ripening. If I'm on my game, they show up at your grocery store looking like something you'd want to peel and eat either on the spot or within a couple of days.

Q: What type of gas do you use?

A: Ethylene, C2H4, which basically is what remains of the alcohol in liquor if you take out the water. Plants naturally produce ethylene, and it causes fruit to ripen. I'm in charge of manipulating the gas vents so the fruit ripens when we want it to ripen. The gas itself doesn't harm the fruit, nor does it harm any people.

Q: Not to sound critical, but how hard is it to open a gas vent?

A: There's more to it than that. Here's the reality: First I have to evaluate where the bananas came from and how long they have been traveling. They go immediately into the chamber, and I have to decide how much gas to give them, what temperature to keep them at, and for how long. A given chamber's bananas need to have the same general characteristics, because one batch of fast-ripeners could cause the whole chamber of bananas to ripen too early.

✳ Chapter 13

Miscellaneous Curios

Does a Curveball Really Curve?

The curveball has been baffling hitters for well more than a hundred years.

✳ ✳ ✳ ✳

HALL-OF-FAMER WILLIAM "CANDY" Cummings is credited with being the first pitcher to master the curve, which he began throwing with great success in 1867 as a member of the Brooklyn Excelsiors. Anyone who has stood in a batter's box and faced a pitcher with an effective curve will tell you that the ball is definitely doing something unusual.

Yet for almost as long as pitchers have been buckling batters' knees with the breaking ball, a small contingent of naysayers has maintained that the ball does not curve, that the whole thing is an optical illusion. Before the availability of advanced photographic technology, various tests were occasionally performed by those curious enough to want to know the truth. Pitchers threw curveballs around boards, through hoops, and parallel to long rods. In 1941, *Life* did a photographic analysis of a curveball and concluded that, no, the ball does not actually curve. The same year, however, *Look* did its own tests and came to the opposite conclusion.

Today's super-slow-motion TV technology catch curveballs in the act all the time, and it sure looks like the ball is curving.

And the truth is, a curveball does curve. Physicists have said so, decisively. They can even tell you the name of the principle that makes it curve: the Magnus effect.

A pitcher who throws a curveball snaps his wrist when he throws, creating a high rate of topspin on the ball. As the ball travels, the air passing on the side against the spin creates drag and higher pressure, while the air passing on the other side has no drag and the pressure is lower. The higher air pressure on one side effectively pushes the ball toward the low-pressure side, sometimes a foot or more. That's the Magnus effect.

The force of gravity also comes into play. Because a curveball is thrown at a slower rate of speed than a fastball, the curve will drop more noticeably than a fastball as it approaches the batter. The combination of the curve and the drop makes it difficult to predict where the ball will be once it reaches home plate and, thus, confounding to hit.

There's even an equation to figure out how much a ball will curve. It looks like this: FMagnus Force = KwVCv. This may seem complicated, but we're willing to bet that for most people, it's still easier to solve the curveball equation than it is to actually hit the thing.

Nauseating Info about That New Car Smell

Actually, the scent we all know and love should be called "volatile organic chemical off-gassed vapor." That's a far more accurate—and unsettling—description.

* * * *

YOU MAY THINK of "new-car smell" as the aroma of cleanliness, but chemically speaking, it's quite the opposite. The distinctive odor is a heady mix of potentially toxic chemicals from the plastics, sealants, adhesives, upholstery, paint, and foam that make up car interiors. Many of the chemicals in these materials are volatile organic compounds. The key word here is "volatile"—these compounds can evaporate at normal temperatures in a process known as "off-gassing." In other words, when the material is new, some of it turns from a solid to a gaseous vapor, and you breathe it all in while you cruise around showing off your new wheels. Heat things up by parking your car in the sun and you get an especially rich chemical cloud.

I Love the Smell of Formaldehyde in the Morning

Don't think the term "organic" means that the chemicals are good for you—it merely signifies that they are carbon-based. In 2006, the non-profit group the Ecology Center released a report that showed that new cars emit potentially dangerous chemicals from manufacturing, and at much higher concentrations than in a new home or building. The study identified phthalates—a class of chemicals used to make PVC plastic more flexible—as one of the biggest problems. Some varieties of phthalates cause liver, reproductive, and learning issues in lab animals, and they may be carcinogenic.

The study also cited polybrominated diphenyl ethers (PBDEs), which are commonly used as fire retardants in cars, as a major concern. Research shows that PDBEs can cause neurodevelopment and liver problems. Other nasty new-car chemicals include formaldehyde (also used as an embalming fluid) and toluene (a noxious chemical that is also found in gasoline and paint thinner). Wonderful, huh?

Car manufacturers (and crafty salesmen) may also spray interiors with a "new-car smell" perfume on top of the existing smell in order to add even more zest to their products. This fragrance may not be as bad for you as the phthalates and PBDEs, but it contains questionable chemicals, too.

The good news is that the Ecology Center report and other studies have spurred carmakers to reduce the use of potentially dangerous materials in new cars. For example, Toyota has developed an alternative car plastic made from sugar cane and corn, while Ford has come up with a soy-based foam for its seats. Don't you just love soy-based smell?

Why Isn't the Whole Plane Made of the Same Stuff as the Black Box?

It's all about weight and sturdiness.

* * * *

PLANES, BELIEVE IT or not, are pretty lightweight. They're built with light metals, such as aluminum. The newer ones are built with even lighter composite materials and plastics. This allows them to be fairly sturdy and resilient without adding too much weight.

If planes were made of the same stuff as the black box, they wouldn't get off the ground. But let's backtrack a bit here. The term "black box" is misleading. What the media refer to as a "black box" is actually two boxes: the Flight Data Recorder (FDR), which records altitude, speed, magnetic heading, and so on; and the Cockpit Voice Recorder (CVR), which records the sounds in the cockpit (presumably ensuring that pilots are on their best behavior). What's more, these boxes are generally bright orange, making them easier to find after a crash. "Black" box either comes from older models that were black or from the charred and/or damaged states of the boxes after a crash.

Whatever the reason behind the name, black boxes are sturdy little things. They carry a bunch of microchips and memory banks encased in protective stainless steel. The protective casing is about a quarter-inch thick, which makes the boxes really heavy. Furthermore, black boxes are not necessarily indestructible—they usually remain intact after a crash partially because they are well placed. They're generally put in the tail of the plane, which often doesn't bear the brunt of a crash.

Even with this extra protection, black boxes sometimes don't survive a plane crash. Still, they typically have been useful, though not so useful that you'd want to build an entire plane with their stainless steel casings.

Why Is the Sky Blue?

What if the sky were some other color? Would a verdant green inspire the same placid happiness that a brilliant blue sky does? Would a pink sky be tedious for everyone except girls under the age of fifteen? What would poets and songwriters make of a sky that was an un-rhymable orange?

* * * *

WE'LL NEVER HAVE to answer these questions, thanks to a serendipitous combination of factors: the nature of sunlight, the makeup of Earth's atmosphere, and the sensitivity of our eyes.

If you have seen sunlight pass through a prism, you know that light, which to the naked eye appears to be white, is actually made up of a rainbow-like spectrum of colors: red, orange, yellow, green, blue, and violet. Light energy travels in waves, and each of these colors has its own wavelength. The red end of the spectrum has the longest wavelength, and the violet end has the shortest.

The waves are scattered when they hit particles, and the size of the particles determines which waves get scattered most effectively. As it happens, the particles that make up the nitrogen and oxygen in the atmosphere scatter shorter wavelengths of light much more effectively than longer wavelengths. The violets and the blues in sunlight are scattered most prominently, and reds and oranges are scattered less prominently.

However, since violet waves are shorter than blue waves, it would seem that violet light would be more prolifically scattered by the atmosphere. So why isn't the sky violet? Because there are variations among colors that make up the spectrum of sunlight—there isn't as much violet as there is blue. And because our eyes are more sensitive to blue light than to violet light, blue is easier for our eyes to detect.

That's why, to us, the sky is blue. And we wouldn't want it any other way.

How Do Fireworks Form Different Shapes?

Fireworks have been delighting people (and on the negative side, blowing off fingers) for more than seven hundred years, and the design hasn't changed much in that time. Getting those fireworks to form complex shapes is a tricky challenge, but the basic idea is still fairly old-school.

* * * *

TO UNDERSTAND WHAT'S involved, it helps to know some fireworks basics. A fireworks shell is a heavy paper container that holds three sections of explosives. The first section is the "lift charge," a packet of black powder (a mixture of potassium nitrate, sulphur, and charcoal) at the bottom of the shell. To prepare the shell for launch, a pyrotechnician places the shell in a mortar (a tube that has the same diameter as the shell), with the lift charge facing downward. A quick-burning fuse runs from the lift charge to the top of the mortar. To fire the shell, an electric trigger lights the quick fuse. It burns down to ignite the black powder at the bottom of the shell, and the resulting explosion propels the shell out of the mortar and high into the air.

The second explosive section is the "bursting charge," a packet of black powder in the middle of the shell. When the electric trigger lights the quick-burning fuse, it also lights a time-delay fuse that runs to the bursting charge. As the shell is hurtling through the air, the time-delay fuse is burning down. Around the time the shell reaches its highest point, the fuse burns down to the bursting charge, and the black powder explodes.

Expanding black powder isn't exactly breathtaking to watch. The vibrant colors you see come from the third section of explosives, known as the "stars." Stars are simply solid clumps of explosive metals that emit colored light when they burn. For example, burning copper salts emit blue light and burning barium nitrate emits green light. The expanding black powder ignites the stars and propels them outward, creating colored streaks in the sky.

The shape of the explosion depends on how the manufacturer positions the stars in the shell. To make a simple ring, it places the stars in a ring around the bursting charge; to make a heart, it positions the stars in a heart shape. Manufacturers can make more complex fireworks patterns, such as a smiley face, by combining multiple compartments with separate bursting charges and stars in a single shell. As the fuse burns, these different "breaks" go off in sequence. In a smiley face shell, the first break that explodes makes a ring, the second creates two dots for the eyes, and the third forms a crescent shape for the mouth. It's hard to produce designs that are more complex than that, since only a few breaks can be set off in quick succession. So if you're hoping to see a fireworks tribute to origami, you're out of luck.

The Cold Realities of a Sunless Earth

If that fiery ball in the sky went out, the future of our planet would not be bright. And that's putting it mildly.

❋ ❋ ❋ ❋

IF THE SUN suddenly stopped emitting light and warmth, Earth would get dark in about eight minutes—the length of time it takes for light to reach us once it escapes the sun—and would gradually become colder. It's been hypothesized that crops would freeze and die within days, rivers would freeze within weeks, and the warming Gulf Stream waters in the Atlantic would freeze within months.

None of us would live much longer than a few weeks, thanks to subtle factors like scarcity of food and water, more drastic factors like severe weather, and absolutely nutso factors like widespread panic. And the only reason anyone would last even a few weeks is that Earth and its atmosphere have some capacity to retain heat, which explains why it doesn't become frigid immediately every night.

Now, if the sun's core stopped undergoing fusion (which is possible but entirely unlikely), up to a million years would pass before we felt the full effect, as that's how long it takes light that's generated by fusion to escape from the plasma-like material that makes up the sun. But long before that, scientists would detect clues—such as the lack of neutrinos (tiny elementary particles) coming from the sun and pulsations on its surface owing to the imbalance between its weight (which produces gravitational force) and the heat-and-pressure force of the fusion, which counteracts the gravity and keeps everything in balance—that would tell us there's something wrong. Slowly the sun would start to shrink.

So you'd rather have the sun die than just suddenly stop emitting light and heat. But either way, the picture isn't pretty.

Why Are Skunks so Stinky?

Oh, so you smell like a bed of roses? But seriously, skunks have earned their odiferous reputation through their marvelous ability to make other things stink to high heaven.

⁂ ⁂ ⁂ ⁂

ALL ELEVEN SPECIES of skunk have stinky spray housed in their anal glands. However, as dog owners can attest, skunks aren't the only animals to have anal glands filled with terrible-smelling substances. Opossums are particularly bad stinkers; an opossum will empty its anal glands when "playing dead" to help it smell like a rotting corpse.

While no animal's anal glands are remotely fragrant, skunks' pack an especially pungent stench. This is because skunks use their spray as a defense mechanism. And they have amazing range: Skunks have strong muscles surrounding the glands, which allow them to spray sixteen feet or more on a good day.

A skunk doesn't want to stink up the place. It does everything in its power to warn predators before it douses its target with *eau de skunk*. A skunk will jump up and down, stomp its feet, hiss, and lift its tail in the air, all in the hope that the predator will realize that it's dealing with a skunk and go away. A skunk only does what it does best when it feels it has no choice. Then it releases the nauseating mix of thiols (chemicals that contain super-stinky sulfur), which makes whatever it hits undateable for the foreseeable future. Skunks have enough "stink juice" stored up for about five or six sprays; after they empty their anal glands, it takes up to ten days to replenish the supply.

Being sprayed by a skunk is an extremely unpleasant experience. Besides the smell, the spray from a skunk can cause nausea and temporary blindness. Bobcats, foxes, coyotes, and badgers usually only hunt skunk if they are really, really hungry. Only the great horned owl makes skunk a regular snack—and the fact that the great horned owl barely has a sense of smell probably has a lot to do with it.

Should you find yourself on the receiving end of a skunk shower, your best deodorizer is alkaline hydrogen peroxide. But unless you startle a skunk (which is possible, since the critter doesn't have keen eyesight), you'll probably have plenty of chances not to get sprayed. You and a skunk have a lot in common: You don't want to get sprayed, and the skunk doesn't want to spray you.

How to Beat a Lie Detector Test

If you're told you need to take a polygraph test before accepting a job or to be cleared of a crime, watch out—you're about to be duped. The polygraph or "lie detector" test is one of the most misunderstood tests used in law enforcement and industry.

⁂ ⁂ ⁂ ⁂

MANY EXPERTS WILL tell you that lie detector tests are based on fallible data—regardless of how scientific the equipment appears, there's no sure way a person can tell whether or not someone is lying. Here are a few suggestions on how to beat one:

- Unless you're applying for a job, refuse to take the polygraph test. There are no laws that can compel anyone to take it.
- Keep your answers short and to the point. Most questions asked of you can be answered with a "yes" or "no."
- During the polygraph test, you'll be asked three types of questions: irrelevant, relevant, and control questions. Irrelevant questions generally take the form of, "Is the color of this room white?" Relevant questions are the areas that get you into trouble.
- Control questions are asked so that the technician can compare the responses to questions against a known entity. The easiest way to beat a lie detector test is to invalidate the control questions. Do this by changing your breathing rate and depth from the normal 15 to 30 breaths per minute to anything faster or slower. You can practice math problems or bite the sides of your tongue until it begins to hurt.

The Route to China Via a Hole

What would you encounter if you tried to dig a hole to China? Hopefully a chiropractor, because severe back pain is about all that your journey would yield. It's obviously impossible to carve out such a tunnel, but for the sake of knowledge, we'll examine this little gem.

⁕ ⁕ ⁕ ⁕

We'll Start in the United States

BEFORE STARTING OUR dig, let's establish that the starting point for our hole is in the United States, where this expression appears to have originated. Nineteenth-century writer/philosopher Henry David Thoreau told the story of a crazy acquaintance who attempted to dig his way to China, and the idea apparently stuck in the American popular mind.

We also need to clear up a common misconception. On a flat map, China appears to be exactly opposite the United States. However, about five hundred years ago or so, humanity established that Earth is round, so we should know not to trust the flat representation. If you attempted to dig a hole straight down from the United States, your journey—about 8,000 miles in all—would actually end somewhere in the Indian Ocean. Therefore, our hole will run diagonally; this will have the added benefit of sparing us from having to dig through some of the really nasty parts of the earth's interior.

Let's Dig

The hole starts with the crust, the outer layer of the planet that we see every day. The earth's crust is anywhere from about three to 25 miles thick, depending on where you are. By the time we jackhammer through this layer, the temperature will be about 1,600 degrees Fahrenheit—hot enough to fry us in an instant. But we digress.

The second layer of the earth is the mantle. The rock here is believed to be slightly softer than that of the crust because of intense heat and pressure. The temperature in the mantle can exceed 4,000 degrees Fahrenheit, but who's counting?

Since our hole is diagonal, we'll probably miss the earth's core. At most, we'll only have to contend with the core's outermost layer. And it's a good thing, too: Whereas the outer core is thought to be liquid, the inner core, which is about 4,000 miles from the earth's surface, is believed to be made of iron and nickel, and is extremely difficult to pierce, particularly with a shovel. But either way, it would be hotter than hot; scientists think the outer core and inner core are 7,000 and 9,000 degrees Fahrenheit, respectively.

And you thought the hot wings night at your local bar took a toll on your body! No, unless fire and brimstone are your thing, the only journey you'll want to take through Earth's center is a hypothetical one.

Running Through the Rain vs. Walking Through the Rain

Which technique keeps you driest? Read on to learn the answer.

⁕ ⁕ ⁕ ⁕

IT MAKES INTUITIVE sense that running through the rain will keep you drier than walking. You will spend less time in the rain, after all. But there's a pervasive old wives' tale that says it won't do any good. So every time there's a downpour and you need to get to your car, you are faced with this confounding question: Should you walk or run?

The argument against running is that more drops hit your chest and legs when you're moving at a quicker pace. If you're walking, the theory goes, the drops are mainly hitting your head. So the proponents of walking say that running exposes you to more drops, not fewer. Several scientists have pondered this possibility (after finishing their work for the day, we hope). In 1987, an Italian physicist determined that sprinting keeps you drier than walking, but only by about 10 percent, which might not be worth the effort and the risk of slipping. In 1995, a British researcher concluded that the increased front-drenching of running effectively cancels out the reduced rain exposure.

These findings didn't seem right to two climatologists at the National Climatic Data Center in Asheville, North Carolina, so they decided to put them to the test. In 1996, they put on identical outfits with plastic bags underneath to keep moisture from seeping out of the clothes and to keep their own sweat from adding to the drenching. One person ran about 330 feet in the rain; the other walked the same distance. They weighed the wet clothes, compared the weights to those when the clothes were dry, and determined that the climatologist who walked got 40 percent wetter than the one who ran.

In other words, run to your car. You're justified—no matter how silly you might look

It's Iron-IC

Popeye credited his trusty can of spinach for his bulging biceps. But his assumptions about spinach were based on a widespread misconception about its iron content.

* * * *

POPEYE DIDN'T START the rumor about the nutritional value of spinach. He simply popularized the widely held belief, based on a study that spinach is a superior source of iron.

Thanks to a Typo

But the leafy greens' reputation wasn't so ironclad. An 1870 German study of spinach claimed it had ten times the iron content of other green leafy vegetables. This claim, uncontested for 70 years, turned out to be based on a misprint—a misplaced decimal point. The iron content of spinach was overestimated by a factor of ten! By the time the error was discovered in 1937, Popeye had already helped spread the myth far and wide—and encouraged several generations of children to tolerate the unappealing vegetable in hopes of developing their hero's brawn.

The hype about spinach didn't end, however, because the error wasn't publicized. It wasn't until an article on the mistake was published in a 1981 issue of the *British Medical Journal* that the public was informed of the true iron content of spinach.

In the 1990s, spinach received another nutritional blow when it was discovered that its oxalic acid content prevents the body from absorbing more than 90 percent of the vegetable's iron. Oxalic acid binds with iron and renders most of it unavailable for absorption.

Still Packs a Wallop

Luckily for Popeye's legacy, however, the spirited sailor was not wrong to think that spinach has abundant nutritional merit. It's a terrific source of vitamins A, B1, B2, B3, B6, C, E, and K, as well as magnesium, calcium, and potassium.

So if it's muscles you're after, you'll need to pump iron rather than consume it.

The Mysterious Blue Hole

State Route 269 hides a roadside attraction of dubious depth and mysterious origin, a supposedly bottomless pool of water that locals simply call the "Blue Hole."

* * * *

EVERY STATE HAS its tourist traps and bizarre little roadside attractions that are just intriguing enough to pull the car over to see. Back in the day, no roadside attraction brought in the Ohio travelers more than a bottomless pond filled with blue water: the mysterious Blue Hole of Castalia.

The Blue Hole's Origins

The Blue Hole is believed to have formed around 1820, when a dam burst and spilled water into a nearby hole. The ground surrounding Castalia is filled with limestone, which does not absorb groundwater well. The water quickly erodes the limestone, forming cave-ins and sinkholes. It wouldn't be until the late 1870s, however, that most people were made aware of the Blue Hole's existence; the hole was in a very isolated location in the woods. Once the Cold Creek Trout Club opened up nearby, however, its members began taking trips out to see the hole, and people all over the area were talking about the mysterious Blue Hole hiding out in Castalia. In 1914, a cave-in resulted in the Blue Hole growing to its current size of almost 75 feet in diameter.

Stop and See the Mystery

The owners of the property where the Blue Hole is situated began promoting it as a tourist stop beginning in the 1920s. It didn't hurt that the entrance to the Blue Hole property was along State Route 269, the same road that people took to get to Cedar Point amusement park. It is estimated that, at the height of its popularity, close to 165,000 people a year came out to take a peek at the Blue Hole.

The Blue Hole was promoted as being bottomless. Other strange stories were often played up as well, including the fact that the water temperature remained at 48 degrees Fahrenheit year-round. Tour guides would point out that regardless of periods of extreme rainfall or even droughtlike conditions, the Blue Hole's water level remained the same throughout.

So What's Up with This Hole, Anyway?

Despite the outlandish claims and theories surrounding the Blue Hole and its origins, the facts themselves are rather mundane. The Blue Hole is really nothing more than a freshwater pond. It isn't even bottomless. Sure, the bright blue surface of the water does indeed make the hole appear infinitely deep, but in fact, it's really only about 45 feet to the bottom at its deepest parts. The blue color of the water is from an extremely high concentration of several elements, including lime, iron, and magnesium. That's the main reason there are no fish in the Blue Hole; they just can't survive with all that stuff in the water.

One Hole or Two?

During the 1990s, the owners of the Blue Hole fell on hard times, forcing them to close the attraction. Families who would show up at the front entrance were forced to stare sadly through a locked gate at the small trail into the woods. That is until several years ago, when the nearby Castalia State Fish Hatchery began clearing land to expand its hatchery. Lo and behold, workers uncovered a second Blue Hole.

Just how this second Blue Hole came to be is still unknown, although the popular belief is that both holes are fed by the same underground water supply. None of that seems to matter to the Blue Hole faithful—they're just thankful to be able to take a gander at a Blue Hole again.

Forensic Dentist

"The remains were identified from dental records." That statement requires the involvement of a forensic dentist, or odontologist. Here's some insight into the job.

* * * *

Q: You could be doing root canals on live, squirming people. But you look at the teeth of the dead and tell us about their final hours.

A: True confession: In dental school, I learned that I loved dentistry but couldn't stand inflicting pain. I got physically ill if I felt I was hurting someone, but I still had a strong interest in the science of teeth.

Q: How long has forensic dentistry been around?

A: In 1849, a Viennese opera house went up in flames, and a number of people were incinerated beyond recognition. Teeth are incredibly durable; the fact that we have the power to break a healthy tooth with our bite proves just how strong our jaws are. The fire victims' teeth enabled families to obtain the correct remains to bury and have closure. Odontology has grown with advancements in dental science and record-keeping.

Q: This is a full-time job?

A: In larger cities it is. In smaller towns, most odontologists operate traditional dental practices and consult on the side as needed by local government. I like being near research universities, because my work also fits into archaeology.

Q: So you get to examine mummies?

A: Not quite. Bites leave patterns. What people eat and the way they chew suggests things about their diets. We can estimate age. All odontologic procedures begin with specific questions. In the case of a tooth from an ancient midden, they might include: "What species was this? From what period of time? How long did he or she live? What did this person usually eat?"

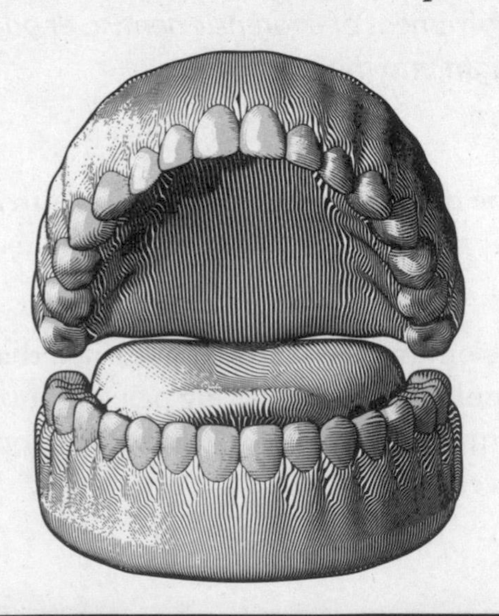

Exposed to Poison

Long a favorite of mystery-novel writers and opportunistic bad guys, poison has an ancient and infamous relationship with people. Some poisons can be found in the natural world while others are manufactured, but all of them spell bad news if you're the unlucky recipient of a dose.

* * * *

Poison Plants

Deadly nightshade, aka belladonna: Every part of this perennial herb is poisonous, but the berries are especially dangerous. The poison attacks the nervous system instantly, causing a rapid pulse, hallucinations, convulsions, ataxia, and coma.

Wolfsbane: This deadly plant was used as an arrow poison by the ancient Chinese, and its name comes from the Greek word meaning "dart." Wolfsbane takes a while to work, but when it does, it causes extreme anxiety, chest pain, and death from respiratory arrest.

Meadow saffron: This tough little plant can be boiled and dried, and it still retains all of its poisonous power. As little as seven milligrams of this stuff could cause colic, paralysis, and heart failure.

Hemlock: This plant is probably the best known of the herbaceous poisons: It was used to knock off the Greek philosopher Socrates. Hemlock is poisonous down to the last leaf and will often send you into a coma before it finishes you for good.

Plans of Attack

There are five ways a person can be exposed to poison: ingestion (through the mouth), inhalation (breathed in through the nose or mouth), ocular (in the eyes), dermal (on the skin), and parenteral (from bites or stings).

Helpful Poison Stats

More than half of poison exposures occur in children under the age of six, and most poisonings involve medications, vitamins, household and chemical personal-care products, and plants. Eighty-nine percent of all poisonings occur at home. If you or someone in your house ingests something poisonous, stay calm and call 911 (if the person has collapsed or is not breathing) or your local poison control center (three-quarters of exposures can be treated over the phone with guidance from an expert).

Good Old Arsenic

Mystery novels are filled with stories of characters choosing to off their enemies with arsenic. Colorless and odorless, this close relative of phosphorous exists in a variety of compounds, not all of which are poisonous. Women in Victorian times used to rub a diluted arsenic compound into their skin to improve their complexions, and some modern medications used to treat cancer contain arsenic. When certain arsenic compounds are concentrated, however, they're deadly; arsenic has been blamed for widespread death through groundwater contamination.

The Dubiously Poisoned

Napoleon Bonaparte: Many historians believe that Napoleon died of arsenic poisoning while imprisoned, because significant traces of arsenic were found in his body by forensics experts 200 years after his death. It has been argued, however, that at that time in history, wallpaper and paint often contained arsenic-laced pigments, and that Napoleon was simply exposed to the poison in his everyday surroundings.

Vincent Van Gogh: Emerald green, a color of paint used by Impressionist painters, contained an arsenic-based pigment. Some historians suggest that Van Gogh's neurological problems had a great deal to do with his use of large quantities of emerald green paint.

Yasser Arafat: Founder of the Palestinian liberation movement, Nobel Peace Prize winner, and politically controversial figure, Yasser Arafat died in 2004 from unknown causes. Leaders of the Islamic Resistance Movement, or Hamas, still accuse Israel of poisoning Arafat with an undetectable toxin, but there's no proof of that so far.

Food Poisoning

Unfortunately, this is a form of poisoning most of us know something about. When food is spoiled or contaminated, bacteria such as salmonella breed quickly. Because we can't see or taste these bacteria, we chomp happily away and don't realize we're about to become really sick. The Centers for Disease Control and Prevention estimates that in the United States alone, food poisoning causes about 76 million illnesses, 325,000 hospitalizations, and up to 5,000 deaths each year.

Blood Poisoning

This form of poisoning occurs when an infectious agent or its toxin spreads through the bloodstream. People actually have a low level of bacteria in their blood most of the time, but if nasty bacteria are introduced, they can cause sepsis, a life-threatening condition. The bacteria can enter the bloodstream through open wounds or from the bite of a parasite.

Snakebites

Because snakes' venom is injected, snakes themselves are considered "venomous" rather than "poisonous." Still, an estimated 8,000 snakebites occur in the United States every year. Poisonous snakes found in North America include rattlesnakes, copperheads, cottonmouths, and coral snakes. While most of these reptiles won't bite unless provoked, if you are bitten you have to take the antivenin fast.

Skull and Crossbones

When pirates sailed the seas, they flew a flag with a skull-and-crossbones symbol. When seafarers saw this Jolly Roger flag, they knew trouble was ahead. Bottles that contain poisons or other toxic substances often bear this symbol to warn anyone against drinking or even touching the contents with bare hands.

Color Me Toxic

Mares eat oats, and does eat oats, and little kids eat crayons. But should Junior wear a HAZMAT suit as he munches on Burnt Sienna or Cadet Blue?

* * * *

ON MAY 23, 2003, the *Seattle Post-Intelligencer* fired a shot heard by parents around the world. In independent tests of eight brands of coloring crayons, three brands (Crayola, Prang, and Rose Art) had colors that contained more than trace levels of asbestos.

The three manufacturers immediately dismissed the findings as wrong, citing their own industry tests. Despite the denials, this report set off a firestorm of fear, criticism, and consumer panic.

"Asbestos" is a general term for several minerals that break easily into fibrous threads, and it has been linked to various forms of cancer, especially when inhaled. In these cases, asbestos is likely mixed with the mineral talc, which was used as a binding agent in crayons. Talc and asbestos are found together in rock formations and frequently are combined in the mining process.

In follow-up tests, the Consumer Product Safety Commission (CPSC) found traces of asbestos in Crayola and Prang crayons, but it assured the public that the amount was insignificant. Similar but non-hazardous "transitional fibers" also appeared in the tests. Although the CPSC wasn't concerned about children ingesting any of these materials and found no airborne fibers even after 30 minutes of simulated scribbling, it requested that the manufacturers reformulate their products. All three companies agreed, and later tests showed all of the crayons to be asbestos-free.

Get Savvy on the Seas

Many people fall (overboard) for some of the most enduring myths about the largest area of our world—the oceans.

* * * *

Myth: Oceans don't freeze solid because of deep currents.

Fact: Although the water around the Arctic and Antarctica is freezing cold, oceans don't freeze solid for several reasons. Oceans contain a lot of water, which circulates around the world. Water from warmer oceans and from underground volcanoes flows into the Arctic, which warms it up a bit. But the main reason oceans don't go into a deep freeze is the salt in the water. The freezing point of saltwater is lower than that of freshwater, and as ocean water reaches the freezing point, the salt crystals interfere with the formation of ice crystals. This water is also more dense and therefore sinks, allowing warmer water to come up to the top—below the surface ice. The surface ice actually insulates the warmer water in the same way an igloo insulates the air inside it. The ice also reflects the sun's rays, and this helps warm the surface and prevents the ice from thickening further. Nearly all the ice in the Antarctic is "seasonal," which means it melts and reforms annually.

Myth: Icebergs are made of frozen seawater.

Fact: This seems like common sense because icebergs float in seawater, but many natural phenomena defy common sense. Oceanographers agree that icebergs are made of freshwater—in the form of snow—that has compacted over hundreds or thousands of years. True icebergs are huge pieces of ice that have broken off from the glaciers that make up the continental ice sheets (as found in Antarctica or Greenland). Seawater, with its salt content, doesn't mix with the freshwater of the iceberg. To test this theory at home, put saltwater and freshwater in the freezer at the same time—and see which solidifies faster.

Anatomical Anomalies

Mother Nature isn't always right. Here are a few examples.

* * * *

But Can It Wag?

Many people don't know that every human embryo actually starts life outfitted with a tail, though we usually lose them before birth. The reason why there are only several dozen known cases of people born with "true human tails" is because almost as soon as the embryonic tail develops, its growth is suppressed by *apoptosis*, or cell death. True human tails, when they are retained by newborns, range from one to more than five inches in length—and, yes, they are able to wag!

Breathe Easily—If You Can

Flying is a strenuous activity requiring vast amounts of oxygen and an efficient respiratory system. Birds have a flow-through system of ventilation, which doesn't let freshly oxygenated air mix with the depleted air leaving the body. Mammals, however, mix fresh air with depleted air when they breathe, which is very inefficient. This is unfortunate for bats, since these flying mammals must take to the sky with inferior ventilation systems. Too bad they can't swap lungs with flightless birds such as emus and penguins, who still have the avian flow-through system that is of little use to them on the ground.

Picasso's Dream Fish

The bony flatfish has evolved to lay flat on the seabed. The most sensible way to flatten itself would be back-to-stomach, like a ray, so that both eyes are on top of its head. However, most bony fish are flat sideways, which means lying down flat would leave one of their eyes pointing uselessly at the ground. But where there's a will, there's a way: The flatfish asymmetrically reshapes its skull during adolescence, so that one of its eyes migrates over the top of the head and winds up on the other side! The result is both comical and creepy.